NMR Basic Principles and Progress 18

Editors: P. Diehl E. Fluck R. Kosfeld

Gérard J. Martin
Maryvonne L. Martin
Jean-Paul Gouesnard

^{15}N-NMR Spectroscopy

With 142 Tables and 11 Figures

Springer-Verlag
Berlin Heidelberg New York 1981

Prof. Dr. Gérard J. Martin
Prof. Dr. Maryvonne L. Martin
Dr. Jean-Paul Gouesnard
Chimie Organique Physique, Faculté des Sciences
Université de Nantes, F-44072 Nantes-Cedex

ISBN 978-3-642-50174-6 ISBN 978-3-642-50172-2 (eBook)
DOI 10.1007/978-3-642-50172-2

Library of Congress Cataloging in Publication Data. Martin, Gérard J. ^{15}N-NMR spectroscopy. (NMR, basic principles and progress; 18), Bibliography: p. Includes indexes. 1. Nuclear magnetic resonance spectroscopy. I. Martin, Maryvonne L., joint author. II. Gouesnard, Jean Paul, 1944- joint author. III. Title. IV. Series. QC490.N2 vol. 18 [QC762] 538.3s 80-28627 ISBN 0-387-10459-3 (U.S.) [543'.0877]

Softcover reprint of the hardcover 1st edition 1981

2152/3140-543210

Contents

1 Introduction

After the proton and carbon, nitrogen is, with oxygen, the most important atom in organic and especially bioorganic molecules. However, the development of nitrogen spectroscopy is indeed very recent. This is due to the fact that nitrogen-14, which is the naturally abundant isotope, suffers, for structural studies, from the disadvantages inherent in nuclei with a quadrupolar moment (Table 1.1). Actually, indirect ^{15}N measurements were reported in the early days of double resonance spectroscopy and the first direct detection of ^{15}N resonance signals at the natural abundance level was realized in 1964 (R 17) at 4.33 MHz (~ 1T) using a 15 mm o.d. cell in the field sweep mode (≃ 0.16 min^{-1}). Signal-to-noise ratios only of 3-4 were obtained for neat liquids and this low sensitivity of the ^{15}N resonance still remains the main disadvantage for ^{15}N spectroscopy (Table 1.1). However, nitrogen-15 has, probably more than any other nucleus, benefited from the advances of NMR technology, i.e. Fourier transformation, multinuclear facilities, wide-bore superconducting solenoids, and, with the new generation of spectrometers, ^{15}N-NMR is entering the field of routine investigation. Nevertheless, in spite of these spectacular improvements, obtaining ^{15}N spectra of diluted species or large biochemical molecules is often not very easy and a good knowledge of the relaxation properties peculiar to ^{15}N may be necessary in order to adjust the pulse sequences and the decoupler duty cycle correctly (Section 2).

The considerable efforts recently devoted to ^{15}N-NMR are certainly justified by the very interesting features of this resonance. First of all, the ^{15}N signals have a relatively narrow natural width and dipolar broadening is smaller for an N-H fragment than for a C-H bond. This property, along with the fact that the number of nitrogen atoms in a molecule is relatively small explains the frequent simplicity of ^{15}N spectra. Another interesting feature is the great range displayed by ^{15}N chemical shifts and illustrated in the accompanying chart. Such a great chemical shift selectivity as compared with proton and even carbon NMR makes ^{15}N spectroscopy especially attractive for the study of bioorganic molecules (Sect. 6.1.3).

It can be seen in Table 1.1 which collects the main nuclear properties of hydrogen, carbon and nitrogen isotopes that the gyromagnetic ratio of ^{15}N is relatively small and negative. It follows that the enhancement factor of the nuclear Overhauser effect, η, may have a high absolute value ($|\eta\ max| \approx 5$). This property is likely to improve the signal-to-noise ratio considerably in a number of cases. However, it must be kept in mind that the fact that η is negative may lead to the cancelling of the ^{15}N signal (Sect. 2).

The aim of this monograph is to present the different aspects of ^{15}N spectroscopy in order to answer two questions of practical interest:

- How to obtain ^{15}N parameters: δ, J, T_1
- How to use the ^{15}N data in structural and physical chemistry.

Since good reviews have previously dealt with the theoretical aspects concerning nitrogen chemical shifts and J_{N-X} coupling constants (W 3, W 23), we shall not attempt to discuss the problem of the semi-empirical determination of these parameters here. Instead, special emphasis will be placed on the relaxation phenomena, solvent effects, and experimental techniques used to obtain ^{15}N spectra.
This monograph also provides an extensive compilation of the ^{15}N parameters, δ, J and T_1, irrespective of whether they have been derived from ^{15}N spectra or not. Rather than discussing qualitatively the changes in δ and J in terms of structural modification within different classes of compounds - an approach which has been widely covered in previously published monographs (L 30, W 23) - we shall present the behaviour of these parameters in terms of the various correlation analyses which have been attempted. Such a correlation analysis of ^{15}N data provides a quantitative basis for explaining and predicting ^{15}N parameters in structural studies. ^{15}N chemical shifts of up to 1400 compounds, and J $^{15}N{\sim}X$ coupling constants concerning approximatively 1200 structures, have been collected in the tables. Thus, the number of ^{15}N parameters presented here amounts to about 5000 and may provide a good basis for a ^{15}N data bank.

Table 1.1. Nuclear properties of hydrogen, carbon and nitrogen isotopes. Spin number: $^{1}_{1}H$, $^{13}_{6}C$, $^{15}_{7}N$: I = 1/2 β_N = 5.0505 10^{-27} JT^{-1} (L 52)

$^{2}_{1}H$, $^{14}_{7}N$: I = 1

Nucleus	$^{1}_{1}H$	$^{2}_{1}H$	$^{13}_{6}C$	$^{14}_{7}N$	$^{15}_{7}N$
Relative Abundance (%)	99.985	0.015	1.108	99.635	0.365
Gyromagnetic Ratio $\gamma \times 10^{-7}$ (T^{-1} s^{-1})	26.752	4.107	6.726	1.9329	- 2.7112
Magnetic Moment					
μ in units of βN	2.79268	0.85738	0.70220	0.40358	- 0.28304
μ (JT^{-1}) x 10^{27}	14.104	4.3302	3.5465	2.0383	- 1.4295
Quadrupolar Magnetic Moment Q ($10^{-28}m^2$)	/	0.00277	/	0.071	/
Overhauser Enhancement Factor $\{^1H\}X$, η_{max}	0.5	/	1.99	/	- 4.93
Frequency ν_0 (MHz)					
Bo = 1 (T)	42.57	6.54	10.71	3.08	4.31
2.114 (T)	90.00	13.82	22.63	6.50	9.12
4.228 (T)	180.00	27.64	45.26	13.00	18.24
6.342 (T)	270.00	41.46	67.89	19.50	27.36
8.456 (T)	360.00	55.28	90.52	26.00	36.48
Relative Sensitivity (a)	1	9.65 10^{-3}	1.59 10^{-2}	1.01 10^{-3}	1.04 10^{-3}
(b)	1	1.45 10^{-6}	1.8 10^{-4}	1. 10^{-3}	3.8 10^{-6}

a) For an equal number of nuclei at constant field.

b) At the natural abundance level.

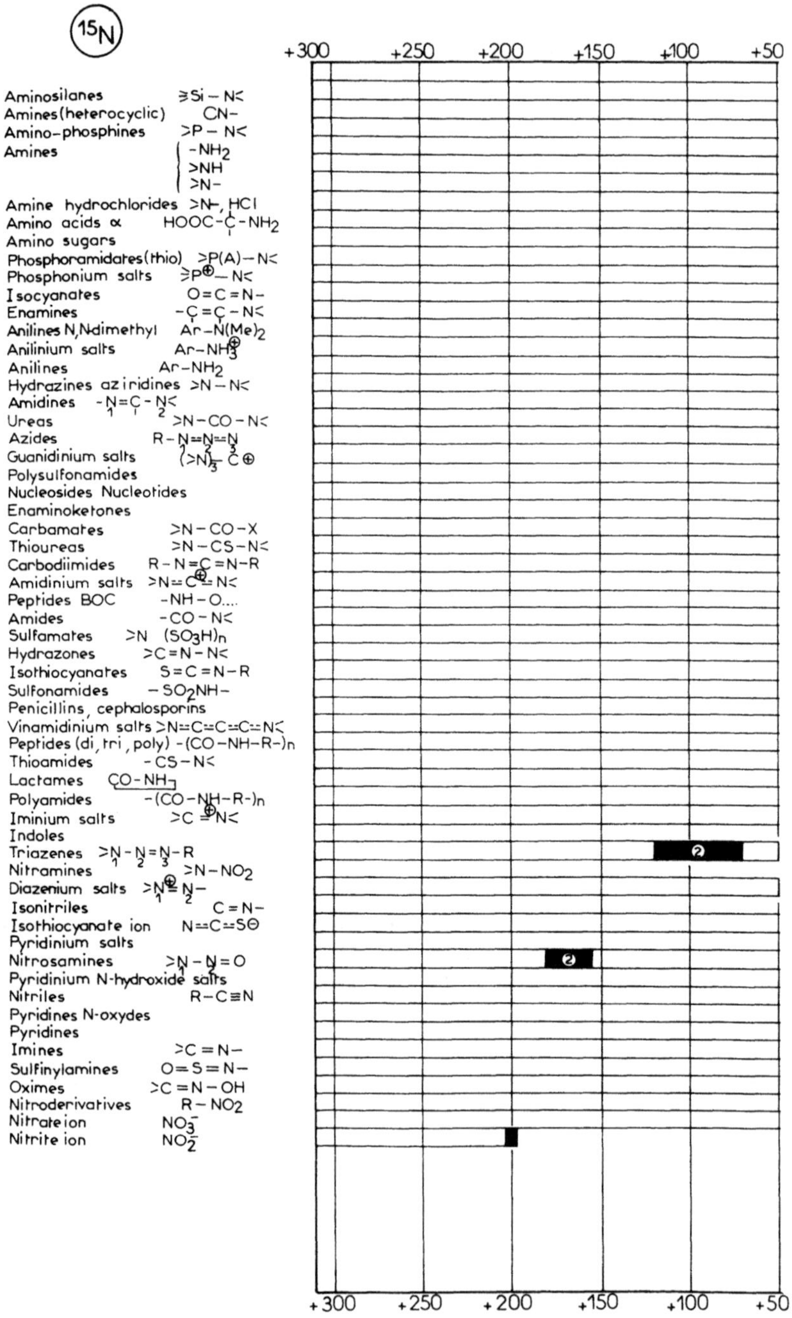
15N
+300 +250 +200 +150 +100 +50
Aminosilanes ⩾Si – N<
Amines (heterocyclic) CN–
Amino-phosphines >P – N<
Amines –NH2 >NH >N–
Amine hydrochlorides >N–, HCl
Amino acids α HOOC–C–NH2
Amino sugars
Phosphoramidates (thio) >P(A)–N<
Phosphonium salts ⩾P⊕–N<
Isocyanates O=C=N–
Enamines –C=C–N<
Anilines N,N-dimethyl Ar–N(Me)2
Anilinium salts Ar–NH3⊕
Anilines Ar–NH2
Hydrazines aziridines >N–N<
Amidines –N=C–N<
Ureas >N–CO–N<
Azides R–N=N=N
Guanidinium salts (>N)3–C⊕
Polysulfonamides
Nucleosides Nucleotides
Enaminoketones
Carbamates >N–CO–X
Thioureas >N–CS–N<
Carbodiimides R–N=C=N–R
Amidinium salts >N=C⊕=N<
Peptides BOC –NH–O....
Amides –CO–N<
Sulfamates >N (SO3H)n
Hydrazones >C=N–N<
Isothiocyanates S=C=N–R
Sulfonamides –SO2NH–
Penicillins, cephalosporins
Vinamidinium salts >N=C=C=C=N<
Peptides (di, tri, poly) –(CO–NH–R–)n
Thioamides –CS–N<
Lactames CO–NH
Polyamides –(CO–NH–R–)n
Iminium salts >C=⊕N<
Indoles
Triazenes >N–N=N–R
Nitramines >N–NO2
Diazenium salts >N⊕=N–
Isonitriles C=N–
Isothiocyanate ion N=C=S⊖
Pyridinium salts
Nitrosamines >N–N=O
Pyridinium N-hydroxide salts
Nitriles R–C≡N
Pyridines N-oxydes
Pyridines
Imines >C=N–
Sulfinylamines O=S=N–
Oximes >C=N–OH
Nitroderivatives R–NO2
Nitrate ion NO3−
Nitrite ion NO2−
+300 +250 +200 +150 +100 +50

Fig. 1-1

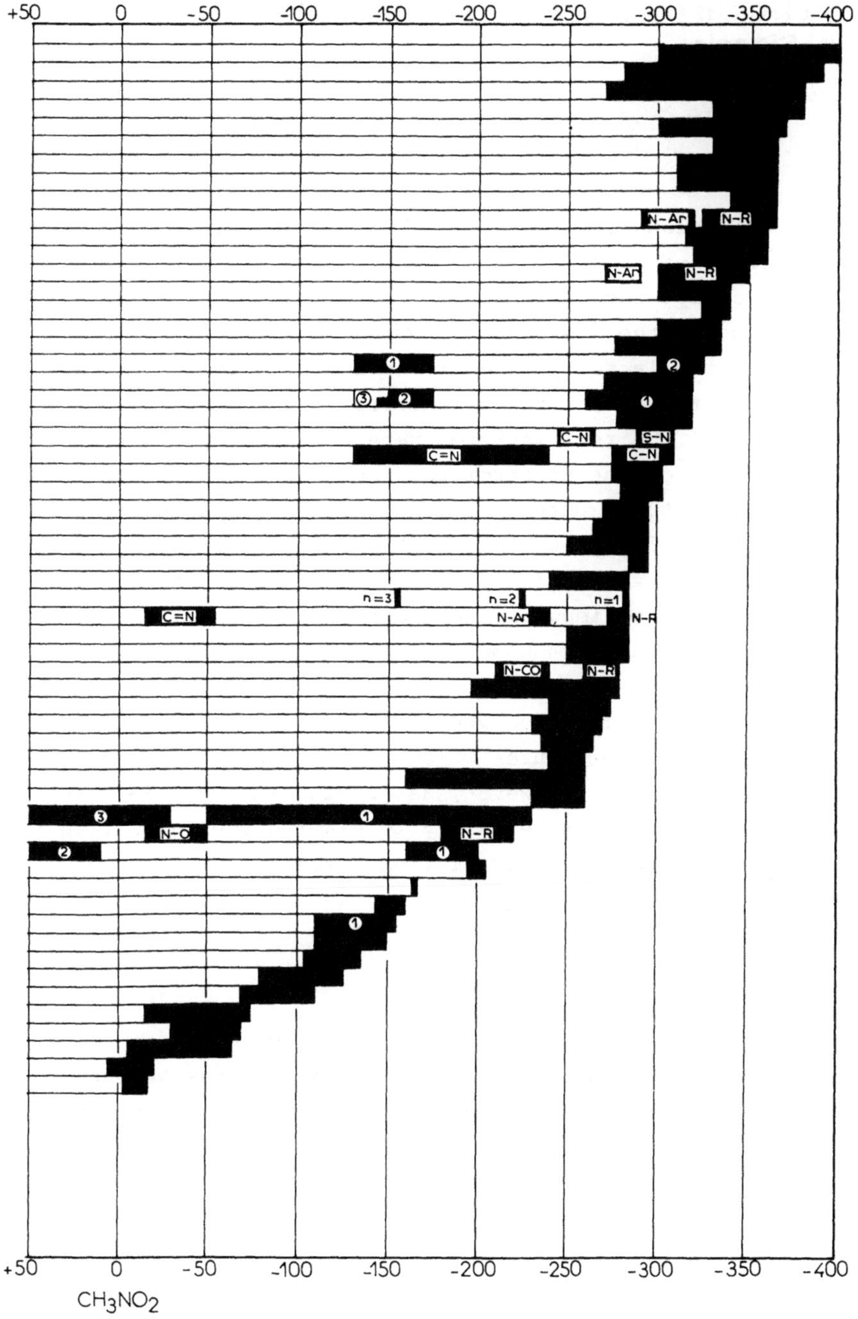
+50
0
-50
-100
-150
-200
-250
-300
-350
-400
N-Ar
N-R
N-Ar
N-R
1
2
3
2
1
C-N
S-N
C=N
C-N
n=3
n=2
n=1
C=N
N-Ar
N-R
N-CO
N-R
3
1
N-O
N-R
2
1
1
CH3NO2

2 Relaxation Phenomena and Nuclear Overhauser Effects. Molecular Dynamics and Observation of the ^{15}N Signals

A knowledge of the relaxation processes which govern nitrogen relaxation is especially useful, not only in order to obtain information on molecular dynamics, but also in order to select the best conditions for the observation of the ^{15}N signals. Owing to the lack of sensitivity of ^{15}N-NMR, the latter problem is indeed of prime importance. It is therefore helpful to be able to anticipate the behaviour of T_1 as a function - of the reorientation rate of the compound - of the molecular structure - of the spectrometer frequency and - of the medium properties and temperature. Determination of the pulse sequence is, in fact, critically conditioned by the values of T_1. Moreover, as the value of the transverse relaxation time T_2^* governs the signal width ($\Delta\upsilon_{1/2} = 1/\pi T_2^*$) it is also important to appreciate the influence of the various experimental parameters upon T_2^*.

Frequently, the discussion of relaxation phenomena is restricted to the extreme narrowing situation where $\omega_o^2\ \tau_c^2 < 1$. It therefore only concerns relatively small molecules which tumble rapidly according to a short correlation time, τ_c. However, as growing interest is directed towards biopolymers it becomes necessary to examine the relaxation behaviour in the whole range of correlation times.

2.1 The Mechanisms of Nitrogen Relaxation

With the exception of quadrupolar interaction, any of the mechanisms discussed in a general interpretation of the relaxation times can be considered when nitrogen-15 relaxation is concerned. Thus, to the extent that a longitudinal relaxation time T_1 may be safely defined, the observed relaxation rate can be written as a sum of several contributions.

$$\frac{1}{T_1} = \frac{1}{T_{1e}} + \frac{1}{T_{1\ DD(intra)}} + \frac{1}{T_{1\ DD(inter)}} + \frac{1}{T_{1\ SR}} + \frac{1}{T_{1\ SC}} + \frac{1}{T_{1\ CSA}}$$

In this equation T_{1e}^{-1} refers to the electronic interaction with a paramagnetic species,

$T_{1\ DD}^{-1}$ (intra) and $T_{1\ DD}^{-1}$ (inter) originate in intra- and inter-molecular nuclear dipole-dipole interactions, $T_{1\ SR}^{-1}$ is the spin rotation contribution, $T_{1\ SC}^{-1}$, the scalar coupling contribution, may result in principle, either from an exchange process of protons bound to nitrogen, or from the quadrupolar relaxation of a coupled nucleus of spin $I > 1/2$, such as deuterium, and $T_{1\ CSA}^{-1}$ is associated with the chemical shift anisotropy at the nitrogen site.

Estimating the various mechanisms which contribute to a given nitrogen relaxation rate is a necessary condition for calculating an effective correlation time. Indiscriminate transformation of T_1 values into τ_c values through the familiar Eq. (2.2) given below may lead to erroneous interpretations and is, at best, of poor conceptual interest.

All the mechanisms considered above have been shown to intervene in carbon-13 relaxation but they may take very different absolute and relative importance in nitrogen NMR (see Tables 2.1 to 2.4).

2.2 Electron Nuclear Relaxation and Influence of Paramagnetic Species

2.2.1 Effect of Paramagnetic Impurities

As nitrogen frequently constitutes a basic site which is likely to interact specifically with the metal site of paramagnetic species, the problem of the contamination by paramagnetic impurities is more crucial in nitrogen NMR than in carbon NMR. Thus, remaining traces of paramagnetic compounds have been shown (I 2) to invalidate the interpretation of the relaxation results in glycine (C 23) (L 36). In fact, the presence of paramagnetic impurities accelerates the longitudinal relaxation, and since it acts as a quencher for dipolar relaxation, it leads to a lowering of the Overhauser effect (see Sect. 2.5). Thus, for a sample of 2-aminoethanol which is characterized before distillation by a nitrogen relaxation time T_1 = 2.1 s, the relaxation becomes essentially dipolar after distillation and T_1 = 4.3 s. On addition of 10^{-4}M of Ni(acac)$_2$, T_1 falls to 0.065 s and η = 0 (L 21). Much higher concentrations (10^{-2} M) of the paramagnetic complex would be required to produce a similar effect in ^{13}C-NMR. In addition, paramagnetic impurities may be responsible for undesirable broadening of the nitrogen lines. Various methods for eliminating paramagnetic impurities are given in Sect. 3.2.

By contrast, the problem of degassing is less crucial. As no effect upon the T_1 values of $HOCH_2CH_2NH_2$ is observed by saturating the solution with pure oxygen it is concluded that the electronic contribution of O_2 to T_1 is higher than 30 s. Therefore, $T_{1e}(O_2)$ is probably higher than 150 s under the usual conditions (L 21).

2.2.2 Use of Relaxation Reagents

The electronic relaxation mechanism may be exploited to facilitate the observation of nitrogen resonances by selecting shiftless reagents which do not introduce too severe a broadening of the lines. Thus, the T_1-reagent $Cr(acac)_3$ (or $Fe(acac)_3$ which is somewhat more efficient but less stable), can be used in non-polar organic solvents to noticeably shorten the longitudinal relaxation times, especially for nitrogens devoid of bound protons. Being coordinatively saturated, $Cr(acac)_3$ acts predominantly through an outer sphere mechanism and produces rather indiscriminate effects. When aqueous solutions or polar solvents are concerned the cryptate $Gd(2:2:1)^{3+}$ has been claimed to exhibit convenient efficiency (G18). Thus,for an aqueous solution of formamide, T_1 = 36 s and 12 s respectively, in the absence and in the presence of cryptate. Therefore, the addition of relaxation reagents enables more efficient pulsing and observation of otherwise hardly detected nitrogens, such as those of 2,6-t-butylpyridines, can be obtained in more reasonable time (D 10).

It should be noted, however, that the addition of noticeable amounts of $Cr(acac)_3$ is likely to introduce non-negligible chemical shift perturbations. These effects occur not only through intermolecular interactions but also through magnetic susceptibility variations since an external reference is usually employed. When quantities up to 0.08 M are added in order to shorten the T_1 values significantly, chemical shift variations of about 0.5 to 1.5 ppm are frequently introduced.

Moreover, as such reagents act as quenchers for the nuclear Overhauser enhancement, the risk of cancelling the nitrogen signal is present (Sect. 2.5) and they should be used with caution.

Paramagnetic relaxation reagents capable of specifically interacting with a basic site of the investigated substrate can also be used as a help for structural elucidation. Since T_1 measurements are frequently prohibitively time-consuming in ^{15}N NMR, the effects of the relaxation reagents can alternatively be observed through line broadening effects and variations of the nuclear Overhauser enhancement factor. These effects are likely to be significant since, owing to the frequent occurence of a lone pair of electrons, nitrogen is a good candidate for specific interactions with a paramagnetic Lewis acid. In fact, the interpretation of line broadening effects in terms of distance parameters should be attempted with caution since it is based on the predominance of a pseudo-contact mechanism. Indeed, it has been shown that transition metal ions, such as Mn^{2+} for example, are the source of predominating contact interactions (M 31a). In favourable cases, the variations of the Overhauser enhancements can be exploited since they are directly related to the variations of the longitudinal relaxation times. On condition that the electron contribution T_{1e}^{-1} is the only competitor to the nuclear dipole-dipole mechanism, the Overhauser factor observed in the presence of a specific relaxation reagent is:

$$\eta_{obs} = \eta_{max} \left(\frac{T_{1e}}{T_{1D} + T_{1e}}\right) \quad (2.1)$$

It follows that: the more effective the electron contribution, that is to say, the shorter the distance to the interacting site, the more affected the signal intensity will be by the paramagnetic substance.

The gadolinium derivative $Gd(dpm)_3$ thus constitutes a good agent for structural elucidation. In acetaldoximes, for example, it alters to a greater extent the nitrogen T_1 relaxation times of the *syn* isomers which are sterically more accessible than the *anti* isomers (L 34). Similarly, the addition of $5 \cdot 10^{-5}$M of $Gd(dpm)_3$ to α-methyltryptamine (Table 2.2) cancels the signal of the basic NH_2 site through NOE quenching whereas the signal of the indole nitrogen is less affected (L 34).

2.3 Nitrogen Relaxation in Small Molecules

To the extent that molecular reorientation may be considered as approximatively isotropic the overall rotation is characterized by a single correlation time τ_c. For small - or medium - size molecules the reorientation is fast, τ_c is short, and the extreme narrowing condition, $\omega_o^2\tau_c^2 \ll 1$ holds. In this section we shall restrict the discussion to this situation for which the frequency dependent terms disappear in the general relaxation equations given in the next section. We are therefore concerned with the left-hand sides of the diagrams given in the Figs. 2.1 to 2.5.

a) Nuclear dipole-dipole interactions quite frequently provide the dominant mechanism for relaxation of ^{13}C nuclei. This contribution may also be of importance in nitrogen NMR even for nuclei devoid of bound protons. Thus, for a NH fragment in the extreme narrowing limit

$$\frac{1}{T_{1\ DD}} \simeq \left(\frac{\mu_o}{4\pi}\right)^2 \gamma_N^2\ \gamma_H^2\ \frac{\hbar^2}{r_{N-H}^6}\ \tau_c = A\tau_c = 5.85.\ 10^9\ \tau_c \qquad (2.2)$$

(assuming $r_{N-H} = 10^{-10}m$)

Indeed, since the magnetic moment of ^{15}N is much lower than that of ^{13}C, the dipolar relaxation is expected to be significantly less efficient for a ^{15}N-H fragment than for a ^{13}C-H fragment. This attenuation effect is somewhat compensated for by the fact that the bond length is slightly shorter for N-H than for C-H [see Eq. (2.2)] but, both factors being taken into account, the $T_{1\ DD}$ value remains about four to five times higher for ^{15}N - ^{1}H than for ^{13}C - ^{1}H (see Fig. 2.1). Consequently, when the ^{15}N and ^{13}C nuclei are bonded to N and N' protons respectively, in a given molecule, the fact that NT_1^{-1} (^{13}C) significantly differs from 4-5 $N'T_1^{-1}$ (^{15}N) can be used as a indication for - non purely dipolar mechanism - anisotropic molecular tumbling - internal rotation - or hydrogen bonding ... In this respect, it should be noted, however, that the T_1 values are very sensitive to the bond lengths which have to be known accurately in order to predict a theoretical ratio. Thus, by assuming different lengths of the N-H bond, relatively large differences between the factors A of Eq. (2.2) or between the $T_{1\ DD}$ values for $\tau_c = 10^{-11}s$ are calculated.

$r_{N-H} \times 10^{10}m$	0.96	0.98	1.0	1.02	1.04	1.06	1.08	1.10
$A \times 10^{-9}$	7.47	6.60	5.85	5.19	4.62	4.12	3.69	3.30
T_{1DD} s	13.4	15.1	17.1	19.2	21.6	24.2	27.1	30.3

Obviously, owing to the r_{N-H}^{-6} dependency, the intramolecular dipolar contribution becomes very small in the absence of directly bonded protons. In fact, other interactions can assist nitrogen relaxation when the intramolecular dipolar mechanism is very ineffective. Thus, the *intermolecular dipolar relaxation* which may generally be ignored for ^{13}C, is more likely to intervene for the less protected ^{15}N basic sites. In this respect, $T_{1\ DD(inter)}^{-1}$ has been shown to reach 85 % of the total dipolar contribution for the nitrogen of pyridine under certain conditions (S 22).

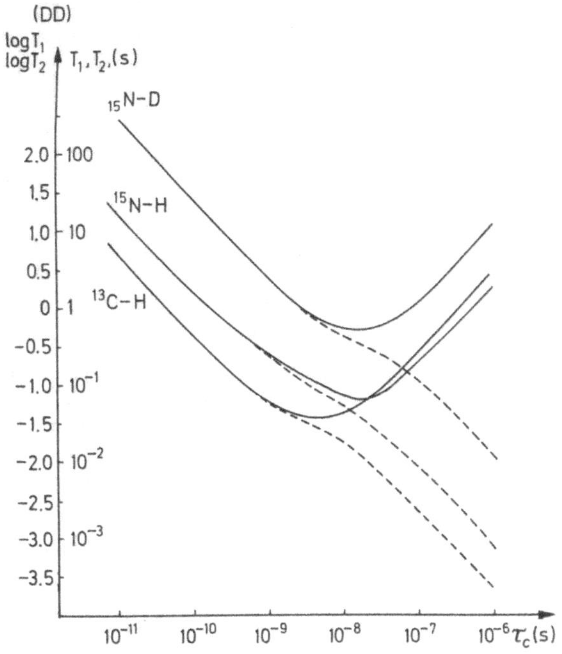

Fig. 2.1 Behaviour of the longitudinal (bold lines) and transverse (dotted lines) nuclear dipole-dipole relaxation times T_1(DD) and T_2(DD) for ^{15}N-H and ^{15}N-D bonds as compared to a ^{13}C-H bond (isotropic reorientation). The bond lengths are assumed to have the following values: $r_{N-H} = 1 \cdot 10^{-10}$m, $r_{C-H} = 1.09\ 10^{-10}$m. The relaxation times are very sensitive to a change in the bond length and the curves can be changed according to the values given in the table (p 10).

b) Spin-rotation interactions can also palliate to some extent inefficient dipolar relaxation. Thus, a shortening of the relaxation times of non-protonated nitrogens is sometimes observed when the temperature is increased. Such a behaviour, which contrasts with that of the dipolar contribution, [Eq. (2.2)] is consistent with the participation of spin-rotation relaxation. As an example, $T_{1\ SR}^{-1}$ (= $7.10^{-3}s^{-1}$) constitutes about 50% of the total relaxation rate for the nitrogens of *trans* azobenzene at 353 K (L 39). From the same point of view the nitrogen of n-butyl nitrite offers a remarkable example of a relaxation entirely governed by a spin-rotation mechanism resulting from internal rotation around the C-O bond (L 37) (Table 2.3).

c) Whereas active participation of the *chemical shift anisotropy* contribution is very infrequent in carbon spectroscopy this type of interaction competes more efficiently with the dipolar mechanism in ^{15}N-NMR. For a system with axial symmetry and assuming isotropic reorientation

$$\frac{1}{T_{1\,CSA}} = \frac{2}{15}\gamma_N^2 \Delta\sigma^2 B_o^2 \frac{\tau_c}{1+\omega_N^2\tau_c^2} \qquad (2.3)$$

where $\Delta\sigma$ represents the anisotropy of the screening tensor: $\Delta\sigma = \sigma_{||} - \sigma_{\perp}$.

In fact, the range of chemical shift anisotopies is about 1 to 4 times larger for nitrogen than for ^{13}C. As an example, $\Delta\sigma = \sigma_{zz} - 1/2(\sigma_{xx} + \sigma_{yy})$ reaches 672 ppm for nitrogen in pyridine whereas $\Delta\sigma$ is only 180 ppm for ^{13}C in benzene (S 22). As illustrated in Fig. 2.2, this mechanism, which depends on Bo^2, becomes much more active when the field strength is increased and it is likely to take more relative importance with the rising use of superconducting apparatus. For small molecules in moderate fields, determining contributions $T_{1\,CSA}^{-1}$, have been detected for nitrobenzene (S 23), pyridine (D 23) (S 22) and the cyanide ion (L 32) for example (Table 2.3). Some values of σ(N) anisotropy have been collected in Ref. (E 6).

Compound	C_5H_5N	N_2	HCN	N_2O	CH_3CN	CH_3NC	NO_3^-	NH_3
$\Delta\sigma$ (ppm)	672	657	577	512 369	452	360	210	39

d) As regards the *scalar relaxation mechanism*, it can originate either in the exchange of protons directly bound to nitrogen, or in the quadrupolar relaxation of deuterium, nitrogen-14 or any other B nucleus with spin I > 1/2 in a N-B fragment:

$$\frac{1}{T_{1\,SC}} = \frac{8\pi^2 J^2}{3} S(S+1) \frac{\tau}{1+(\omega_I-\omega_S)^2\tau^2} \qquad (2.4)$$

In this equation J is the ^{15}N-B coupling constant, S is the nuclear spin quantum number of B and τ is the correlation time for the scalar coupling mechanism. When relaxation is modulated by chemical ex-

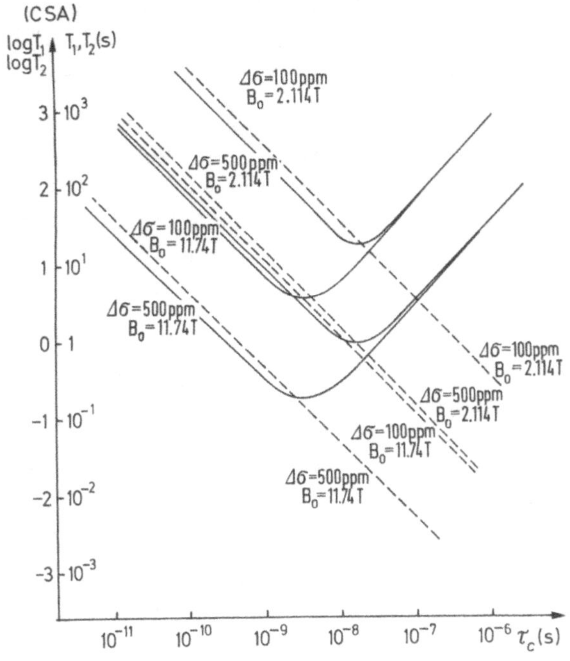

Fig. 2.2 Behaviour of the ^{15}N longitudinal (bold lines) and transverse (dotted lines) chemical shift anisotropy relaxation times T_1(CSA) and T_2(CSA) for two field strengths (2.114 and 11.74 T) and two values of the chemical shift anisotropy (100 and 500 ppm) (axial symmetry - isotropic reorientation)

change of B = ^{1}H, τ is the lifetime which governs the exchange process. When B is a rapidly relaxing quadrupolar nucleus, τ is the quadrupolar relaxation time T_{2a} of this species.

Actually, although such a mechanism is sometimes considered (L 56), it can be calculated that chemical exchange as well as ^{2}H or ^{14}N relaxation are usually ineffective in significantly accelerating longitudinal relaxation of ^{15}N. By contrast, contributions to T_2 can be observed, as will be discussed further in the next section.

e) Once the contributing mechanisms have been elucidated, nitrogen relaxation can be exploited in the same way as carbon relaxation, in order to derive *information about molecular dynamics*. In practice, it is generally convenient to examine the nitrogen and carbon relaxation parameters in conjunction with each other. Thus, when the dipolar nature of the relaxation parameters can be ascertained in both ^{13}C and ^{15}N-NMR, the T_1 values of carbons pertaining to the molecular skeleton allow an overall correlation time for isotropic reorientation to be determined. Measurement of the ^{15}N relaxation times then offers a means for investigating internal rotational mobilities of NH_2 or NH_3^+ groups. Such complementary studies give rise to a better insight into the dynamics of anilines (L 35) acetamides (G 16) or oxytocin (L 22) for example.

Investigation of nitrogen relaxation may also be of interest, in deriving information about anisotropic molecular motion. In fact the determination of a single τ_c value cannot be considered as a proof of isotropic motion, if independent internuclear vectors have not been examined. In this respect, results derived from N ~ H dipolar interactions may provide determining complementary information about eventual motional anisotropy.

f) *To conclude* it can be emphasized that, in numerous cases, inefficiency of the dipolar mechanism is hardly compensated for and renders the nitrogen *longitudinal relaxation rate much slower than the carbon relaxation rate*. Thus, as shown in Tables 2.1 to 2.3, T_1 values up to 20 s are common for NH or NH_2 groups and non-protonated nitrogen may be characterized by T_1 values higher than 50 s. Such slow relaxation adds a further difficulty to the observing of the already poorly sensitive ^{15}N resonances. In routine one pulse sequences, it frequently requires the use of small pulse angles and/or appropriate delays between the pulses to allow for convenient recovery of the magnetization. This situation also explains the extensive use of relaxation reagents (Sect. 3.2).

2.4 Nitrogen Relaxation in Large Molecules. The Isotopic Substitution Method

Nitrogen NMR acquires its structural dimension with the investigations of large molecules containing several nitrogen atoms, and much work has been done to improve this field.

Fortunately, the inconvenience resulting from the poor efficiency of the N-H dipolar relaxation is rapidly attenuated as the molecular reorientation becomes slower, i.e. as the molecular weight of the investigated compounds increases. As shown in Fig.2.3, *T_{1DD} rapidly decreases towards a minimum* as τ_c increases. This minimum is displaced towards longer correlation times for ^{15}N-H as compared to ^{13}C-H, but it appears at progressively shorter τ_c when the magnetic field increases. For example, when measured with a spectrometer of nominal frequency 180 MHz for proton, the relaxation times of NH and NH_2 groups in molecules such as lysozyme ($\tau_c \simeq 5 \cdot 10^{-9}$s) or protamines containing about 30 aminoacid residues are found in the sub-second range (G 17). It is this behaviour which permits fast signal accumulation of the free induction decays and therefore renders the observation of resonances in many biopolymers possible in spite of the low nitrogen content in each site. However, for molecules of very high molecular weight and long correlation times, such as nucleic acids, the relaxation times again increase with τ_c and the use of progressively higher magnetic fields becomes more and more unfavourable from this point of view (Fig. 2.3a).

Another severe limitation to the investigation of biopolymers is caused by the *dipolar broadening* which accompanies the increase in τ_c (Fig. 2.3b). Fortunately, owing to the lower value of γ_N with respect to γ_C, the ^{15}N signals are less liable to dipolar broadening than the ^{13}C signals (Fig. 2.1). Moreover, it should be noted that in a certain range of correlation times, the lines are expected to be sharper in superconducting spectrometers (Fig. 2.3b). However, the dipolar effect may be a severe impediment to the resolution of nitrogen lines of proteins, for example.

A good method for reducing the dipolar contribution to T_2^{-1} consists of *exchanging the hydrogen atoms of the NH and NH_2 groups with deuterium atoms.* Since the gyromagnetic ratio of ^2H is much smaller than that of ^1H, the dipolar contribution becomes much less efficient for N-D than for N-H (Fig. 2.1). As a result, the dipolar half-height line-width is expected to be 8 to 16 times smaller for ^{15}N-D. However, due to the existence of a quadrupolar moment for ^2H, contribution of the scalar coupling mechanism to the transverse relaxation rate has to be considered.

$$\frac{1}{T_{2\,SC}} = \frac{8\pi^2 J^2}{3} \left(T_{1Q} + \frac{T_{2Q}}{1 + (\omega_H - \omega_N)^2 T_{2Q}^2}\right) \qquad (2.5)$$

Noticeable contributions of this mechanism are in fact predicted (Fig. 2.4). However as long as a high field superconducting spectrometer is not used, the substitution of ^1H by ^2H in slowly tumbling molecules, such as proteins characterized by correlation times in a range $5 \cdot 10^{-9}$ to $5 \cdot 10^{-8}$s, remains advantageous.

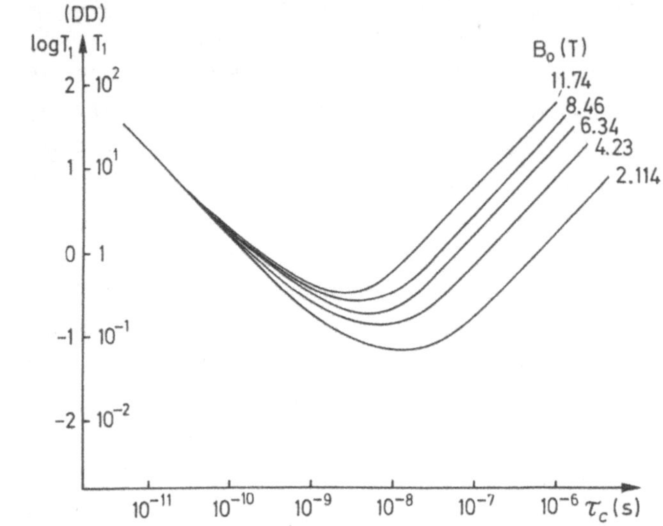

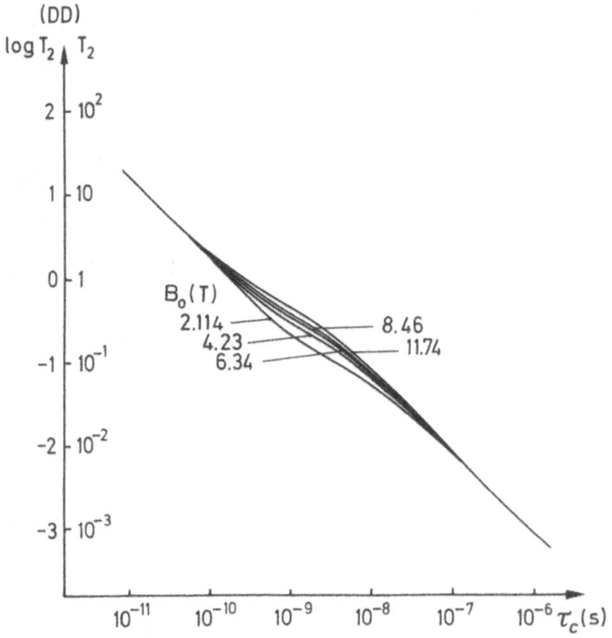

Fig. 2.3 Behaviour of the longitudinal (*a*) and transverse (*b*) nuclear dipole-dipole relaxation times T_1(DD) and T_2(DD) for different field strengths B_o (2.114, 4.23, 6.34, 8.46 and 11.74 T)

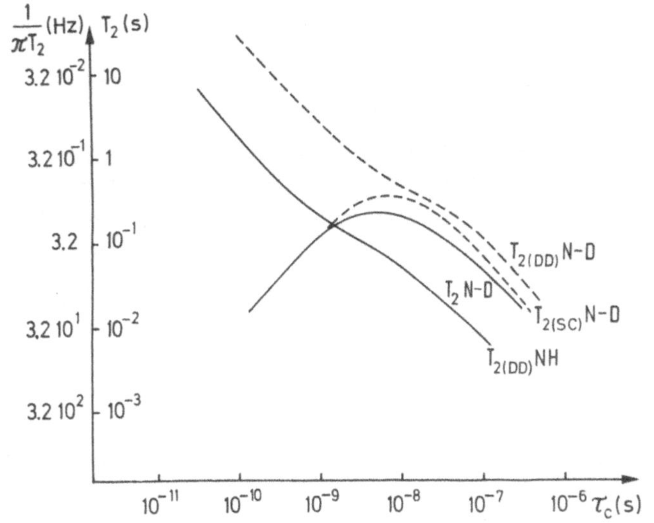

Fig. 2.4 Contributions of the dipole-dipole and scalar relaxation mechanisms to the overall transverse relaxation time T_2 of a N-D bond as compared to the corresponding relaxation time of N-H. B_0 = 2.114 T, J^{15}N-H = 90 Hz, J^{15}N-D = 13.82 Hz, e^2qQ/h = 0.2 MHz.

As illustrated in Figure 2.4., the sum of the dipolar and scalar coupling contributions to the nitrogen linewidth in N-D is then smaller than the dipolar contribution to the width of the nitrogen signal of N-H. Thus, whereas only three poorely resolved glycyl resonances are observed in the 9.12 MHz coupled spectrum of a H_2O solution of (^{15}N - Gly) hemoglobin, up to seven lines are clearly resolved in D_2O after exchange of the labile amide hydrogens with deuterium (L 19). Such an improvement in the resolution enables the investigation of, for example, the effect of denaturation upon the degree of random coil transformation of proteins.

2.5 The Nuclear Overhauser Effect

In nitrogen-15 NMR, advantage can be taken of the very favourable Overhauser factors, η. Indeed, in the extreme narrowing limit, the ratio of the signal intensities in the presence and in the absence of Overhauser effect, $I/I_0 = 1 + \eta$, is found to be between -3.93 and +1, since the maximum Overhauser factor η_{max} reaches the value $\gamma_H/2\gamma_N = -4.93$. The signal enhancement is therefore higher than in $^{13}C\ \{^1H\}$ NMR where I/I_0 varies between +1 and +2.99. The observed η value is determined by the proportion of dipolar relaxation in the total relaxation.

$$\eta_{obs} = -4.93 \frac{T_{1\ obs}}{T_{1\ DD}} \tag{2.6}$$

and it attains its maximum value only on condition that the mechanism is entirely dipolar in nature.

From this point of view, nitrogen-15 exhibits a peculiarity which is characteristic of nuclei with a negative gyromagnetic ratio and is frequently emphasized. Indeed, it is possible for the relative contribution of the dipolar mechanism to be reduced to about 20% and therefore to give values of $\eta = -1$ and $I/I_0 = 0$. The signal is then cancelled by the Overhauser effect. This phenomenon should be kept in mind, in particular, when using relaxation reagents, since the electronuclear interaction may compete with the nuclear dipolar effect to yield approximatively $\eta_{obs} = -1$ [Eq. (2.6)] (H 13).

Even when the relaxation is fully dipolar the *Overhauser effect decreases as a function of* τ_c, and the intensity conditions therefore become unfavourable for the study of large molecules. As seen in Fig. 2.5, η_{max} reaches the value -0.123 for long correlation times and I/I_0 cannot exceed + 0.88. The higher the magnetic field used, the earlier the loss of Overhauser enhancement, with respect to increasing τ_c. Thus, for molecules tumbling at about $\tau_c = 10^{-9}$s and observed at 50.67 MHz, the nitrogen signal is cancelled in a $^{15}N\sim\{^1H\}$ experiment. In fact, for a range of τ_c values comprised between about $3\cdot10^{-10}$s and $3\cdot10^{-9}$s, the use of superconducting spectrometers is unfavourable from the point of view of the Overhauser effect.

In practice, in order to overcome the disadvantages resulting from the cancelling or reduction of the signal through Overhauser effect, the nitrogen spectra must frequently be recorded either in an inverse gated decoupling mode (M 31c) or in an undecoupled mode. Moreover, care should be taken in selecting the decoupler off-time of the inverse gated sequence since efficient suppression of the Overhauser effects requires a waiting period well suited to the T_1 values of the considered nitrogens (K 25).

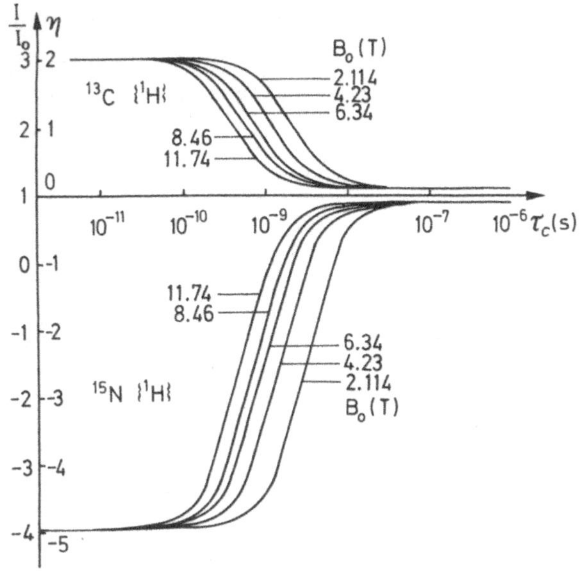

Fig. 2.5 Comparison of the NOE enhancement factor η and intensity ratio $\frac{I}{I_o} = 1+\eta$ for a ^{13}C-H and a ^{15}N-H bond at different field strengths B_o (2.114, 4.23, 6.34, 8.46 and 11.74 T).

Interestingly, the behaviour of the Overhauser factor can be exploited to provide information about the spectral assignment and molecular dynamics of complex biological species. Thus, the intensity variations displayed by the decoupled and coupled spectra of whole cells isotopically labelled is very illustrative (L 40). Whereas the resonances of the protein backbone amides are clearly observed as a broad signal at about 255 ppm in a proton non-decoupled spectrum measured at 9.22 MHz, the envelope of these resonances is nulled by proton broad-band decoupling. This result indicates either a reorientation rate of the cellular proteins corresponding to a τ_c value of about 5.10^{-9}s (Fig. 2.5) or an averaging out over inverted and non-inverted signals pertaining to faster and slower tumbling proteins. By contrast, segmental motion in side chains allows sharp inverted signals to be observed in proton-decoupled spectra (L 40). For very slowly tumbling molecules such as nuclei acids (τ_c >10^{-8}s) decoupling is expected to produce only a small signal attenuation ($I/I_o \simeq 0.9$) but saturation effects associated with increased T_1 values, and dipolar broadening (Fig. 2.1), are severe impediments. Similarly, whereas the NOE effect, $I/I_o \simeq 0.5$, measured at 18.25 MHz for the backbone amide resonances of lysozyme suggests an overall reorientation characterized by $\tau_c \geq 5.10^{-9}$s, the narrow inverted lines of the guanidino and the ammonium group indicate a strong negative NOE and therefore shorter correlation times resulting from a considerable degree of internal motion (G 17).

Table 2.1. Longitudinal relaxation times T_1 and nuclear Overhauser enhancement factor η of ^{15}N atoms in compounds containing an amino group. (See also Table 2.2)

ν_o spectrometer frequency for the resonance of ^{15}N

T temperature in Kelvin

The Overhauser factor η is defined as $(I-I_o)/I_o$ where I and I_o are the signal intensities in the presence and in the absence of Overhauser effect respectively.

$X-NH_2$ (*)		T_1 (s)	η (a)	Experimental Conditions (b) Solvent		ν_0 (MHz)	T(K)	Ref.
$CH_3(CH_2)_3NH_2$		70	-3.9	neat	(b_2)	27.4	299	L34
$C_6H_5NH_2$		13.0	-5.0	neat	(b_2)	27.4	299	L34
		22.8	-4.2	neat	(b_2)	27.4	313	L21
		31.5	-4.6	$CDCl_3$ 1M	(b_2) (c)	27.4	297	L35
$C_6H_5-NH_2$, CF_3COOH		1.6	-5.1	$CDCl_3$ 1M	(b_2) (c)	27.4	297	L35
1,2-$C_6H_4(COOH)NH_2$		0.57	-4.3	DMSO d_6 4M	(b_2) (c)	27.4	313	L35
	cation	0.39	-3.8	CF_3COO^-,DMSO d_6 4M	(b_2) (c)	27.4	313	L35
	anion	3.67	-4.8	Na^+ , D_2O 4M	(b_2) (c)	27.4	313	L35
1,3-$C_6H_4(COOH)NH_2$		0.68	-3.8	DMSO-DMSO d_6 4M	(b_2 (c)	27.4	313	L21-L35
	cation	0.43	-3.8	CF_3COO^-,DMSO d_6 4M	(b_2 (c)	27.4	313	L35
	anion	3.90	-4.1	Na^+, D_2O 4M	(b_2) (c)	27.4	313	L35
1,4-$C_6H_4(COOH)NH_2$		0.54	-3.1	DMSO d_6 4M	(b_2) (c)	27.4	313	L35
	cation	0.30	-3.7	CF_3COO^-,DMSO d_6 4M	(b_2) (c)	27.4	313	L35
	anion	3.54	-4.1	Na^+, D_2O 4M	(b_2) (c)	27.4	313	L35
1,4-$C_6H_4(NO_2)NH_2$		1.7	-4.8	DMSO-DMSO d_6 3M	(b_2) (c)	27.4	303	L21
$HOCO-CH_2NH_2$		12.7	-4.85	H_2O(EDTA)pH<-0.5	(b_4)	9.12	315	L36,C23
		1.1	-0.5 (O)	(d,e) pH 8.1	(b_4)	9.12	315	L36
$HOCO-CH_2NH-CO-CH_2\underline{N}H_2$		0.58 (h)	0 (h)	H_2O(resin) pH 5	(b_2)	9.12	(g)	I2
		4.6 (i)	-4.5 (i)		(b_2) (f)	9.12	(g)	I2

Table 2.1 (continued)

$X-NH_2$ (*)	T_1 (s)	η (a)	Experimental Conditions (b) Solvent		ν_o (MHz)	T (K)	Ref.
$HO-CH_2CH_2-NH_2$	4.3	-4.8	neat	(b_2) (b_4) (c)	27.4	303	L21
$HO-CH_2CH_2CH_2-NH_2$	3.6	-4.8	neat	(b_2) (b_4) (c)	27.4	303	L21
$HCONH_2$	14.5	-5.0	acetone d_6 90%	(b_4) (c)	27.4	303	L21
	21.0	-5.0	neat	(b_4)	18.24	305	L21
CH_3CONH_2	11.9	-5.1	H_2O 6M pH 5.4	(b_3) (c)	6.08	303	G16
	14.2	-5.1	5.3M	(b_3)	6.08	303	G16
	21.6	-3.6	0.6M pH 5.1	(b_3)	6.08	303	G16
$(H_2N)_2^+C-NH-(CH_2)_3-CH(COOH)-NH_3^+$	4.6		H_2O 3M pH 5.8	(b_3)	18.25	306	G17

(a) The relaxation mechanism is mainly dipolar (a_0) unless otherwise stated

(a.1) SR Spin Rotation

(a.2) CSA Chemical Shift Anisotropy

(a.3) SC Scalar Coupling

(b) Method used to measure T_1

(b.1) SR Saturation Recovery

(b.2) IRFT Inversion-Recovery Fourier Transform

(b.3) PSFT Progressive Saturation Fourier Transform

(b.4) DOE Dynamic Overhauser Effect

(b.5) ARP Adiabatic Rapid Passage

(c) The sample has been degassed

(d) Undegassed

(e) EDTA was claimed not to entirely remove paramagnetic impurities (I 2)

(f) Treated with resin for eliminating paramagnetic impurities

(g) Room temperature

(h) Paramagnetic impurities are present

(i) Paramagnetic impurities have been eliminated

(j) NH_2^+ guanidino

(k) NH guanidino

(l) 0.55 is for asparginine and cysteine and 0.60 is for leucine

(m) Probable contribution of paramagnetic impurities

(n) Both nitrogens averaged

(o) Not specified

(*) Light and heavy ammonia have also been studied (L33) at 6.07 MHz, using the SR method:

NH_3 T_1 = 186 s (302.6 K) DD-SR
117 s (213 K)

ND_3 T_1 = 413 s (302.6 K) SR
695 s (207 K)

Table 2.2. Longitudinal relaxation times T_1 and nuclear Overhauser enhancement factor η of ^{15}N atoms in molecules containing a >NH group.

For explanation of abreviations see caption of Table 2.1.

X-NH-Y	T_1 (s)	η (a)	Experimental Conditions (b) Solvent	ν_0 (MHz)	T(K)	Ref.
$H_2N-CH_2CO-\underline{NH}-CH_2$ COOH	5.26 (h)	-4.9 (h)	H_2O(resin) pH5 (b_3)	9.12	(g)	I2
	4.7 (i)	-4.7 (i)	(f) (b_3)	9.12	(g)	I2
$CH_3-NH-CHO$	23		(b_2)	27.4	299	L34
	40	-4.3	neat (b_2)	27.4	299	L34
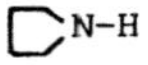	58	-4.6	neat (b_2)	27.4	299	L34
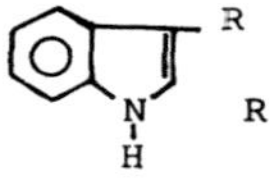 R = H	3.0	-4.7	DMSO 8M (b_2)	27.4	299	L34
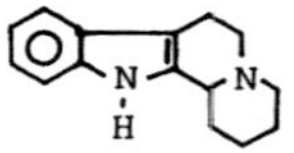	3.5	-5.0	$CDCl_3$ 1M (b_2,b_4)	18.2	301	L21
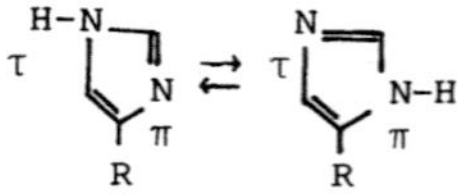						
R = H	7.6	-4.7	DMSO 8M (b_2)	27.4	299	L34
R = $CH_2-CH(NH_2)-COOH$	τ 9, π 13	τ -4.9, π -4.8	H_2O pH 1.4 (b_2)	9.12	311	B35
	τ 4.7, π 13	τ -2.8	pH 6.6 (b_2)	9.12	311	B35
	τ 7, π 25	τ -2.7, π -2.3	pH 7.8 (b_2)	9.12	311	B35
	τ 7.7, π 24	τ -2.0, π -1.8	pH 11.0 (b_2)	9.12	311	B35
$H_2N\overset{...}{-}C(=\overset{+}{N}H_2)-NH-(CH_2)_3-CH(\overset{+}{N}H_3)-COOH$	NH 4.5		H_2O 3M pH 5.8 (b_3)	18.25	306	G17
	NH_2^+ 2.8		H_2O 3M pH 5.8 (b_3)	18.25	306	G17

Table 2.3. Longitudinal relaxation times T_1 and nuclear Overhauser enhancement factor η of ^{15}N atoms in molecules containing a >N- group.

For explanation of abreviations see caption of Table 2.1.

X - N<Y,Z	T_1 (s)	η (a)	Experimental Conditions (b) Solvent			ν_0(MHz)	T(K)	Ref.
N_2	13.0	(a_1)	neat	(b_2,b_3)		3	77	I 7
	14.9	(a_1)	neat	(b_3)		4.4	77.3	K 27
	1.5	(a_1)	neat	(b_3)		4.4	126	K 27
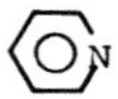	85	-0.4(a_2)	neat	(b_2)		27.4	299	L 34
	50	(a_o,a_1,a_2)	neat		(c)	30	290	S 22
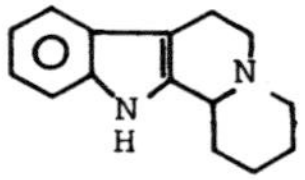	25	-5.0	$CDCl_3$ 1M	(b_2,b_4)		18.2	301	L 21
CH_3CN	90					27.4	299	L 34
	63.2						303	L 32
⊖CN	21.2	0(a_1,a_2)	D_2O 7.9 M	(b_3)		6.08	303	L 32
	14.5	0(a_1,a_2)		(b_3)		18.23	303	L 32
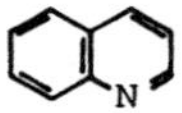	≃ 200	(a_o, a_1)	neat	(b_5)	(c)	6.08	310	L 38
$C_6H_5N{=}N(\rightarrow O){-}C_6H_5$	≃ 150	(a_o, a_1)	neat	(b_5)	(c)	6.08	310	L 38
$C_6H_5N(CH_3)_2$	≃ 300							L 32
quinuclidine	≃ 1300							L 32
$C_6H_5N{=}N{-}C_6H_5$ trans	56.1	(a_o,a_1,a_2)	$CDCl_3$ 2.7 M	(b_2)	(c)	9.12	301	L 39
	72.0	(a_o,a_1,a_2)		(b_2)		9.12	343	L 39

Table 2.3 (continued)

$X - N\begin{smallmatrix}Y\\Z\end{smallmatrix}$	T_1 (s)	η (a)	Experimental Conditions (b) Solvent			ν_o (MHz)	T(K)	Ref.
$_2(HOCH_2)_2C(CH_3)NO_2$	36.5	- 1.0	H_2O-D_2O 5M	(b_2,b_4)		27.4	303	L 21
$C_6H_5NO_2$	{400	(a, a_1)	neat	(b_5)	(c)	6.08	305	L 38
	{180	(a_1,a_2)	neat	(b_5)	(c)	14	310	S 23
$C_6D_5NO_2$	≃ 550	(a, a_1)	neat	(b_5)	(c)	6.08	310	L 38
$1,4-C_6H_4(NH_2)NO_2$	48.0	$0(a_1,a_2)$	DMSO-DMSO d_6 3M	(b_2)	(f)	27.4	303	L 21
$CH_3(CH_2)_3-O-NO$	{24.4	(a_1)	neat	(b_2)	(c)		303	L 39
	{12.1		neat	(b_2)	(c)	9.12	328	L 37
$CH_3CH=NOH$ syn 62 %	54	-2.5 (m)	neat	(b_2)		27.4	301	L 20
anti 38 %	35	-2.5 (m)		(b_2)				L 20
$CH_3CH_2CH=NOH$ syn 44 %	25	-1.4 (m)	neat	(b_2)		27.4	301	L 20
anti 56 %	39	-2.3 (m)		(b_2)				L 20
$(CH_3)_2CH-CH=NOH$ anti 75 %	26	-1.9 (m)	neat	(b_2)		27.4	301	L 20
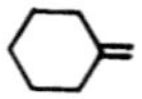 NOH	46	-2.7 (m)	neat	(b_2)		27.4	301	L 20
$CH_3CH_2-CH(CH_3)-C(CH_3)=NOH$	18.8	-2.9	neat	(b_2)	(c)	27.4	303	L 21
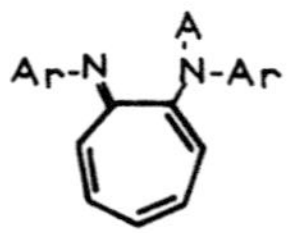 A = H	4.17	-4.35	$CDBr_3$ 0.3M	(o)		(o)	299	J 11
D	24.8							

Table 2.4. Longitudinal relaxation times T_1 and nuclear Overhauser enhancement factor η of ^{15}N atoms in large biological molecules.

For explanation of abreviations see caption of Table 2.1.

Large Biological Molecules	T_1 (s)	η (a)	Experimental Conditions (b) Solvent	ν_0 (MHz)	T(K)	Ref.
Lysozyme (guanidino NH_2^+)	0.33		H_2O 9.4 m M pH 5.0 (0.1 M citrate Buffer)(b_3)	18.25	306	G 17
Protamine from herring (clupeine sulfate)	0.47 (j) 0.86 (k)		H_2O 7 g/18 ml pH 0.67 (b_3)	18.25	308	G 17
Oxytocin	0.55 0.60 (l)	-4.93	H_2O 60 mg/ml pH 4.0 (b_3)	9.12		L 22
Tetraphenylporphyrin dication	1.9	-5.1	$CHCl_3$ 0.016 M + CF_3COOH 5/1 (b_3)	18.25	311	G 8
Tetraphenylporphyrin Zn	56	≃ -1 (m)	$CHCl_3$ $5.6.10^{-3}$M (b_3)	18.25	311	G 8
6-Deoxy-1,2:3,5-di-O-isopropylidene-6-phthalimido-α-D-glucopyranose	31 37 71		$CDCl_3/C_6F_6$ 9/1 0.85 M 0.85 M (b_2) (c) 0.28 M (b_2)	9.12 9.12 9.12		C 19 C 19 C 19
6-Deoxy-1,2:3,4-di-O-isopropylidene-6-phthalimido-α-D-galactopyranose	20 20	-4.4	$CDCl_3/C_6F_6$ 9/1 (b_2) (b_2) (c)	9.12		C 19 C 19
Alumichrome gly	0.58 to 0.64	-4.7 to -4.9	DMSOd^6 0.07 M (b_2)	10.13	318	L 48
Drn	0.49 to 0.57	-4.7 to -5.0	DMSOd^6 0.07 M (b_2)	10.13	318	L 48

3 Experimental Techniques in ^{15}N Spectroscopy

3.1 Experimental Methods for Recording ^{15}N Spectra

3.1.1 Indirect Detection: CW INDOR Spectra and FT Double Resonance Experiments

Although the vast majority of ^{15}N spectra is now directly recorded by pulse FT spectroscopy, indirect detection may still be of interest. Indeed, a lot of ^{15}N parameters have been obtained in the past through CW double resonance experiments. In this area, the various techniques of $^{1}H\{^{15}N\}$ double irradiation can be exploited and information about the ^{15}N resonances is then obtained via proton responses to pertubations applied, more or less selectively, to the ^{15}N transitions (M 31d). The INDOR method, in particular, is well suited to indirect detection of the ^{15}N spectrum (M 28) (M 33). In this method a proton line corresponding to a given $^{15}N{\sim}H$ scalar coupling is continuously monitored while the double irradiation field $\vec{B_2}$ is swept with an amplitude $\gamma B_2 \simeq \Delta\upsilon_{1/2}(Hz)$ through the ^{15}N transitions. Responses are then obtained in the ^{1}H spectrum each time that the double irradiation frequency encounters a ^{15}N transition connected with the considered proton line. This technique allows a determination of the ^{15}N coupled spectrum which benefits from the sensitivity of the corresponding proton resonances. However, it requires the existence of detectable $^{15}N{\sim}H$ coupling constants and lacks general applicability. Moreover, owing to the difficulty of detecting $^{1}\underline{H}{\sim}^{15}N$ satellites in natural ^{15}N abundance, the experiments are performed on ^{15}N enriched samples.

$^{1}H\{^{15}N\}$ double resonance can also be exploited in the FT mode either in a selective or in a non-selective way. Thus, selective double resonance methods, analogous to those commonly used in the CW mode, can be applied to FT detection and assignment of nitrogen resonances in compounds isotopically enriched with nitrogen-15 (R 11). For example, collapse of a given multiplet is obtained in the FT proton spectrum when the double irradiation field is set, with a magnitude appropriate to selective N~H decoupling, just at the resonance frequency of the corresponding coupled nitrogen. The other $^{1}H{\sim}^{15}N$ splittings are then reduced according to a proximity of their ^{15}N frequencies to that of the decoupled ^{15}N multiplet. This off-resonance technique (M 31e) allows proton and ^{15}N spectra to be interconnected and has been successfully applied to the spectral assignment of a polypeptide structure, for example (L 16).

A difference method can also be combined with double irradiation to obtain the $^{15}N{\sim}H$ coupling constants of unenriched samples by means of the the proton spectrum (L 16)(M 4). Actually, easy observation of the $^{1}H{\sim}^{15}N$ satellite spectrum is precluded by the unfavourable dynamic range resulting from the low natural abundance of ^{15}N with respect to ^{14}N. However, subtraction of the intense proton spectrum corresponding to the molecules which contain ^{14}N can be achieved by successively accumulating pairs of free induction decays (FID) in the following way. In each pair, a first FID is acquired while the double irradiation field is applied to the ^{15}N spectrum and this FID is coadded to the computer memory. Then a second FID, which is acquired while $\vec{B}_2$ is set well outside the range of ^{15}N frequencies, is co-subtrated from the memory. In the difference spectrum thus obtained, the proton lines associated with the ^{14}N species are, in principle, eliminated and the resulting pattern depends on the characteristics of the double irradiation field. If broad-band decoupling is achieved, both the coupled $^{1}H{\sim}^{15}N$ multiplets and the corresponding decoupled proton lines, which are inverted with respect to the undecoupled spectrum, are simultaneously observed. Alternatively, with a moderately strong CW $\vec{B}_2$ field, off resonance effects, similar to those discussed above, can be obtained which allow the ^{1}H and ^{15}N spectra to be connected. Altough short-term drift of the spectrometer field is somewhat compensated for by the alternate procedure employed, the method requires good instrumental stability for the strong resonances associated to the ^{14}N species to be efficiently cancelled.

3.1.2 The Conventional Single Pulse Method

The spectra of ^{15}N in natural abundance are now routinely obtained by the classical single pulse method in the presence of broad-band proton decoupling. In principle all the variants developed in ^{13}C spectroscopy are applicable to ^{15}N NMR. Thus undecoupled spectra benefiting from Overhauser enhancement (M 31f) or partially decoupled spectra resulting from off-resonance irradiation (M 31e) can be obtained. However, such experiments are in most cases drastically compromised by the low sensitivity of 15-nitrogen and even the recording of completely decoupled spectra requires much attention in order to avoid wasting too much time or undue failure of the experiment.

As already mentioned, proper adjustment of the pulse interval and pulse angle, and working out of the decoupling field are critically conditioned by the relaxation behaviour and Overhauser effects, which are themselves controlled by the size and the structural properties of the molecule, by the nature of the medium, by the temperature and by the spectrometer frequency. It is thus of prime importance to bear in mind all the properties discussed in the section devoted to relaxation and especially the possible cancelling of the ^{15}N signal resulting from unfavourable NOE, when starting a pulse sequence. Indeed, as the relaxation properties exhibit considerable differences among the samples, one is easily convinced that general experimental parameters cannot be proposed. Some typical experimental conditions are given in Table 3.1. In cases where both the longitudinal and transverse relaxation times are long, it may be advantageous to allow for accumulation of transverse components of magnetization in a steady state regime (M 31g). In spite of the high T_1, T_2 values, relatively short pulse intervals and pulse angles larger than the Ernst value are then selected. Noticeable intensity improvement may be achieved in such conditions but this procedure is likely to produce phase anomalies which cannot be corrected for. An absorption-like shape is nevertheless obtainable by recording the magnitude spectrum (M 31h).

Finally, it should be noticed that full power noise decoupling may have an important influence on the sample temperature. Depending on the probe configuration, an increase in temperature of 10 to 20K can be observed when a 5W power of the B_2 field is applied in the probe.

3.1.3 Methods of Polarization Transfer

A number of methods can be used to enhance the signals of nuclei with low gyromagnetic ratios through polarization transfer. These methods exploit the nuclear spin polarization of protons which are characterized by much larger Boltzmann population differences than nitrogen. (See also atoms (B 71)).

Selective population inversion experiments

The sensitivity limitations inherent in the observation of proton decoupled ^{15}N signals become more significant if undecoupled spectra are desired. The method of selective population inversion (M 31i), which is capable of noticeably enhancing the line intensities of coupled spectra, therefore seems very attractive. In this method a π pulse is applied selectively to a given component (or an appropriate sub-multiplet) of the satellite proton spectrum pertaining to the molecules which contain the ^{15}N isotope. Just after this selective inversion, a non selective pulse is applied in order to observe the ^{15}N spectrum (Fig. 3.1A)

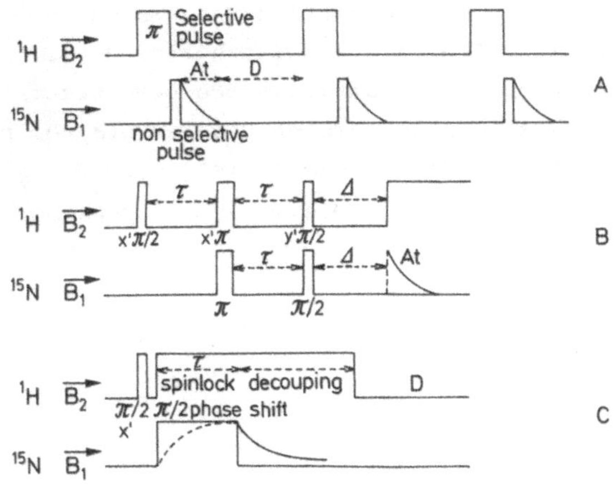

Fig. 3.1. Characteristics of the principal methods used to increase the signal to noise ratio in ^{15}N spectroscopy.

A Basic sequence for an SPI experiment (M 31i). At is the acquisition time. D, the delay for magnetization recovery, is conditioned by the proton relaxation times.

B Pulse sequence for the INEPT method (M 38). For an ^{15}N-H doublet $\tau = 1/4\ J_{NH}$ and $\Delta = 1/2\ J_{NH}$. For an ^{15}N-H_2 triplet $\tau = 1/4\ J_{NH}$ and $\Delta = 1/4\ J_{NH}$. In order to obtain an undecoupled spectrum the $\vec{B}_2$ field is switched off during data acquisition.

C Diagram of a J-cross polarization experiment (B 57), τ is the cross polarization time.

Due to population inversion of the energy levels involved in the perturbed lines, intensity changes are produced in the connected ^{15}N transitions. These intensity variations depend on the ratio of the gyromagnetic ratios γ_H/γ_N, which is especially favourable in the $^{15}N\{^1H\}$ case since it reaches a value of about 10 (J 9) (M 31j). Thus, for an ^{15}N~H fragment, uncoupled with the rest of the molecule, inversion of one component of the satellite proton doublet leads in principle to regressively and progressively connected ^{15}N components with intensities of -9 and +11. By subtraction of the normal undecoupled ^{15}N spectrum a difference spectrum with transition intensities -10 and +10 is expected. Still higher sensitivity enhancement accompanies the irradiation of a degenerate ^1H transition. As an example, the new population distribution, brought about by inversion of a compoment of the proton doublet of

$NH_4^+ Cl^-$, leads to a ^{15}N pattern with intensities -39 -76 +6 +84 +41 instead of the normal 1 4 6 4 1 multiplet (M 31j). Moreover, the ^{15}N spectrum may now be acquired with a pulse repetition rate which is determined by the proton relaxation times instead of the frequently much longer nitrogen relaxation times. In addition, decoupled spectra benefiting from still higher intensity enhancements can be obtained by appropriately gating the double irradiation field (H 19) (C 29). The SPI method is useful for determining the relative signs of coupling constants as illustrated in Fig. 3.2 for compound $C_6H_5CH_2NHCHO$ (T 8).

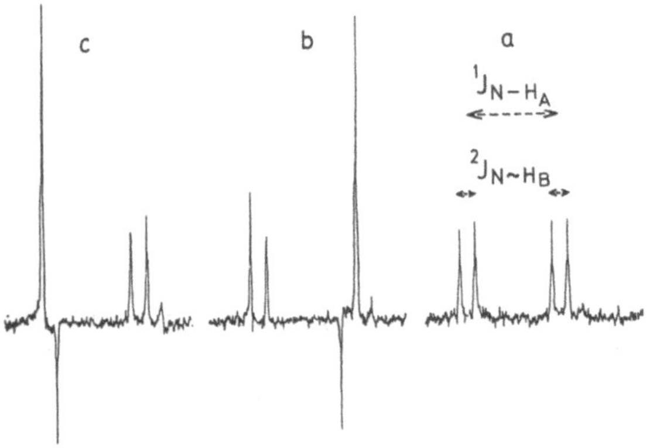

Fig. 3.2 ^{15}N spectra of the amide $C_6H_5CH_2NH\text{-}CHO$
(C) (A) (B)

obtained in selective population inversion experiments. Spectra a, b and c are obtained in the same experimental conditions except for the selective irradiation. In spite of overlapping between the proton spectra of H_A and H_B, experiments run at several frequencies allow the four lines of the satellite spectrum of H_B for example, to be precisely localized. These components are denoted a b c d when going from high frequencies towards low frequencies.

a) normal undecoupled spectrum: $|J_{NH_A}|$ = 93 Hz, $|J_{NH_B}|$ = 15 Hz, $|J_{NH_C}| \simeq$ 2 Hz
b) the frequency component b of the 2×2 multiplets pertaining to the H_B satellite spectrum associated with the species containing the ^{15}N isotope has been inverted.
c) as in b) but the frequency component c has been inverted.

The parameters used in the SPI experiments were: At = 3.4s, pulse repetition time = 3.9s, lenght of the $\pi(B_2)$ pulse = 0.5s

These experiments indicate that J_{NH_A} and $J_{H_AH_B}$ have different signs. The same type of experiment performed on one of the 2×3×2 satellite transitions of H_A indicates that J_{NH_B} and $J_{H_AH_B}$ have different signs.

Similarly, a negative sign of $^{2}J^{15}N-^{19}F$ in 2-fluoropyridine has been determined by this technique (J 1). However, the usefulness of selective population transfer experiments for routinely obtaining coupled ^{15}N spectra is rather limited. A first requirement consists in precisely positioning the double irradiation frequency upon usually unobservable satellite proton transitions associated with the nitrogen-15 isotope. Unless the J_{N-H} coupling constants have already been obtained by recording a normal coupled ^{15}N spectrum (!) it is necessary to estimate, a priori, these coupling constants. In a number of cases, the tables given in Sect. 7 may probably be used rather successfully to this aim. However, a second requirement, which concerns the necessity of the existence of $^{15}N{\sim}H$ coupling constants large enough for selective inversion to be performed easily, is even more stringent. Indeed, intensity enhancements are more specially desired for non-protonated nitrogens which are frequently characterized by long relaxation times, but, unfortunately, the $^{15}N{\sim}^{1}H$ coupling constants across more than one bond are often relatively small, as discussed in Sect. 7, and this situation renders the selective inversion method impracticable in a number of cases.

3.1.3.1 The INEPT Pulse Sequence

In contrast to the SPI sequence, the INEPT method (insensitive nuclei enhanced by polarisation transfer) (M 38) (M 39) is capable of inverting simultaneously more than one proton transition. Consider a two spin system, N-H subjected to the pulse sequence shown in Fig. 3.1B. The first two pulses excite an ^{15}N modulated spin echo at time 2τ for the proton spins (M 31e). As a π pulse has been applied to the nitrogen spectrum at time τ, the slow and fast components of the proton doublet have been interchanged and have in fact continued to diverge. If $\tau = 1/4\ J_{NH}$, they are found in the +x' and -x' directions at time 2τ, whereas the chemical shift effects are refocused. A $\pi/2\ \vec{B}_2$ pulse, applied in the y' direction then bring the proton components into the +z and -z directions. One transition of the proton doublet is hence inverted and leads to polarisation transfer by a factor $\gamma_H/\gamma_N \approx 10$ into the ^{15}N system. A $\pi/2(x')$ pulse applied at the ^{15}N resonance frequency then allows the ^{15}N doublet to appear as enhanced positive and negative lines. Alternatively, if a delay $\Delta = 1/2\ J_{NH}$ is left before applying proton decoupling while simultaneously acquiring the ^{15}N spectrum (Fig. 3.1B), an enhanced proton decoupled spectrum may be obtained. A delay $\Delta = 1/4\ J_{NH}$ should be inserted if triplets of NH_2 groups are concerned. Apart from the sensitivity enhancement brought about by polarization transfer, a further gain in intensity may result from the possibility of repeating the pulse sequence at a rate governed by the relatively short proton relaxation times instead of the often long nitrogen relaxation times.

Since the range of one-bond N-H coupling constants is relatively small (see Tables 7.1 to 7.35), the INEPT sequence performed for a given τ value is capable of rather conveniently inverting ^{15}N satellite lines of all protons directly coupled to nitrogen and good intensity enhancements are then obtained for all the protonated nitrogens. The method is also applicable to the observation of nitrogens which exhibit only long-range couplings, provided that these couplings are resolvable, but it then requires long τ delays and is less appropriate (M 39).

3.1.3.2 J-Cross-Polarization Experiments

The technique of cross-polarization is familiar to the users of solid state NMR (H 20) (P 15). It allows the strong spin polarization of protons to be transferred to low abundance nuclei such as ^{13}C and involves the dipolar interactions. This method can be applied to the liquid state but the scalar coupling J is now responsible for the polarization transfer. As proton polarization is about ten times higher than ^{15}N polarization, a ten-fold intensity gain may be anticipated, all things being identical otherwise. The sequence for J-cross-polarization is shown in Fig. 3.1C. The proton magnetization vectors are spinlocked (M 31k) along the y' ($\vec{B}_2$) direction by means of $\pi/2$ $\vec{B}_2$ pulse followed by a $\pi/2$ phase shift of the $\vec{B}_2$ field. The proton magnetic moments then precess about the $\vec{B}_2$(y') direction at an angular frequency $\gamma_H B_2$. Simultaneously to proton spin locking, the $\vec{B}_1$ field is applied for a time τ to the nitrogen magnetic moments which then precess at the angular frequency $\gamma_N B_1$. If the HARTMANN-HAHN condition

$$\gamma_H B_2 = \gamma_N B_1 \qquad (3.1)$$

is satisfied, polarization can be transferred from protons to nitrogens (B 57) and the enhanced nitrogen free induction decay is acquired following switching off of the $\vec{B}_1$ field. If $\vec{B}_2$ is maintained during signal aquisition, an enhanced decoupled spectrum is obtained. In addition to the sensitivity improvement which results from the favourable ratio γ_H/γ_N, further gain can be obtained by taking advantage of the fact that the pulse repetition time is now conditioned by the proton relaxation times rather than by the usually longer nitrogen relaxation times. Thus, in the case of pyridine, an overall time advantage of about 350 is predicted in favour of a J-cross polarization experiment as compared to the conventional acquisition technique (B 57). It should be noted, however, that the method suffers from several limitations. In fact, the signal amplitudes are modulated by the J_{N-H} frequencies and, for a two spin system N~H for example, the nitrogen polarization takes the form:

$$M_N = M_H \sin^2 \pi J \tau / 2 \qquad (3.2)$$

Thus, for a given nitrogen the full polarization transfer can be reached only to the extent that the cross polarization time τ is given the appropriate value. Moreover the HARTMANN-HAHN condition [Eq. (3.1)] is less easily fullfilled in the liquid state than for solids since the tolerable mismatch is now of the order of the scalar coupling constants which are at least two orders of magnitude smaller than the dipolar ones. The homogeneity of the rf fields is also a critical condition and in practice probe coils doubly tuned for ^{15}N and protons have been especially constructed. (See also (C 28)).

Finally, it should be noted that the modulation of signal amplitudes by frequencies related to the scalar coupling constants can be exploited to measure the $J_{N\sim H}$ values (B 65) (M 40) (M 41). Alternatively, if the ^{15}N spectrum is observed without decoupling for an appropriate cross-polarization time τ, the spin coupling constants can be directly measured provided that an additional π/2 pulse is conveniently applied at the end of the cross-polarization sequence in order to remove the phase anomalies (C 25).

3.2. Sample Preparation

In general, a careful choice of the sampling conditions is certainly of great importance for the quality of the NMR results but, although recent advances in technology have rendered ^{15}N spectra more easily available, this problem of sample preparation remains crucial in ^{15}N spectroscopy at the natural abundance level.

Among the different experimental procedures which are routinely used in NMR spectroscopy, attention will be focussed on some specific topics concerning the NMR cell, the purification of the sample, the chemical reagents and the labelling experiments. We shall discuss in Sects. 4 and 5, two other points of practical importance, the referencing of ^{15}N shifts and solvent effects on ^{15}N spectra.

3.2.1 The NMR Cell

As discussed below in Sect. 4, the procedure which is the most widely adopted for locking the spectrometer and referencing ^{15}N shifts makes use of a co-axial cell assembly which involves a "capillary" mounted inside the main cell. The "capillary", which contains both the reference and the lock substance, has an outer diameter, Ø of 3, 4, or 5 mm depending on the size of the main cell which is filled with the sample. In order to obtain optimum sensitivity, probes capable of accomodating large cells have been developed: the outer diameters of the cells usually range from 10 to 25 mm and for a given concentration, the maximum sensitivity is obviously obtained with the largest cell. However, it

Ø (mm)	10	12	15	20	25
vol (ml)	2	2.5	3.5 to 4	8 to 12	17 to 22

should be kept in mind that such cells make great demands on sample quantity. Moreover, large diameter cells lead to the formation of intense vortices which must be reduced by using Teflon plugs in order to force the sample into the coil space. As their axes are drilled, these plugs may act as apacers for the co-axial "capillary" which contains the reference and the lock substances. Another advantage of using large diameter sample tubes lies in the fact that the filling factors of such cells can be easily kept to a maximum value (M 31m).

3.2.2 Purification of the Sample

For purposes of structure elucidation, it is generally not necessary to eliminate paramagnetic impurities since most spectra are obtained in the proton decoupled mode which is usually the source of signal broadening (M 31n). However, filtration of the sample solution may be advisable to avoid the field inhomogeneities arising from the fluctuation of small solid particles in the liquid.

When relaxation times, NOE enhancement, η, or exchange mechanisms are to be investigated, it is then strongly recommended (see Sect. 2.2) to eliminate paramagnetic impurities, namely paramagnetic metal ions, which are often found in solutions of nitrogen-containing molecules and may be very tightly bound. Thus, the anomalous pH dependence of the NOE η factor of glycine, which was attributed to scalar and spin-rotation relaxation (C 23, L 36), arises, as a matter of fact, from contamination by paramagnetic impurities (I 2). In this respect, atomic absorption spectroscopy may be used for detecting the presence of small amounts of metal ions before attempting purification. Addition of EDTA appears to be inefficient in fully removing transition metals which are strongly bound to the basic nitrogen atoms and Chelex 100 chelating resin was claimed (I 2) to be a more convenient method. Thus 0.25 ml of activated Chelex 100 (B 68) with NH_4OH, H_2O, and HCl, H_2O is shaken in a plastic vial with 1 ml of 1 M solution of the nitrogen-containing molecule. Selectivity of Chelex 100 decreases in the series: $Cu^{2+} \gg Fe^{3+} > Cr^{3+} > Ni^{2+} > Co^{2+} > Mn^{2+} \gg Na^{+}$.

Finally degassing under a vacuum may be useful to eliminate the small quantities of paramagnetic oxygen dissolved in the sample; the freeze-pump-thaw cycle technique appears to give the best results for eliminating oxygen in purified organic solutions.

3.2.3 Chemically Assisted ^{15}N Spectroscopy

Since the relaxation times of non-protonated nitrogens or of ^{15}N nuclei pertaining to small molecules may be very long (see Sect. 2.3) the *"relaxation reagents"* (M 31 o), which are capable of considerably shortening the T_1 values without introducing impedient broadening, are very useful in ^{15}N spectroscopy. The chromium tris-acetylacetonate complex, $Cr(arac)_3$, is the most widely used reagent and concentrations of up to 0.05-0.1 M generally give good results. However as this reagent acts as a quencher for the nuclear Overhauser effect (Sect. 2), a careful adjustment of the decoupler duty cycle is necessary to avoid eventual cancelling of the ^{15}N signal. Moreover, it should be recalled that shift variation may be as high as 1.5 ppm when 0.1 M concentrations are used, depending on the structure of the nitrogen-containing molecule. It is therefore recommended not to add T_1 reagent to the sample when accurate changes in ^{15}N values are to be detected. Other organic ligands, such as dipivaloylmethane, were used to make Cr^{3+} ion soluble in some organic solvents, but the efficiency of the quencher is reduced. Table 3.2 gives illustrative examples of the efficiency of Cr^{3+} and Gd^{3+} bound to acac and dpm (L 34). Moreover, it must be kept in mind that Gd chelates appear to induce selective decrease in the T_1 values of basic and unhindered nitrogen atoms.

As anticipated from the results of ^{14}N spectroscopy (W 27), shift reagents such as $Eu(dpm)_3$ or $Yb(dpm)_3$ (M 31p) may induce important shift variations since nitrogen is often able to directly coordinate to paramagnetic reagents which act as Lewis acids. Shifts even higher than 500-1500 ppm, and characteristic of the type of nitrogen concerned, are expected. The first results obtained in ^{15}N spectroscopy show that the induced shifts take smaller values when preferential complexation occurs at the nucleophilic O atom of diazine mono-N-oxides for example (S 25). Several measurements, taken in the concentration range of 0 - 0.1 M of the $Yb(fod)_3$ reagent, enable the extrapolation to 1:1 molar ratios of reagent and substrate to be conveniently performed. Thus for pyrimidine N-oxide, bound shifts of +31.6 ppm and -4.7 ppm have been measured for the N-oxide nitrogen and the azine nitrogen respectively. Finally, it should be noticed that shift reagents may also act as more or less selective quenchers for the nuclear Overhauser effect and this property provides additional structural information (S 25).

3.2.4 ^{15}N Labelling Experiments

Although progress in the technology of superconducting solenoids has made possible the recording of natural abundance ^{15}N spectra of solutions containing ≃ 0.1 atom l^{-1}, ^{15}N data concerning for example i) chemical shifts of tertiary nitrogen atoms in diluted solutions or in large molecules of biological interest, ii) relaxation times and iv) chemical mechanisms are often more readily available from labelling experiments. Many ^{15}N-enriched simple molecules are commercially available from most of the fine chemicals or isotope dealers: Merck Sharp and Dohme (Canada), Spectrometric Spin et Techniques, Commissariat à l'Energie Atomique (France), Isocommerz (East Germany), MSD, Merck (West Germany), Junta de Energia Nuclear (Spain), Fluka (Swiss), the Norrell Chemical Co, Prochem. (United Kingdom), Aldrich, Bio-Rad Lab., Intern. Chem. Nuclear Corp., Stohler Isotope Chemicals (U.S.A.) ... More sophisticated ^{15}N labelled molecules are easily synthetized from cheap raw materials such as ammonia, ammonium salts, nitric acid and nitrates, nitrites, phthalimide, using well documented procedures. Various references, in which methods for synthesis of ^{15}N labelled molecules can be found, are summarized in Table 3.3. It is also possible to obtain biosynthetic ^{15}N enriched molecules from simple ^{15}N labelled compounds (nitrates, nitrites, aminoacids ...). Thus, adenosine (K 4) Chlorophyll

(B 32) Nybomycin (N 1) Oxytocin (B 38, L 22) Valinomycin (S 5), Alumichrome (L 16) and whole cell proteins (L 40) have been biosynthetized. The list is certainly not exhaustive and other interesting ^{15}N labelling experiments are described in special literature, namely in the Journal of Labelled Compounds.

Table 3.1. Practical conditions (sample preparation and spectrometer settings) used for obtaining ^{15}N spectra of typical compounds at the natural abundance level of the ^{15}N isotope.

Sampling condition:

T_1R: $Cr(acac)_3$ r.t.: room temperature

Spectrometer settings: all spectrometers were operated under quadrature (or single side-band) detection, PW pulse width expressed in term of flip angle

τ: pulse repetition time

decoupler: ND noise decoupling, CS coupled spectrum, IGD inverse gate decoupling

cell: outer diameters of the cell containing the sample and of the coaxial cell containing the lock and reference substances

NS: number of scans

t: duration of an experiment.

COMPOUNDS	Sampling Conditions		Spectrometer Settings							Ref.
	Solvent Concentration	T(K)	ν_0(MHz)	PW(°)	τ(s)	Decoupler	Cell(mm)	NS x 10^3	t(h)	
Amines NH_2	C_6H_{12} 20% mol	r.t	18.25	25°	2 to 5	ND	25/5	0.5 to 1	0.6 to 1.4	D12
>N-	C_6H_{12} 20% mol	r.t	18.25	25°	10 to 30	ND	25/5	0.5 to 1	1.4 to 8.3	D12
Amino-acids	H_2O 0.1M	r.t	9.12		0.4	ND	10/5	4	1.1	S5
Amino-sugars	H_2O 25%	308	18.25	20	2	ND	25/5	1.8 to 3.8	1 to 2	B45
Enamines	C_6H_{12} 20% mol	300	18.25	20	4.5	ND	25/5	2	2.5	W16
Amides (-NHCO)	$CHCl_3$ 40% mol	r.t	18.25	20	4.5	ND	25/4	2	2.5	W17
	neat	301	9.117	30	1.4	ND	15/4	1 to 5	0.4 to 2	D18
Isocyanates		303	18.25	45	30	IGD	25/5			Y11
Isothiocyanates	neat 0.08MT_1R	301	9.117	45	2	ND	15/4			M20
Ureas NH_2-CO-	DMF 4M	r.t	10.09	20 to 25	2.0	ND	10/3	2 to 2.5	1.1 to 1.4	S16
	H_2O EtOH 3M	301	9.117	30	1.0	ND	15/4	2	0.6	M9
>N-CO-	DMF 4M + 1510^{-3}gT_1R	r.t	10.09	20 to 25	2.0	ND	10/3	20	11	S16
	neat 0.08MT_1R	301	9.117	45	2.0	ND	15/4	10 to 15	6 to 9	M9

Azines (ring nitrogen)	TFA 0.5M	313	10.1	50	10	ND	20/	2	5.5	S25
	neat T_1R	r.t	10.09	30	2	ND	10/3	4.5	2.5	D16
	various 14% mol	301	18.25	70	100	IGD	25/5	0.1 to 0.4	2.8 to 11	D17
Imines, oximes	$CHCl_3$ 20 to 36% mol	298	18.25	70	30	IGD	25/5	0.2	4.1	W19
	$CHCl_3$ 50% $^v/v$ T_1R	301	9.117	40	1	ND	15/5	20 to 80	6 to 22	N5
Nucleosides	H_2O	r.t	9.12	30	0.8	ND	10			H10
	H_2O 0.5 to 1M	r.t	18.25	30	2 to 5	ND	25/5	4 to 6	3 to 6	M21
	H_2O	r.t	18.25	90	> 10	IGD	25/5			H10
Pterins	H_2O 0.7M	r.t	10.09	40	4.4	ND	20/	0.8	1(e)	S15
Polypeptides	TFA 1.4 g in 7 ml solv.	306	9.12	50	0.5	ND	20/	30	4.2	K19
	TFA 4 g in 20 ml solv.	313	18.24	30	1.0	ND	25/5	30 to 50	8 to 14	K19
	TFA 1 g in 5 ml solv.	308	36.48	50	1.2	ND	15/	3.5	1.2	K19
	DMSO 5 g in 25 ml solv.	313	18.24	45	7.7	IGD	25/5	25 to 40	53 to 85	K25
Lysozyme (a)	H_2O 9 10^{-3}M pH 3.9	r.t	18.25	90	0.82	ND	25/5	93	21.2	G17
tRNA (b)	H_2O 11 10^{-3}M pH 5.6	303	18.25	90	0.41	ND	25/5	302	34.4	G17
	H_2O 11 10^{-3}M pH 5.4	353	18.25	90	0.82	ND	25/5	82	18.7	G17
Haemoglobin (c)	H_2O 95% enriched ^{15}N	r.t	9.12	90	0.7	CS	10	120	23.3	L19
Whole cells (d)	H_2O 95% enriched ^{15}N	300	9.12	90	0.7	ND	10	30	5.8	L40

(a) From hen egg-white (natural abundance ^{15}N spectra)

(b) From yeast (natural abundance ^{15}N spectra)

(c) From 95% $[^{15}N\text{-Gly}]$ enriched

(d) From 95% enriched materials (escherichia coli, baker yeast, ...)

(e) For protonated N atoms.

Table 3.2. Electron-nuclear relaxation times in amines measured after addition of T_1 reagent. (Values normalized to a viscosity of 1cP and expressed in cPs units (L 34))

	$Cr(acac)_3$	$Cr(dpm)_3$	$Gd(acac)_3$	$Gd(dmp)_3$	No T_1R
$C(T_1R)$ (M)	0.05	0.05	5.10^{-4}	5.10^{-4}	0
Pyrrolidine	2.7	5.4	6.9	3.2	58 (a)
Pyridine	3.7	5.2	7.4	8.6	85 (a)
Pyrrole	1.8	4.1	66	38	40 (a)

(a) Diamagnetic T_1 values

Table 3.3. List of ^{15}N enriched compounds obtained from organic syntheses. The syntheses are described or quoted in the references.

Compounds	Ref.	Compounds	Ref.
Amines (saturated)	B24,M22,M42 R6	imines	B24,F3
Amines (acetylenic)	B22	oximes (ethers)	B13,B17,C10,D1 F3
Amino acids (a)	B24,F5,L6 M42	nitriles,cyanamide	B24,H3,M42
Amino sugars	C19	thiocyanates	R19,M42
Amino alcohols	M42		
Enamines	A2	hydrazines	L28,M42
Anilines	A14,A21,N1,M42	oxazines	R9
Pyrroles	G2	nitrosamines	S3
Pyridines	O 11	nitro derivatives,nitrates	B24,M42
Imidazoles, pyrazoles Pyrazolinones	A27,L24,W26 M42,H4	azides	M10,M11,M42 M24,W26
		nitramines,nitrourethans	S1,I1
Tetrazoles	T7,W26		
Thiazoles	M42	formazanes	F9
Pyrimidines,pyrimidinones	R4,R20		
Indoles	M42	nitrites,nitroso derivatives	S1,L39
Quinolines	A21,K10,M42		
		diazo derivatives	A12,B24,L2,L39 M42,P7
Quinolines N-oxides	T2		
		azoxybenzenes	S9
Quinolines,quinazolones	W5,N1		
		triazenes	A11,M42
Benzopurines,purines	M42,W5,Y2		
		triazines	B14,R19
Enamino ketones	K6		
		aminophosphines	C13
Amides	B24,D5,M22,M42 O11,R6,S4 N1	phosphazenes,phosphazides	M10,T4,T5
Ureas,thioureas	B24,B66,E2,M42 R19,R20,S6	silylamines	C13
		porphyrins	D22,K14a,M12
Dipeptides	K15		

(a) Syntheses of an exhaustive series of amino acids are described in M42.

4 Reference for ^{15}N Chemical Shifts

Chemical shifts are relative parameters which can be measured with a high degree of reproductibility providing that the same reference substance is used in the same experimental conditions. Unfortunately, this goal was not achieved in the last decade, which saw the sudden rise of ^{15}N spectroscopy, since at least thirteen different molecules in a variety of solutions were used as standards for ^{15}N chemical shifts. Indeed, this is not really surprising if we bear in mind that about fifteen years were required for the TMS δ-scale to be universally accepted in ^{1}H spectroscopy!

According to the IUPAC recommendations (I 8) the δ parameter is defined by Eq. 4.1

$$\delta_i = \sigma_{ref} - \sigma_i = \frac{B_i - B_{ref}}{B_0} = \frac{\nu_i - \nu_{ref}}{\nu_0} \tag{4.1}$$

and a smaller screening constant σ_i with respect to the reference corresponds to a positive δ value

low fields positive values of δ		high fields negative values of δ
δ ←		
high frequencies paramagnetic shifts	0(reference)	low frequencies diamagnetic shifts

4.1 External and Internal Locking and Referencing

In fact as we must consider both problems of locking the field-frequency ratio and referencing the chemical shifts, a multitude of situations arise from the combination of various sample configurations:

		Lock	Reference
External		The lock substance is not in the same RF coil as the sample	The reference substance is in the same coil but not in the same cell as the sample
Internal	a)	The lock substance is in the same coil but not in the same cell as the sample	The reference and the sample are in the same coil and in the same cell
	b)	The lock substance and the sample are in the same cell	

Owing to its minor importance in ^{15}N spectroscopy, the external lock configuration will not be considered here. It should be emphasized that, in this case, the sample and the lock substance are not influenced by the same field. As a result the choice of ν_O in Eq. (4.1) to compute the δ values is critical.

As regards the internal lock configuration, two situations should be examined. *Internal referencing*, which is almost exclusively used in proton and carbon NMR, suffers from a number of disadvantages in ^{15}N spectroscopy. First of all, intermolecular effects on $\delta^{15}N$ are very important (see Sect. 5) due to the presence of a lone pair on nitrogen and to the fact that nitrogen-containing bonds more often exhibit a high polarity. Nitromethane, which is soluble in most organic solvents, (S), shows a solvent dependence which extends over a ~10 ppm range when using reasonably diluted solutions (0.3 M CH_3NO_2 in S) (W 22);

S	$\delta^{14}N$	S	$\delta^{14}N$	S	$\delta^{14}N$
DMSO	2.0	$(CH_3)_2CO$	-0.8	CH_2Br_2	-3.2
H_2O	2.0	dioxane	-1.8	$CHCl_3$	-3.8
DMF	0.7	CH_3OH	-2.0	$(Et)_2O$	-3.9
none	0	C_2H_5OH	-2.7	C_6H_6	-4.4
CH_3CN	-0.2	CH_2Cl_2	-3.2	CCl_4	-7.1

^{15}N measurements of more concentrated samples (25 % v/v)show the same trend, but the solvent effect range is limited to ≃ 4 ppm (S 24). Inorganic ions, such as nitrate or ammonium ions, are also extremely sensitive to medium effects, and are sparingly soluble in organic solvents. The second main disadvantage of internal referencing is the problem of sensitivity, or, which comes to the same thing, the cost. In order to get the optimum signal-to-noise ratio for the sample, in a given period of time, it is necessary to keep the volume of reference -and lock-substance to a minimum. Practically, this condition calls for the use of a ^{15}N-enriched reference, and it seems unrealistic to waste money or time in throwing away or recovering labelled material.

As a consequence, the simplest and more economical way of suiting both requirements of chemical inertness and sensitivity is to mix the enriched reference and the lock substance - usually a deuterated molecule - in a thin coaxial cell, fitted in the tube containing the sample (Fig. 4.1).

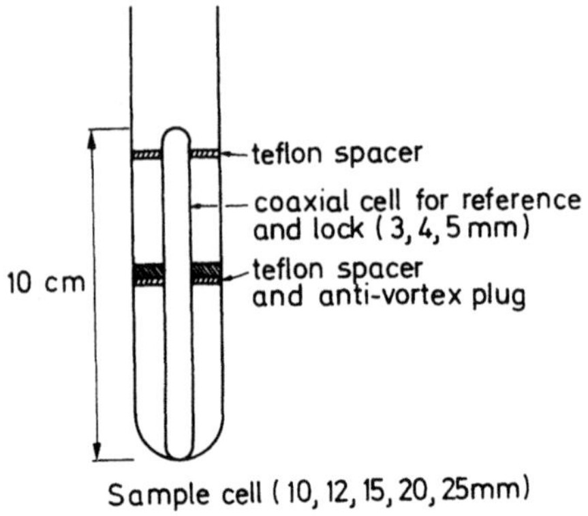

Fig. 4.1. Coaxial assembly in an NMR cell with an external referencing-internal lock configuration

Such an arrangement constitutes an *external reference and internal lock* configuration and now introduces the problem of differential shielding, since the field experienced by the reference and the sample depends on a shape factor (Sf) and on the difference in magnetic susceptibility ($\Delta\chi$) between the reference and the sample (M 31b). Equations have been theoretically derived which enable the chemical shift to be corrected for the susceptibility effect. The correction factor depends on the disposition of the sample with respect to the magnetic field direction. For studies involving coaxial cylinders placed in a permanent or electromagnet, the sample is perpendicular to the field direction and the intrinsic chemical shift, δ_{int}, is given by:

$$\delta_{int} - \delta_{obs}^{\perp} = -\frac{\Delta\chi}{6} \tag{4.2}$$

For a superconducting solenoid, the sample is parallel to the field direction and

$$\delta_{int} - \delta^{\parallel}_{obs} = + \frac{\Delta\chi}{3} \qquad (4.3)$$

Equations (4.2) and (4.3) are given in the SI system of units, and consequently, caution should be exercised when using tables of magnetic susceptibilities. (The conversion factor between SI and CGS χ units is 4π, i.e. $\chi_{SI} = 4\pi\chi_{CGS}$.)

The χ values (in SI units) of the principal organic solvents are given in Table 5.1. Interestingly, nitromethane has the lowest χ value among the main organic solvents, and the $\Delta\chi$ differences are therefore usually positive (Table 4.1). If we consider the largest $\Delta\chi$ value, i.e. $\chi_{CH_3NO_2}-\chi_{CH_2Br_2}$, the corrections deduced from Eqs. (4.2) and (4.3) are -1.1 and +2.2 ppm respectively. Thus, it should be noted that working with a high field superconducting system induces a shift correction which is twice as high as that produced by an electromagnet. Moreover, the corrections are opposite in sign and we think that this consideration may explain a number of discrepancies observed in the literature (see Tables of chemical shifts, Sect. 6).

In principle, a refined procedure developed to eliminate the susceptibiliy contribution is based on the use of two spectrometers. The combination of Eqs. (4.2) and (4.3) gives Eq. (4.4).

$$\delta_{int} = \frac{i}{3} (\delta^{\parallel}_{obs} + 2\delta^{\perp}_{obs}) \qquad (4.4)$$

When a series of measurements is performed at a given field configuration, say 5.87 T, with samples of nearly equal magnetic susceptibilities containing the same external reference, it is advisable in principle to repeat one measurement at another field configuration, say 2.11 T, in order to appreciate the correction factor, and then to correct the other values of the series. Note that the correction factor does not depend on the nucleus considered and can be measured in proton spectroscopy. Unfortunately, the problem of the shape factor is very critical and it seems difficult to obtain reproducible results when the cells are changed.

4.2 The Various References Used in ^{15}N spectroscopy. Conversion to the External Nitromethane Scale

When examining the literature data on nitrogen shifts, the chemist is faced with the same difficulties that arose in the sixties in ^{1}H spectroscopy. Owing to the great number of combinations involving standard substance, solvent, pH, concentration, direct comparison of data is frequently made unpracticable. Tables 4.1 and 4.2 give the different substances used to reference ^{15}N chemical shifts. Two kinds of standards emerge which correspond to *low frequency* and *high frequency* references. The high frequency (or low field) references include mainly nitromethane and the nitrate ion (Table 4.1). The resonances of both CH_3NO_2 and NO_3^- in Na^+ or NH_4^+ salts depend on the solvent and on the gegenion (see Sect. 4.1). The situation is worse for the nitric acid, since the concentration dependence of $\delta^{15}N$ spreads over a 25 ppm range. The low frequency references are shown in Table 4.2 where it is also apparent that solvent effects are important for the ammonium ion. Moreover, the ammonium salts suffer from the disadvantage of yielding spectral patterns which depend on the rates of the proton exchange reactions. In this respect, a multiplet resonance structure may be observed when the reference is dissolved in heavy water. The tetraalkylammonium salts (B 63) do not present such drawbacks but they are not easily accessible to the ^{15}N enriched level. A solution of ^{15}N enriched nitromethane in trideuterio nitromethane appears to present most of the qualities required for a good external standard (L 45). It gives a strong signal, is readily available at the enriched ^{15}N or ^{2}H level and has a resonance which does not interfere with the majority of other ^{15}N signals. In practice, a solution of 0.01 ml of $^{15}NO_2\ CH_3$ (95 % enriched) in 0.39 ml of CD_3NO_2, introduced in an inner coaxial 4 mm cell, gives a signal which falls in the same dynamic range as neat $(CH_3)_2NCHO$ at the natural abundance level in a 15 mm cell. According to the IUPAC recommendations (I 8), the ^{15}N chemical shifts of nitrogen containing compounds will appear with a minus or plus sign depending on whether they are shielded or deshielded with respect to nitromethane. This situation is not unique in NMR spectroscopy, since ^{31}P chemical shifts are also referenced with respect to a standard, 85 % H_3PO_4, the resonance of which lies in the middle of the $\delta^{31}P$ range. Thus, the literature results will be converted to the nitromethane scale and presented with respect to this primary standard. Since it may be anticipated that, in the future, ^{15}N measurements will be referenced with respect to CH_3NO_2 , the values quoted in the Tables of this book will be directly comparable.

Unfortunately, the conversion of the published data to the CH_3NO_2 scale is not straightforward for a number of papers since the authors have not described clearly the composition of the used standard (solvent and concentration). Tables 4.1 and 4.2 (column 4) provide a series of conversion factors which can be used to convert $\delta^{15}N$ values referenced with respect to a given standard to the nitromethane scale. Thus, for NH_4Cl (5.6 M in H_2O, saturated)

$$\delta Ni/_{CH_3NO_2} = \delta Ni/_{NH_4Cl} - 352.9 \text{ ppm}$$

In some particular cases, the conversion factors used to correct $\delta^{15}N$ values in Tables 6.1 to 6.61 are different from those quoted in Tables 4.2 and 4.3. Therefore, to avoid confusion, the factors used are always specified in the captions.
In order to facilitate the comparison of $\delta^{15}N$ values referenced to a given secondary standard such as NH_4Cl, with data presented with respect to another secondary standard, say NO_3NH_4 (saturated in H_2O), we have given the relevant conversion factors in columns 6 of Tables 4.1 and 4.2. Also listed in these Tables are the values of the magnetic susceptibilities of the main standards taken from Ref. W 22. However, it should be emphasized that, owing to the imperfect shape of most coaxial cells, the corrections computed using Eqs. (4.2) and (4.3) give an estimate rather than an accurate determination of susceptibility effects.

4.3 Universal Scalės of Nitrogen Chemical Shifts

An alternative to the use of standard substances is to measure the absolute frequency given by the computer in FT spectroscopy. If samples of nearly equal susceptibilities are investigated (same solvent), the computer procedure enables us to compare the chemical shifts satisfactorily, provided that the same offset value of the RF pulse and the same lock system are used. The sensitivity is then enhanced since the space which would have been taken up by the coaxial cell can now be filled with sample.

An attractive possibility is also offered by multinuclear spectroscopy. Taking advantage of the technical facilities of modern FT spectrometers, the nuclei of TMS can be used as internal standards for ^{19}F, ^{31}P... and ^{15}N resonances (B 39). Since the ^{13}C frequency lies in the middle of the frequency range of heteronuclei, the ^{13}C signal of TMS may indeed constitute a universal internal reference.

It has also been proposed to refer the resonance of a given nitrogen atom to that of proton assuming that TMS resonates at exactly 100 MHz (M 33). A detailed procedure of such a method is given in reference H 16. Thus, nitromethane resonates at 10 136 719 Hz (0.1M $CDCl_3$) and 10 136 783 Hz (0.1M DMSO-d_6) (B 69); since $(CH_3)_4 N^+I^-$ (in DMSO-d_6) is observed at 10 133 351 Hz (B5) the difference between both signals is calculated to be -3432 Hz (for DMSO solutions) which correspond to -338.7 ppm (The chemical shift of $(CH_3)_4 N^+I^-$ in H_2O, is -337.3 with respect to neat CH_3NO_2, Table 4.2).

4.4 The Absolute Nitrogen Shieldings

From a theoretical point of view, it is interesting to gain a good knowledge of the absolute nitrogen shielding. Unfortunately, direct experimental determination is not actually available and the total screening σN must be obtained as the sum of two contributions σ_{para} and σ_{dia} (W 23). σ_{para} is usually computed (F 11, K 28) from spin-rotation constants, which have been obtained, for NH_3 gas for example, from molecular beam maser measurements (σ_p = -91 ppm). σ_{dia} , can be calculated in the ab initio formalism (H 15) and was found, for NH_3 gas, to be equal to ~ +355 ppm. It is deduced that $\sigma_N(NH_3$ gas) equals about +264 ppm (M 32) and this value can be converted to the different scales of ^{15}N chemical shifts. Since $\delta^{15}N$ (NH_3 liq.) was found to be 18 ppm high-frequency from NH_3 gas and $\delta(NH_4Cl, H_2O)$ 26 ppm high frequency from NH_3 liq. (A 5, A 23), the absolute shielding of CH_3NO_2 is computed to be equal to ≃ -135 ppm and that of NH_4Cl, H_2O ≃ +220 ppm. However, as discussed in reference W 24, these numbers give a trend rather than precise information about the absolute nitrogen shielding. Other absolute screening values have also been compiled in reference A 24, W 24.

$\sigma(N_2)$ = -102 ppm, $\sigma(HCN)$ = -37 ppm, $\sigma(ClCN)$ = -306 ppm (W 24)
$\sigma(CH_3NO_2)$ = -141 ppm (A 24)
$\Delta(\sigma$ absolute): 366 ppm (N_2) 301 (HCN) 570 (ClCN) (computed values)
$\Delta(\sigma$ absolute): 314 ppm (N_2) 225 (HCN) 240 (ClCN) (observed values)
with $\Delta(\sigma \text{ absolute}) = |\sigma_{absolute}(\text{compound}) - \sigma_{absolute}(NH_3)|$

Table 4.1. High frequency references for ^{15}N chemical shifts

Reference	Conditions	χ (10^6) SI Unit (b)	Conversion (a) Factor to CH_3NO_2 Scale	Conversion Factor (a) to Other Scales		Ref.
$CH_3\underline{N}O_2$	neat	- 4.91	0	Absolute scale	~135	M 32
			0			W 22
	0.3 M in C_6H_6		-4.4			W 22
	0.3 M in DMSO		+2.0			W 22
	0.3 M in $CHCl_3$		-3.8			W 22
$H\underline{N}O_3$	1 M in D_2O	-8.98	-6.2	$N(Me)_4Cl$(2 M, H_2O)	332.8	Y9, D12
				$\underline{N}H_4NO_3$(2M, H_2O)	355.0	Y9, D12
				H_2NCONH_2 (2M, H_2O)	298.7	Y9, D12
			-4.4			W 22
	10 M in H_2O			$H_3\overset{+}{N}CH_2COOH$	346.1	B39,C23
			-6.1	$N(Me)_4Cl$(12 M, D_2O)	330.6	W 21
				$N(Me)_4I$ (0.3 M, H_2O)	331.7	W 21
				NH_4Cl (H_2O)	348.7	W 21
			-18.2			W 22
	9 M in H_2O		-14.4			S 24
	15 M in H_2O	-7.77	-31.3			W 22
$Na\underline{N}O_3$	1 M in D_2O			$C_6H_5NH_2$ neat	320.5	M 10
				$\underline{N}H_4NO_3$ sat.1M H^+	355.3	M 10
				$NO_3\underline{N}H_4$ sat.1 M H^+	0.25	M 10

Table 4.1 (continued)

Reference	Conditions	χ (10^6) SI Unit (b)	Conversion (a) Factor to CH_3NO_2 Scale	Conversion Factor (a) to Other Scales		Ref.
	2 M in D_2O, 1M HNO_3		- 1.5	NH_4Cl (1 M H_2O, 1M H^+)	353.3	M 20d
				NH_2CHO (neat)	266.1	M 20d
				H_2N-CO-NH_2 (2M, H_2O)	301.4	M 20d
				CH_3NO_2 + 0.08 M $Cr(acac)_3$	-0.8	M 20d
	2 M in D_2O, 2 M HNO_3		- 3.1			M 20d
	7.9 M in H_2O (sat.)	-9.16	- 3.7	NH_4Cl (sat. in H_2O)	349.2	W 22
				$N(Me)_4Cl$ (sat. in H_2O)	333.0	W 22
$NH_4\underline{N}O_3$	5 M in 2 M HNO_3, D_2O			NH_4Cl (H_2O)	356.2	J 7
	5 M in 2 M HNO_3		- 4.6	$\underline{N}H_4NO_3$ (5M in 2M HNO_3)	354.4	W 22
					354.1	B 58
	5 M in 2M HCl		- 5.2	$\underline{N}H_4NO_3$ (5M in 2M HCl)	352.8	W 22
	12.3 M in H_2O (sat.)	-9.07	- 4.0	NH_4Cl (sat. in H_2O)	348.9	W 22
				$N(Me)_4Cl$ (sat. in H_2O)	332.7	W 22
K $\underline{N}O_3$	0.3 M in H_2O		- 3.5			W 22
K $\underline{N}O_2$	in H_2O		+237.1	NH_4Cl (sat. H_2O)	590.7	F 1
Na$\underline{N}O_2$	0.3 M in H_2O		+227.6	NH_4Cl (5.6 M in H_2O)	580.5	W 22

(a) To convert a given ^{15}N resonance δ_i, measured with respect to the reference i, to the CH_3NO_2 scale or to a secondary scale, algebraically add the conversion factor to the δ_i value: $\delta^{15}N(CH_3NO_2) = \delta_i$ + conversion factor.

(b) The susceptibility values of the references are given in SI units: $\chi_i(SI) = 4\pi\chi_i(CGS)$

Table 4.2. Low frequency references for ^{15}N chemical shifts

Reference	Conditions	(10^6) SI Unit	Conversion Factor to CH_3NO_2 Scale (a)	Conversion Factor to Other Scales (a)		Ref.
NH_3	gas			absolute scale -264		M 32
	liquid		-380.2			S 24
			-381.9			W 22
			-379.6(b)			S 24
NH_4Cl	5.6 M in H_2O (sat.)	-9.66	-352.9	$N(Me)_4Cl$ (6MH_2O)	- 16.2	W 22
				NH_4NO_3 (Sat.H_2O)	6.6	W 22
				NO_3NH_4 (Sat.H_2O)	-348.9	W 22
				NO_3Na (Sat.H_2O)	-349.2	W 22
	5.6 M in 1 M HCl,H_2O		-353.0(d)			C21,C1
			-355.3			L 41
	5 M in 2 M HCl, H_2O		-352.5			W 22
	2.9 M in 1M HCl,H_2O		-351.8_5(c)			S 14
	2 M in 1 M HCl, H_2O		-355.3			S24,L41
	8 mol % in H_2O,H^+		-358 (d)	$N(Et)_4Cl$ (1M H_2O)	- 42.0	C1,C21
$N(Me)_4Cl$	6.03 M in H_2O (Sat.)		-336.7	NH_4Cl (Sat.H_2O)	16.2	W 22
				NH_4Cl (H_2O)	18.0	L21,W21
				NH_4NO_3(5 M H_2O)	22.3	W 22
	12 M in H_2O		-336.7	NH_4Cl (H_2O)	18.1	W 21
$N(Et)_4Cl$	in H_2O		-316(d)			C 1
	4.6 M (Sat.)		-315.8			W 22
NH_4NO_3	in HNO_3 H_2O		-358.6			S 24
	in 2M HNO_3, H_2O			NO_3 NH_4 (2M H_2O)	-354.1	B 58
	5 M in 2 M HCl		-358.0	NO_3 NH_4 (5M H_2O)	-352.8	W 22
	5 M in 2 M HNO_3		-359.0	NO_3 NH_4 (5M H_2O)	-354.4	W 22
				HNO_3 (1M H_2O)	-355.0	D 12
	12.3 M in H_2O (Sat.)	-9.07	-359.5			W 22
	15 M in 2N HCl		-357.4			B 46
$N(Me)_4I$	0.3 M in H_2O		-337.3	NH_4Cl(5M in 2M HCl)	15.2	W 22
	0.3 M in H_2O		-337.8	NH_4Cl (H_2O)	17.0	W 21
				$N(Me)_4Cl$(12 M H_2O)	1.1	

(a) See caption of Table 4.1

(b) CH_3NO_2 in $CDCl_3$ (1/1) containing $Cr(acac)_3$ at 298 K

(c) CH_3NO_2 in C_6D_6

(d) CH_3NO_2 + 0.03 g $Cr(acac)_3$

5 Medium Effects in ^{15}N Spectroscopy

Due to the fact that a nitrogen atom incorporated in a molecule, exhibits a Lewis basicity which depends on the degree of lone-pair delocalisation, solvent effects are expected to be greater in ^{15}N spectroscopy than in carbon NMR. Examination of Table 5.1 shows that amines, amides (or ureas), nitriles and nitro derivatives are characterized by different values of polarity parameters indicatory of their electron-donating or -accepting abilities.

As the exchange reactions which involve the nitrogen atom of the various amino derivatives strongly depend on the medium, the borderline between intermolecular effects and chemical reactions may become blurred. This is particularly so for proton exchange reactions which occur as a function of pH in >N-H derivatives. However, such processes will be considered in Sect. 8, and only conventional medium effects along with protonation effects will be dealt with now.

For the sake of brevity, certain abbreviations for the relevant solvents have been used throughout the book.

DMC	$CO(OCH_3)_2$	DCE	CH_2ClCH_2Cl
DMF	$(CH_3)_2NCHO$	TFE	CF_3CH_2OH
DMSO	$(CH_3)_2SO$	TCE	$CHCl_2CHCl_2$
HMPT	$[(CH_3)_2N]_3PO$	THF	$(CH_2)_4$-O
ODCB	$C_6H_4Cl_2o$	TMB	1,3,5-$C_6H_3(CH_3)_3$
TFA	CF_3COOH	TMU	$[(CH_3)_2N)]_2CO$

5.1 Physical Properties and Polarity Factors of Solvents

In order to understand the solvent induced ^{15}N shifts better, it is worthwhile briefly recalling some of the physical properties and parameters of solvent polarity which are frequently discussed in papers dealing with intermolecular effects.

5.1.1 Physical Properties of Solvents

The physical properties of the solvents which are concerned in NMR determinations are: density, melting point, boiling point, refraction index, dielectric constant, permanent electric moment and magnetic volume susceptibilities. The first five properties have been compiled for 911 organic solvents (S 26) and the list is available on request. Dipole moments are found in ref. (M 34) and volume susceptibilities in ref. (L 43).
Limited tables containing the physical properties of selected solvents are presented in specialized monographs dealing with solvent effect (R 18a) or NMR (M 31).

5.1.2 Empirical Parameters of Solvent Polarity

A number of empirical parameters have been recommended for expressing the polarity of organic solvents (K 33). Most of them originate from spectroscopic determination or thermodynamic and kinetic measurements. Among these parameters only those which have been discussed in relation with NMR spectroscopy will be dealt with here. First af all we shall consider solvent parameters which have been derived from direct NMR measurements

i) The acidity of the solvent has been characterized by the acceptor number AN (G 19) determined from ^{31}P measurements of Et_3PO in various solvents: $\Delta\delta^{31}P = \delta^{31}P(\text{hexane}) - \delta^{31}P(S)$

$$AN = 2.35\ \Delta\delta^{31}P$$

ii) Another "acidity" factor P is defined as a ^{19}F chemical shift difference of $pF\text{-}C_6H_4\text{-}NO$ in various solvents (quoted in R 21)

$$P = \delta^{19}F(S) - \delta^{19}F(C_6H_{12})$$

iii) The Lewis basicity may be represented in suitable conditions by the value $\Delta\delta_\infty$ (M 35) defined in ^1H spectroscopy as $\Delta\delta_\infty = \delta_i - \delta_o$ where δ_i and δ_o are the chemical shifts of $CHCl_3$ diluted in a given solvent S_i and in cyclohexane (7.06 ppm).

Other solvent parameters, defined on a thermodynamic basis or from electronic absorption spectra are also commonly used to explain the solvent induced NMR shifts.

i) The basicity of the solvent, is represented by the donor number of Gutman DN (G 19) deduced from equilibrium data.

ii) The molar electronic transition energy of a solvent dependent phenomenon has been proposed, in a number of cases, as a measure of the solvent polarity.
Two parameters are used most often in NMR research:
-the Kosower Z factor (K 30) which is related to a charge-transfer electronic absorption of a substituted pyridinium iodide.
-the Dimroth-Reichardt E_T factor (D 25, R 18b) which represents the solvatochromic shift ($\Pi \rightarrow \Pi^*$) of a pyridine derivative.
In fact, both scales are comparable but the E_T factors have been determined for a greater number of solvents than the Z factors.

iii) An infrared wave number difference between the gas phase and the solution, $B = \nu^{(gas)} - \nu^{(solvent)}$, has also been propsed (B 60) (K 33). ν is the stretching absorption of the O-H(O-D) vibration in phenol or deuterated methanol.

Since these different scales do not represent the same kind of solvent properties they are not always comparable. However, the different solvent parameters are not actually physically independent quantities. An interesting approach to the solvent polarity problem is a statistical evaluation, mainly founded on a factor analysis, of the different scales (C 24). Two main factors F_1, F_2, which represent $\simeq$85% of the whole variance for the various scales, seem to be appropriate for representing solvent effects in NMR. In Table 5.1, $\Delta\delta_\infty$, A_N, D_N, E_T, Z, B, F_1, F_2, F_3, are given for the main organic solvents.

5.2 Solvent Effects on ^{15}N Chemical Shifts

The solvent induced ^{15}N shifts are very high and have been shown in a great number of compounds. Only studies specifically devoted to solvent effects will be discussed here:

5.2.1 Qualitative Discussion of Solvent-Induced ^{15}N Shifts

5.2.1.1 Amines and Hydrochlorides

Tertiary amines undergo downfield shifts in polar or acid solvents, which have been explained on the basis of a decrease in electron density due to hydrogen bonding to the nitrogen lone pair or protonation (P 2). The situation is complicated when the nitrogen atom is bound to one or two protons and complex self-association or solvation processes occur (D 26). If basic solvents only are considered, a downfield shift may also be observed (P 2).

5.2.1.2 Amino Acids (K 31, B 39)

Increasing acidity of the medium results in an increasing upfield shift of the nitrogen of α-amino acids. Thus, a change of about -3 ppm may be observed when going from aqueous solutions to 90% H_2SO_4 solutions. This shift results largely from the indirect effect of protonation of the carboxy group rather than from a direct solvent interaction. Since such an indirect effect is not transmitted in ω-amino acids, these compounds are almost insensitive to solvent effects. Attention has also been drawn to the fact that the shape of the titration curves actually depends on the acid used for titration.

5.2.1.3 Amides (M 36) and Related Compounds

The solvent effects on amides are great and can be either diamagnetic or paramagnetic. When formamide *1* or N-methylformamide *2* are dissolved in dioxane, large diamagnetic shifts associated with the breaking of self associations are observed (Fig. 5.1). By contrast, proton donating solvents capable of introducing strong hydrogen bonding may induce high frequency shifts. Thus important paramagnetic shifts occur as the concentration of the amides in water or ethylene glycol decreases. In this respect, it should be noted that the relative dispositions of the resonance signals of *1* and *2* depend on the nature and on the concentration of the solvent (Fig. 5.1). It is therefore necessary to possess a good appraisal of such solvent effects when using ^{15}N chemical shifts for structural elucidations.
The protonation shifts, $\Delta\delta^{15}N = \delta^{15}N(\text{neat}) - \delta^{15}N(\text{protonated species})$,

of amides (M 20b), nitrosamines (G 14) and ureas (M 9) are paramagnetic and their magnitude increases in the series

	amides	nitrosamines (N_I)	ureas
$\Delta\delta^{15}N$ =	3 to 12 ppm	≃ 25 ppm	60 to 80 ppm

On the other hand, the N = O nitrogen atom of nitrosamines moves towards low frequencies by about -5 ppm.

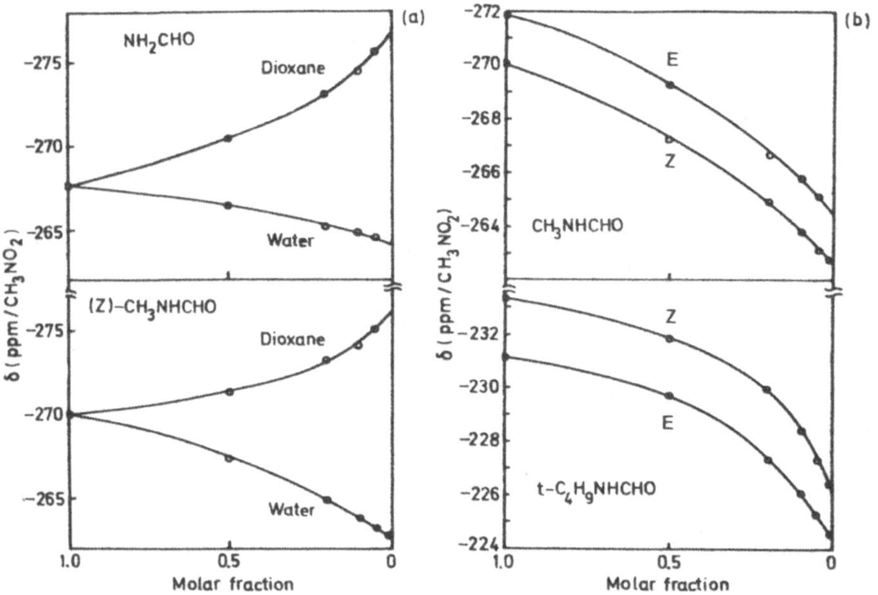

Fig. 5.1. **Concentration dependence of $\delta^{15}N$ of: *a)* Z-MeNHCHO and NH_2CHO in water and dioxane; *b)* Z-E MeNHCHO and Z-E t -C_4H_9NHCHO in water (conversion factor to the CH_3NO_2 scale: -1.5 ppm)**

5.2.1.4 *Sulfonamides*

The case of sulfonamides ($R\text{-}SO_2NH_2$) is interesting to discuss in comparison with that of amides, since in the former, H-bond interactions occur mainly with the nitrogen atom, rather than with the oxygen. Thus diamagnetic shifts are observed ($\Delta\delta^{15}N$):(K 29)

	DMSO	acetone	methanol	formic acid	TFA
$CH_3SO_2NH_2$ (K 29)	0	-4.3	-5.8	-5.0	-7.7 ppm
$CH_3NH\text{-}COCH_3$ (M 36, M 20b)	0	-2.3	+1.5	/	+5.0 ppm

However, a paramagnetic shift occurs when the pH value of an aqueous solution of methanesulfonamide increases

pH	1	4	6.5	8.5	10	11.1	NaOH
$\delta(^{15}N)$	0	0	-0.2	-0.1	1.2	10.9	12.6

5.2.1.5 Peptides and Polypeptides

Small peptides have different kinds of nitrogen atoms which can be readily identified from their solvent - and protonation - shifts (G 13, K 17). Indeed the high-frequency shifts observed from DMSO- to water- or acid-solutions are not of the same magnitude for both nitrogen atoms

Peptide		DMSO	Pyridine	TFE	Acetic acid	Water	TFA	Ref.
Ac-Gly-Gly (1 2)	N_1	0				3.8	9.8	G 13
	N_2	0				4.4	3.8	
BOC-LAla-LAla (1 2)	N_1	0 (a)	0.4	0.3	1.2	3.0	-51.9 (b)	K 17
	N_2	0 (a)	0.9	2.4	3.8	10.1	5.7	
BOC-LAla-DAla (1 2)	N_1	0 (a)	0.4	0.3	1.2	3.0	-51.9 (b)	K 17
	N_2	0 (a)	0.25	2.8	2.2	10.5	6.2	

(a) DMSO/acetone 1/1 (b) the BOC group is hydrolysed in TFA solutions and the shift reported here is that of the ammonium nitrogen atom.

of acetyldiglycine and BOC-dialanine. Moreover, slight differences can be observed between the two diastereomeric species of BOC-dialanine. Polypeptides and polyamides are also distinguishable by their solvent induced shifts (K 22) ($\Delta\delta^{15}N$/ppm)

Polypeptide	DMSO	TFE	H_2O	H-COOH	TFA	FSO_3H
Polyglycine	0	1.4	/	4.5	5.1	19.2
Polysarcosine	0	/	5.8	6.0	7.8	/
Poly(Tau-ε-Aca) (SO_2-NH)	0	/	/	-2.2	-3.1	/

Again, the sulfonamide function behaves abnormally since the protonation shifts of the polymer of taurine and ε-aminocaproic acid is upfield.

From a general point of view, it is important to notice that such pH and solvent effects on peptides units may reach a magnitude which masks the sequence effects on the chemical shifts.

5.2.1.6 Compounds containing on Acyclic Carbon-Nitrogen Double Bond

In imines (W 19), oximes (B 47), thiocyanates (V 1) and carbodiimides (Y 10), the nitrogen lone pair may be engaged in hydrogen bonding. H-bond formation is important for oximes or imines and a diamagnetic shift occurs when going from DMSO to methanol solutions. However, the two compounds exhibit extremely different behaviour in chloroform solutions, which has been tentatively explained on the basis of relationships between Hammett constants and chemical shifts (see Sect. 5.2.2).

(25% solutions)	DMSO	Benzene	Methanol	Chloroform	TFE
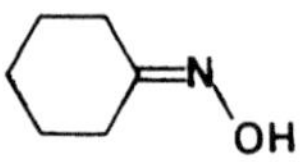	0	-12.2	-12.5	-14.3	/
p-CH_3O-Ar-CH=N-C_6H_5	0	/	-10.0	-0.8	/
i-C_3H_7N=C=NC_3H_7-i	0	/	/	-0.4	+2.2

5.2.1.7 Azines and Related Derivatives

Low-frequency solvent induced shifts are observed for pyridine or quinoline (P 2) (D 17). Thus when comparing the gas phase or a cyclohexane solution to an aqueous or a TFE solution, a -40 ppm shift is measured for pyridine.

gas	C_6H_{12}	CCl_4	C_6H_6	neat	DMSO	CH_2Cl_2	$CHCl_3$	CH_3OH	H_2O	TFE	
0	-1.5	-4.3	-4.9	-6.3	-6.9	-9.1	-12.5	-24.9	-28.1	-39.9	ppm

(These values are corrected from susceptibility effects and extrapolated to infinitely diluted solutions). The large upfield shifts are explained by the preferential stabilization of the electronic ground state through hydrogen bonding to the nitrogen unshared pair. The increase in the n→Π* excitation energy is associated with a substantial decrease of the paramagnetic screening contribution to $\delta^{15}N$ (P 2). The same behaviour is observed for the azine nitrogen of 1-methylimidazole (N III) which is likely to be engaged in strong hydrogen bonding with water for example. By contrast, a small paramagnetic shift is induced from CH_2Cl_2 to water solutions for the sp^3 nitrogen atom (N I) (A 25) (S 28).

	CH_2Cl_2	H_2O
N (I)	0	+4 ppm
N (III)	0	-16 ppm

5.2.1.8 Solvent Effects and Diastereoisomerism

We have already indicated that small but significant ^{15}N differential solvent shifts may be observed for oligopeptides. In monomeric units, the Z and E configurations of the amide bond are also subjected to different changes.

		Neat	Dioxane	DMSO	Water	Methanol
N-methylformamide	Z	0			0	
(M 20)	E	-1.6			-1.9	
N-t-butylformamide	Z	0	0	0	0	0
(M 36)	E	+2.2	+2.7	+1.5	+1.8	+2.2
Ala-Pro; Ac-Lpro	Z			0 (b)	0 (a)	0 (b)
(a) (B 20) (b) (H 21)	E			-0.3	+0.8	+0.6
i-butyraldehyde oxime	Z			0		0
(B 47)	E			-0.7		+1.8

It can be seen that, in protic solvents (H_2O, CH_3OH), a reversal of the assignment of Z and E isomers, with respect to that in aprotic solvents, may be necessary.

5.2.1.9 Solvents Effects and Electronic Paramagnetism

When electronic paramagnetism originates from the coordination state of a metal atom, solvent effects may induce condiderable shifts when the ligand nucleophilicity is high. Thus, the cyanide resonance of ^{15}N-enriched dicyanoprotohaemin complexes moves 250 ppm when passing from DMSO to methanol solutions (M 29)

DMSO	DMSO + D_2O	DMSO + D_2O	Pyridine + D_2O	Pyridine + D_2O	MeOD	MeOD + D_2O
	18/1	18/4	20/4	20/7		1/1
0	-16	-61	-75	-95	-226	-252

The $^{15}N{\equiv}C^-$ groups coordinated in the axial position to the iron atom of the porphyrin fragment undergo a decrease in the spin transfer from the iron in a H-bonding solvent, and the paramagnetic shift is reduced.

5.2.2 Quantitative Treatment of Solvent-Induced ^{15}N Shifts

5.2.2.1 Investigations of the factors which govern the ^{15}N solvent shifts

Several kinds of approaches have been applied to the explanation of solvent effects on ^{15}N shifts and conversely the variations of $\delta^{15}N$ may be used to obtain information about intermolecular interactions.

The differential shifts have been correlated with various solvent polarity parameters. We have already mentioned that many solvent effects have been explained on the basis of hydrogen bonding with the solvent. However, for methylamine for example, a lack of correlation with the acceptor strength of the solvent is noted (P 2) (L 44). Moreover, no simple correlation with the dielectric constant ε or a function of ε: $(\varepsilon-1)/(2\varepsilon+2.5)$, seems to hold in amines or pyridine (P 2) (D 17). However satisfactory one parameter correlations may be found in specific cases. Thus, the pyridine ^{15}N shifts have been shown to correlate ($R = 0.944$) with the Kosower Z value (Table 5.1) (D 17) and this relationship supports the interpretation of the solvent effects on the basis of an increase in the excitation energies through hydrogen bond formation.

A more general analysis concerning the variations introduced by 22 solvents on the ^{15}N and ^{29}Si chemical shifts of methylsilatrane,

$CH_3Si(O\ CH_2CH_2)_3\ N$, has demonstrated the effect of several parameters (P 12).

S	C_6H_{12}	CS_2	CCl_4	C_6H_6	C_6H_5Cl	THF	CH_3COOEt	$(Me)_3COH$	Acetone	CH_2Cl_2	DCE
$\Delta\delta^{15}N$	0	1.0	1.3	2.7	4.0	4.4	4.6	5.6	6.0	6.1	6.2
S	$CHCl_3$	Quinoline	HMPT	Pyridine	$C_6H_5NO_2$	$CH_3CONHCH_3$	DMF	CH_3NO_2	CD_3OD		
$\Delta\delta^{15}N$	6.3	6.3	6.5	6.7	6.8	7.2	7.3	7.8	8.1		
S	DMSO	D_2O									
$\Delta\delta^{15}N$	8.4	11.3									

It is interesting to note that the ^{15}N and ^{29}Si shifts are opposite in direction but linearly correlated

$$-\Delta\delta(^{29}Si) = 0.17 + 1.31\ \Delta\delta(^{15}N) \quad (R = 0.922,\ s = 0.55\ \text{ppm},\ 11\ \text{pts})$$

Three solvent parameters are shown to contribute to these variations: the polarity Y, the polarisability P, and the acidity E (K 33)

$$\Delta\delta^{15}N = -7.45 + 20.74\ Y + 11.34\ P + 0.229\ E \quad (R = 0.976,\ 21\ \text{pts})$$

Evidence for solvent interactions is also seen in their influence on the slopes (ρ) (and interceps, a) of the linear correlations observed between $\delta^{15}N$ of parasubstituted aryl-imines and -oximes and the Hammett substituent parameters σ_H (W 19).

$$\delta^{15}N = a + \rho\sigma_H \quad (R > 0.98 \text{ for I and III},\ R > 0.93 \text{ for II})$$

	DMSO a	DMSO ρ	$CHCl_3$ a	$CHCl_3$ ρ	CH_3OH a	CH_3OH ρ
(I) $X\text{-}C_6H_4\text{-}CH{=}N\text{-}C_6H_{11}$	/		-37.7	+21.3	-48.2	+26.3
(II) $X\text{-}C_6H_4\text{-}CH{=}N\text{-}C_6H_5$	-55.4	+16.5	-56.8	+21.8	-62.8	+28.3
(III) $X\text{-}C_6H_4\text{-}CH{=}N\text{-}OH$	-12.0	+17.3	-27.0	+17.7	/	

The slopes ρ increase for I and II with the hydrogen bonding ability of the solvent. Clearly the hydrogen bond interactions depend on the substituents and different shift variations are produced in different media.

Finally, it should be noted that, from a theoretical point of view, several trends are shown to be satisfactorily reproduced by sum-over-states calculations using INDO/S parameters (A 24). These calculations make use of both the solvaton model to describe the effect of changes in the dielectric constant of the medium, and a dimer model to represent hydrogen bonding interactions. Only the solvaton model seems necessary in a finite pertubation approach.

In fact, the correlations of solvent induced shifts with physical parameters sometimes present a biased approach since the variables are not always independent. To obviate, at least in part, this drawback, a statistical analysis of the significant factors which govern the solvent effects was performed with several kinds of amides: amides presenting an NH-bond (H_2N-CHO, CH_3NH-CHO, tBu-NH-CHO), NN disubstituted amides ($(CH_3)_2$N-CHO, $(CH_3)_2$N-CO-CH_3) and thioamides ($(CH_3)_2$NCSH) (M 36) (M 20b). It was shown that two independent factors explain 99% of the total variance. For the series of amides and solvents used, the most important factor involves the Lewis basicity of the amides since ^{15}N shifts are correlated well with the Dimroth E_T parameter which is a measure of the acid character of the solvent. The second factor should be related to the proton donating ability of the amides. A similar dependence of δ^{15}N on the Er parameter has also been recently demonstrated for a series of merocyanines (R 23).

5.2.2.2 Investigations of the Equilibria Involving Intermolecular Interactions

To conclude this section, it should be emphasized that the concentration dependence of δ^{15}N in a given solvent may provide information about the equilibria which take place in the medium. A linear dependence of δ^{15}N on the concentration, C, has been observed for various solutions of $(CH_3)_3$N (A 26) in water, methanol, ethanol, carbon tetrachloride ... and for pyridine in aqueous solution (D 17). This behaviour enables the extrapolation for a zero concentration of the solute to be carried out with reasonable accuracy. However, in the general case no simple linear concentration dependence is expected and the shape of the curves $\delta^{15}N = f(C)$ depends on the nature of the equilibria involved and on the equilibrium constants K. In fact deviations with respect to linearity have been observed for different solutions of ammonia (L 44), amides (M 36) and electrolytes (G 22). From a more quantitative point of view, a theoretical analysis of the curve shapes has been applied to the study of ionic interactions in lithium thiocyanate solutions (V 1). In dimethylformamide for example, the δ^{15}N variations characterize ion-pairing interactions which involve the nitrogen atom of SCN^- and therefore are of the isothiocyanate type. For a simple association equilibrium, the observed chemical shift is a weighted mean over the chemical shifts of the exchanging species A and AM

$$A + M \rightleftharpoons AM$$

$\delta_A \quad K \quad \delta_{AM}$ x fraction of A which exists in the AM form.

$$\delta_{obs} = \delta_A + x(\delta_{AM} - \delta_A)$$

The apparent equilibrium constant of the process, K, can be computed from a least square fitting of δ vs x with

$$x = \frac{1 + 2\ KC - (1 + 4\ KC)^{1/2}}{2\ KC} \qquad \text{(C concentration of A)}$$

The calculated value for the ion pairing of Li SCN in DMF is K = 2.1 $\pm$ 0.2 M^{-1}. Dimerization equilibria or the formation of S bonded species are also identified in other solvents.

5.3 Solvent effects on the $^{n}J^{15}N{\sim}X$ coupling constants

From a general point of view, coupling constants are *intrinsically* less sensitive than chemical shifts to solvent effects. In fact, it may be anticipated that significant variations in $J^{15}N{\sim}X$ will be principally found when either the nitrogen or a directly bonded hydrogen atom is engaged in specific interactions. Although strictly speaking protonation does not constitute a solvent effect, we shall also consider here the variations of the coupling constants in protonating acid media.

Obviously, the $^{n}J^{15}N{\sim}X$ coupling constants may also be the subject of relatively large changes as a function of the solvent if the measured value actually represents a weighted mean over species in conformational or in *tautomeric equilibria*. These indirect medium effects then reflect the influence of the solvent on the proportions of the exchanging forms.

5.3.1 Direct Solvent Effects on $^{n}J^{15}N{\sim}X$ Coupling Constants

5.3.1.1 $^{1}J_{N-H}$

The one-bond coupling constant $^{1}J\ ^{15}{N-H}$ in aniline, for example, exhibits a sensitivity to solvent effects which is comparable to that of $^{1}J\ ^{13}{C-H}$ of chloroform.

		C_6D_{12}	$CDCl_3$	Dioxane	Pyridine	Acetone	DMF	DMSO
$C_6H_5\underline{N}H_2$	(P2)	-78.0	-78.0	-80.6	-81.4	-82.1	-82.3	-82.3
$\underline{C}HCl_3$	(E4)	208.1	209.5	-	-	215.2	217.4	217.7

Thus, a noticeable increase in the absolute value of the coupling constant accompanies the increase in the hydrogen bonding capability of the solvent. The same trend is observed in a number of ring substituted aniline derivatives for which $|^1J_{N-H}|$ is found to be about 3 to 5 Hz larger in DMSO solutions than in $CDCl_3$ solutions (A 4). This difference in the values of the J_{N-H} coupling constant, measured in $CDCl_3$ and in DMSO, $\Delta J = |J_{NH}(DMSO)| - |J_{NH}(CDCl_3)|$, has been used to propose an order of intramolecular hydrogen bonding abilities of the substituents in ortho substituted anilines (A 6). Indeed, ΔJ is expected to be small if the strength of the intramolecular hydrogen bonding, which exists in the relatively inert $CDCl_3$ medium, is comparable with that of the intermolecular hydrogen bonding taking place between N-H and DMSO. This situation is found for example with the strongly electron-withdrawing nitro group of o-nitroaniline (ΔJ = 0.7 Hz). By contrast, owing probably to its small size which prevents the close approach of the N-H proton, the ortho fluorine atom introduces only weak hydrogen bonding in $CDCl_3$ and ΔJ reaches 3.4 Hz. On this basis, the following order of intramolecular hydrogen bonding abilities is proposed $NO_2 >$ C = O, > CF_3, Br, OCH_3 > F

5.3.1.2 $^{n>1}J_{N-H}$

For oximes in aprotic media, significant induced changes of the $|^2J_{N-H}|$ coupling constant are observed and these changes are found to be of opposite signs in the Z and E configurations (C 9), (K 7), (T_2).

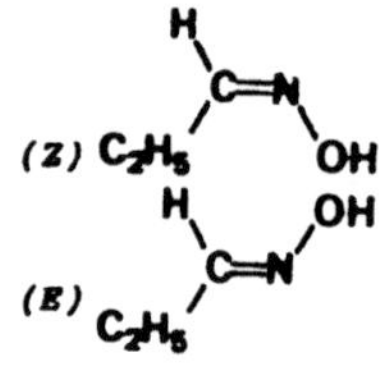

	Pentane	CH_3CN	H_2O	CH_3COOH
(Z) ^2J(Hz)	-15.5	-16.7	-15.8	-14.6
(E) ^2J(Hz)	+3.0	+2.7	+2.9	+3.1

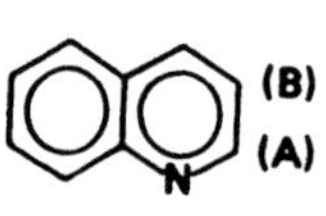

	Pentane	CH_3CN	CH_3COOD	CF_3COOD
$^2J_{NH(A)}$	-10.8	-11.0	-4.6	-2.0
$^3J_{NH(B)}$	-1.3	/	-3.1	-4.5

Such a behaviour may be used as an indication that the considered coupling constants are of opposite sign, Similarly, significant solvent effects are observed for the $^2J_{N-H}$ *parameter* in 2-(α-naphthyl)aziridine. Thus, $|^2J_{N-H_A}|$ varies between 3.6 Hz in $CDCl_3$ and 4.9 Hz in DMSO whereas $|^2J_{N-H_B}|$ and $|^2J_{N-H_C}|$ are equal respectively to 1.4, 1.7 Hz in $CDCl_3$, or 0.3 and 0.7 Hz in DMSO (O 3). The changes in the populations of the configurations as a function of solvent possibly act in addition to the intrinsic solvent effects.

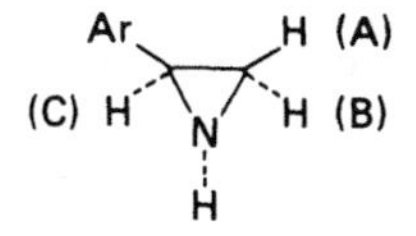

Hydrogen bonding to the oxygen of amides which is expected to enhance the C-N Π bond order also affects the $^2J_{N-CHO}$ coupling constant. This parameter algebraically increases from -15.6 to -14.1 Hz on progressive dilution of N-methylformamide in water (B 26). Although small values of $|^2J_{N-H}|$ are measured in amines, variations reaching 25% are detected in $(CH_3)_3N$ since $|^2J_{N-H}|$ decreases from 0.8 Hz in $CDCl_3$ to 0.60 Hz in CD_3OD (P 2).

5.3.1.3 J_{N-C}

The noticeable solvent effects exhibited by the $^1J_{C-N}$ coupling constant of aniline (W 10) or formamide (H 6) are associated with changes in the C-N Π bond order. Indeed, this bond order increases due to the hydrogen bonding of aniline to DMSO or of formamide to water. For aniline the behaviour of J_{C-N} parallels that of the $^1J_{N-H}$ coupling constant discussed above.

		Neat	$CDCl_3$	DMSO	DCl	Ref.
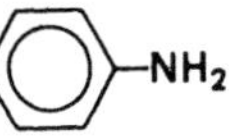	$^1J_{N-C_1}$	-10.9	-10.9	-12.1	-8.9	(W 11)
		neat	H_2O(0.3 m.f)	H_2O		
H_2N-CHO	$^1J_{N-C}$	-13.9	-14.5	-14.8		(H 6)
		$CDCl_3$	DMSO			
pCl-C_6H_4CH=N-OH	$^1J_{N-C}$	(-)4.5	(-)3.9			(D 1)
	$^2J_{N-C}$	(-)7.6	(-)8.0			(D 1)

In the case of formamide the dependence upon the C-N Π bond order is corroborated by the increase in the rotational barrier which accompanies dilution in water (K 23). It is also shown that, whereas the magnitude of $|^1J_{N-C}|$ of benzaldoximes decreases from $CDCl_3$ to DMSO, $|^2J_{N-C}|$ increases. Smaller and more irregular variations are exhibited by the $^3J_{N-C}$ coupling constants (D 1). Indeed, it may be assumed in this case that the formation of complexes between the oxime and DMSO enhances the positive lone-pair contribution to $^1J_{C-N}$, making the net coupling less negative.

5.3.1.4 J_{N-F}

A very large solvent effect is exhibited by the $^4J_{N-F}$ coupling constant of 2-fluorobenzamide.

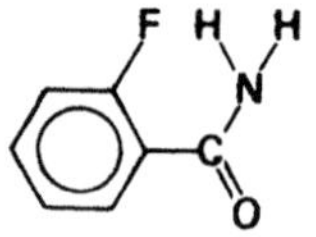

The algebraic increase from -7.0 Hz in $CDCl_3$ to -3.2 Hz in DMSO $-d_6$ corroborates the existence of a through space coupling mechanism (F 2). In fact, inhibition of this contribution due to the rupture of intramolecular hydrogen bonds between fluorine and the N-H proton is produced by DMSO and a concomitant decrease in the absolute value of $^4J_{N-F}$ is observed.

5.3.2 Protonation Effects on $^nJ\,^{15}N{\sim}X$ Coupling Constants

5.3.2.1 J_{N-H}

An algebraic decrease in $^1J_{N-H}$ accompanies the protonation of amines (Tables 7.1, 7.2). Thus, $^1J_{N-H}$ = -64.5 Hz in CH_3NH_2 (gas phase) and -75.4 Hz in $CH_3NH_3^+\ Cl^-$ (A 5) (A 23). The opposite behaviour is found for anilines since ^1J, which equals -82.6 Hz and -89.4 Hz for aniline and p-nitroaniline (A 4) in DMSO, drops to -76.0 Hz for both protonated species following the loss in trigonal hybridization (A 7). For amides which are primarily protonated on the oxygen, a substantial increase in $|^1J_{N-H}|$, associated with an enhanced sp^2 character, is observed. Thus, for acetamide, $^1J_{N-H}$ = -88.4 Hz (*cis* to the oxygen) and -90.9 Hz (*trans* to the oxygen) in water whereas the corresponding values are

$^1J_{N-H}$ = -93.2 Hz and -96.6 Hz in FSO_3H (L 13).

Protonation may also have an influence on the $^2J_{N-C-H}$ coupling constants. Thus the magnitude of $|^2J_{N-C-H}|$ of amino acids such as glycine or alanine is shown to decrease from values of about 1.5 - 2.2 Hz at high pH values to less than 0.5 Hz when the pH is about 6 (L 47). Interestingly, a strong decrease of $|^2J|$ may be observed when the coupled proton is *cis* with respect to a lone-pair at the nitrogen (see data p. 66). This behaviour shows that the orientation of the lone pair is the main factor controlling the difference in magnitude of the $^2J_{N-H}$ coupling constants in the Z and E configurations of oximes (K 7). From a more general point of view, it is observed that both the negative and positive ^2J coupling constants in oximes (C 10), quinoline (T 2) or pyridine (L 8) increase algebraically on protonation (Table 7.13). By contrast, the $^3J_{N-H}$ coupling constants which are negative then decrease (see data p. 66 and Table 7.13).

5.3.2.2 J_{N-C}

As regards the behaviour of the coupling constants between nitrogen and carbon, the loss in planar (sp^2) character of the $C-NH_2$ fragment upon protonation of anilines for example is shown to be associated with an algebraic increase in the magnitude of $^1J_{N-C}$ (W 10). By contrast, the absolute value of $^1J_{N-C}$ of amines (B 11), (B 19), quinuclidine (B 11) and amino acids such as glycine, alanine (L 8), proline (B 20) or guanidino arginine (L 6) more or less increases upon protonation. For amides O-protonation enhances $|^1J_{N-C(O)}|$ as expected from the increase in the N-C(O) double bond character but a concomitant

	$^1J_{N-C}$		$^2J_{N-C}$		$^3J_{N-C}$		
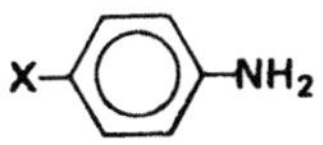	DMSO(1M)	$HFSO_3$(1M)	$CDCl_3$	CF_3COOD	$CDCl_3$	CF_3COOD	
X = H	-12.1	-8.6	-2.8	(-)1.5	-1.2	(-)2.1	(A 14)
X = NO_2	-14.7	-9.2					
	$CDCl_3$	HCl,$CDCl_3$	$CDCl_3$	HCl,$CDCl_3$	$CDCl_3$	HCl,$CDCl_3$	
$CH_3CH_2CH_2NH_2$	3.9	4.2	1.2	<0.2	1.4	1.3	(B 11)
quinuclidine	2.1	4.8	<0.2	<0.2	2.8	6.7	(B 11)
	0.45	12.0	2.4	2.1	3.6	5.3	(L 8)

lowering of the one bond coupling involving an N-alkyl substituent is observed (B 19). Thus, for the amide $(CH_3)_3\underset{2}{C}-\underset{1}{C}O-NH-\underset{3}{C}H(CH_3)_2$ $|^1J_{N-C_1}|$ = 13.2 Hz, $|^1J_{N-C_3}|$ = 9.5 Hz, $|^2J_{N-C_2}|$ = 6.5 Hz in $CDCl_3$ whereas the corresponding values for the hydrochlorides in $CDCl_3$ are 18.8 Hz, 6.3 Hz and 2.3 Hz, respectively. From a similiar point of view, deprotonation of the carboxylate group (pK_a = 3.1) of glycylglycine causes the $|^1J_{N-C(O)}|$ interaction to increase from 17.7 Hz to 18.9 Hz as a result of an increase in the double bond character of the peptide bond (I 4). Such behaviours may prove useful in conformational studies of the peptides, since it is shown that intramolecular H-bonding enhances $|^1J_{C-N}|$ in the β turn of a cyclic peptide and that this coupling is further increased by protonation of the solvent-exposed carbonyls (W 12).

5.3.3 Indirect Solvent Effects on $^nJ_{N\sim X}$ Coupling Constants (Tautomeric Equilibria)

In fact, noticeable variations of coupling constants involving nitrogen may originate from an influence of the solvent upon the relative concentration of exchanging species. As tautomeric equilibria frequently occur in conjugated molecules containing nitrogen, such indirect solvent effects have been observed and discussed in a number of cases. If two tautomers I and II are mutually interconverting, the observed coupling constant is the weighted mean over the J values characteristic of the I and II species: $J_{obs} = p_I\ J_I + p_{II}\ J_{II}$. To the extent that J_I and J_{II} exhibit negligeable solvent dependency, variations in J_{obs} reflect the influence of the solvent upon the populations p_I and p_{II}. Thus, by assuming $J_{N-H} \simeq 0$ in the oxygen tautomer and $^1J\,N-H \simeq 90$ Hz in the nitrogen tautomer of Schiff bases the percentage of the *keto-amine and enol-imine* forms can be estimated and it is shown for example that basic solvents such as DMSO or hexamethylphosphorictriamide stabilize the amine form *1* (≃ 1/3) whereas the imine form is almost exclu-

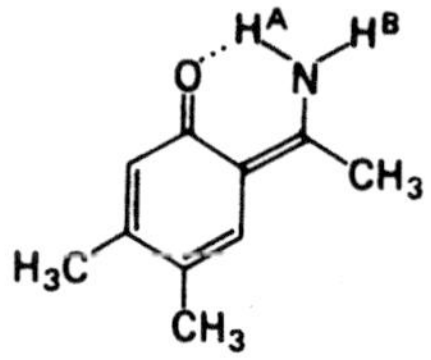

1

sively present in $CDCl_3$ (>90%) (D 20). These conclusions clearly result from the values $|^{J}N\text{-}H_A| = 31.2$ Hz and $|^{J}N\text{-}H_B| = 75.4$ Hz measured in DMSO for the associated and free hydrogen, respectively, and $|^{J}N\text{-}H_B| = 62.8$ Hz measured in $CDCl_3$. Such a drastic influence of the nature of the solvent on the proportion of amine and imine forms has been noticed in a number of other cases (B 52) (D 11) (Tables 7.5, 7.6). It has also been shown by means of ^{15}N labelling, that the derivatives, obtained via reaction of amines with β-chlorovinyl ketones, exist predominantly in the enaminoketone form R-CO-CH = CH-NHR'. Thus, $^{1}J^{15}N\text{-}H = 91$ Hz for the compound with R = R' = C_6H_5 investigated in an inert medium (F 15). However, a small amount of the enolimine form of C_6H_5-NH-CH=CH-CS-C_6H_5 probably intervenes for the homologous sulfurated compound since $^{1}J^{15}N\text{-}H = 87$ Hz. Whereas a *cis* configuration stabilized by chelation between N-H and C=O or C=S is exclusively present in CCl_4 or $CDCl_3$ solutions, a *trans* configuration is formed in basic solvents such as hexamethylphosphorictriamide ($^{J15}N\text{-}H = 91.2$ Hz and 91.5 Hz for the free N-H of the ketone and thioketone, respectively).

Similar analyses have been carried out on several *hydrazono* ⇄ *azo* tautomeric equilibria (P 19) (Y 1). Thus, a phenylazo -3 pyrazolone and a phenylazo -4 pyrazolone are shown to exist in the azo form in $CDCl_3$, DMSO and C_5H_5N, whereas a shift from the hydrazono form to the azo form is detected for 2-phenylazothianapthen-3-one when going from DMSO to pyridine (Y 1). Changes in the percentage of the tautomeric isomers of compound *2* as a function of solvent are also proposed to interpret the

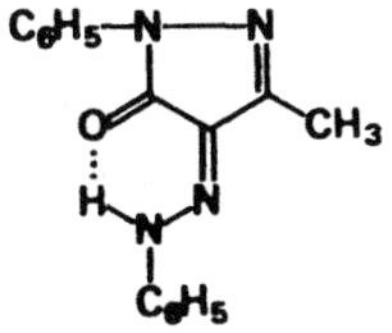

2

behaviour of the N-H spectral pattern. In fact, measurement of $^{J}N\text{-}H = 96$ Hz in $CHCl_3$ indicates a nearly pure hydrazono form but exchange processes may intervene in other media (L 23). More quantitative results could be derived involving the azo-hydrazone equilibrium in o-hydroxy-azo compounds such as *3* (B 43). Assuming that $^{1}J^{15}N\text{-}H = 88$ Hz in the hydrazone structure

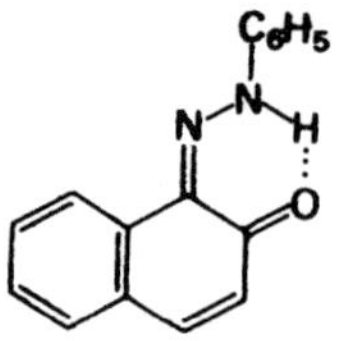

3

the equilibrium constant K = [azo]/[hydrazo] is calculated to vary from 0.5 in CH_2Cl_2 (at 302 K) to 0.35 in $CDCl_3$ (293 K), 0.47 in DMSO (293 K) and 0.72 in CCl_4 (293 K).

In practice, coupling constants other than ^{J}N-H may provide information on tautomeric equilibria. This is illustrated by the use of $^{2}J_{N\sim H}$ coupling constants involving the deprotonated imidazole ring of histidine to estimate the relative amounts of tautomeric species above pH 6.2 (B 35).

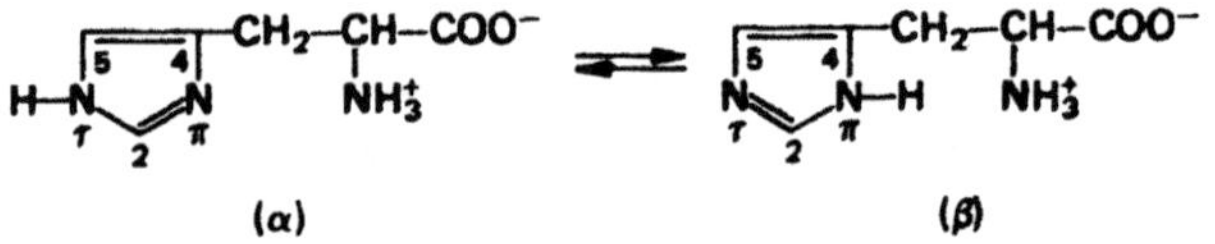

By assuming that $^{2}J_{N_{\tau}H_{5}}$ takes values of -14.4 Hz and -4.8 Hz in the β and α forms respectively the proportion of the β tautomer is calculated to amount to 12% at pH 8.2. This proportion increases to 20% at pH 9.3 as indicated by the fact that $^{2}J_{N_{\tau}H_{5}}$ becomes more negative and $^{2}J_{N_{\pi}H_{2}}$ less negative following respectively the increase and decrease in proportion of a structure with a nitrogen lone-pair in a *cisoid* relationship to the coupled proton.

Table 5.1. Magnetic susceptibility and polarity parameters of solvents, ordered by increasing molecular weights;
χ is expressed in SI units, $\Delta\delta_\infty$ (M35), An (G19), F_1, F_2, F_3 (C24) are dimensionless parameters, Dn (G19) in $kcal \cdot mol^{-1}$, B (K33) in cm^{-1}.

Solvent	$\chi 10^6$	$\Delta\delta_\infty$	An	Dn	Z	E_T	B	F_1	F_2	F_3
Water	- 9.03		54.8	18.0	94.6	63.1	123	4.0	-2.35	-0.48
Methanol	- 6.66		41.3		83.6	55.5		2.6	-1.67	-0.50
Acetonitrile	- 6.71	0.56	18.9	14.1	71.3	46.0	101	1.06	-0.33	0.48
Formamide	- 6.92		39.8		83.3	56.6				
Ethanol	- 7.23	0.78	37.1		79.6	51.9		1.94	-1.30	0.41
Acetone	- 5.78	0.94	12.5	17.0	65.5	42.2	116	0.16	-0.03	0.10
Acetic acid	- 6.92		52.9		79.2	51.2	98			
Nitromethane	- 4.91	0.46	20.5	2.7	71.2	46.3	58	0.86	-0.91	1.31
THF		0.80	8.0	20.0	58.8	37.4	142	-0.98	-0.08	-0.52
DMF			16.0	26.6	68.4	43.8	159	0.75	0.95	-0.28
Diethyl ether	- 6.56	0.78	3.9	19.2		34.6	129	-1.87	-0.23	-0.72
Carbon disulfide	- 6.68					32.6				
Benzene	- 8.78		8.2	0.1	54.0	34.5	52	-2.98	-1.76	0.23
DMSO		1.32	19.3	29.8	71.1	45.0	193	1.08	1.05	-0.52
Pyridine	- 7.68	1.56	14.2	33.1	64.0	40.2	220	-0.11	0.83	-1.29
Cyclohexane	- 7.88	0			60.1	31.2				
Dichloromethane	- 9.21		20.4		64.7	41.1	40			
Hexane	- 7.01	0	0			30.9		-3.95	-1.44	0.21
DMA			13.6	27.8	66.9	43.7	165	0.52	1.29	-0.30
Morpholine	- 7.93	1.20								
Dioxane	- 7.62	0.64	10.8	14.8	36.0	48.4	129	-2.29	-1.34	-0.95
Toluene	- 7.67					33.9	54			
Dichloroethylene Z	- 8.53									
Dichloroethylene E	- 8.02									
DCE			16.7	0	63.4	41.9	54	-0.50	-1.09	1.15
TMU	- 7.97					41.0				
Chloroform	- 9.30				63.2	39.1	35			
Freon 12	- 8.07									
Quinoline	- 9.16					39.4	267			
Trichlorethylene	- 9.22									
Decaline	- 8.56									
O Dichlorobenzene	- 9.40						41			
Carbon tetrachloride	- 8.68	0.18	8.6			32.5	31			

Table 5.1 (continued)

Solvent	$\chi 10^6$	$\Delta\delta_\infty$	An	Dn	Z	E_T	B	F_1	F_2	F_3
Tetrachloroethane	-10.75									
Methylene bromide	-11.75									
HMPT		2.05	10.2	38.8	62.8	40.9	234	-0.25	3.80	-0.39

6 ^{15}N Chemical Shifts

Compilation until 1979 of the literature on ^{15}N parameters, namely chemical shifts and coupling constants, has provided a large amount of data. Examination of Fig. 1.1 shows that most ^{15}N resonances lie in a 500 ppm range, although some compounds, such as nitroso derivatives, extend the field of δ^{15}N to 800 ppm.

Extensive tables of δ^{15}N corresponding to many families of compounds will be given in this section but no specific comments concerning the trends observed in each group will be developed. The latter aspect is in fact one of those which have been the most widely reviewed in the past and we shall instead focus our attention on the analyses of the chemical shifts which have been attempted in terms of various correlations.

6.1 Tables of ^{15}N Chemical Shifts

6.1.1 Accuracy of ^{15}N chemical Shift determination

Before considering the whole of ^{15}N shifts, we shall recall here some of the problems connected with the accuracy of the reported values. Indeed the intrinsic accuracy of the published δ^{15}N values depends mainly on the digital resolution, i.e. the reciprocal of the acquisition time, which is usually of the order of 0.1 ppm (see Sect. 2). However, the degree of uncertainty which is attached to the comparison of chemical shift values measured in different experimental conditions, is much higher. A first source of discrepancy comes from the fact that the magnetic susceptibilities of the solutions corresponding to the compared ^{15}N values are not the same. As discussed in Sect. 4.1, the correction may be as high as 1 ppm, but usually remains of the order of 0.6 ppm. A second, but minor, cause of error concerns the nature of the lock substance. Since the deuterium resonance signals of heavy water, perdeuterated nitromethane ... which are used to lock the field-frequency ratio, may have half-height line-widths of the order of 0.5 to 1.5 Hz, i.e. usually one address of the computer memory, an additional 0.1 ppm contribution to the total uncertainty should be taken into account. A third difficulty arises if we consider that the frequency ν_0 used in Eq. (4.1) (see p.44) to compute the ^{15}N values is not always accurately known by the users, since it depends not only on the lock substances but also on the specifications of the spectrometer; another 0.1 ppm error may then be introduced. It should also be kept in mind that the temperature coefficients of most associated liquids are of the order of -2 to +4 10^{-2}ppm K^{-1} (A 5) (L 57). A 10 K difference between two experiments may therefore lead to a change of about 0.4 ppm in chemical shift. Finally, two important sources of errors, which have already been mentioned, concern, firstly, the nature of the standard substance used as a reference and the estimation of the conversion factor (see Sect. 4.2) and, secondly, whether a relaxation reagent such as $Cr(acac)_3$ was added to the sample or not (see Sect. 2.2).

Connected to this problem of accuracy of the ^{15}N shifts, it is useful to be aware of the possibility of rather high isotope effects when the ^{15}N resonance of a nitrogen atom bound to a ^{14}N isotope is compared to the resonance of a ^{15}N atom bound to another ^{15}N isotope (S 1). This situation occurs when using enriched materials and $\Delta\delta$, defined as $\Delta\delta = \delta(^{15}N - {}^{14}N) - \delta(^{15}N - {}^{15}N)$, may be as high as 3.4 ppm

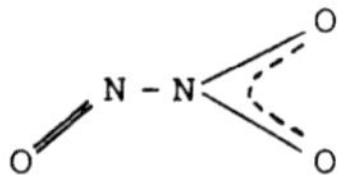

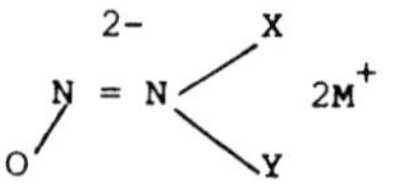

{ X = O Y = O, SO_3
{ Y = O X = $CH_2N(O)N(O)$

Δδ (ppm) -3.4 -0.3   1.9/2.2 0.9/1.3

To conclude, it can be said that the ^{15}N shifts are measured with a high degree of precision but that the comparison of the literature data is frequently only accurate within a 1 ppm range. Moreover, when the reported $\delta^{15}N$ values concern molecules dissolved in different solvents, a critical consideration of the possible solvent shifts is indispensable (see Sect. 5). The data presented in Tables 6.1 to 6.61 always include, where available, the solvent and the concentration of the solute, the frequency ν_0 and the temperature, the reference substance and the conversion factor used to convert the ^{15}N shifts to the CH_3NO_2 scale.

6.1.2 List of Tables of $\delta^{15}N$ Data

6.1.3 Bioorganic Molecules

Since nitrogen atoms are present in a number of molecules of great biological importance, this field of ^{15}N spectroscopy is rapidly growing in spite of the technical difficulties. At present, the ^{15}N spectra of a large number of nitrogen-containing biomolecules have already been published and numerous data are given in the tables. In addition, we shall present below some rather more specific results concerning typical biomolecules.

6.1.3.1 Small Polypeptides

Gramicidin S (H 21) is a cyclic decapeptide (M = 1120) which shows six ^{15}N signals:

```
   ---- -255.7/-254.3/-252.5 ------------------------
D - Phe - L - Pro - L - Val - L - Orn - L - Leu
        -245.0         -265.6
L - Leu - L - Orn - L - Val - L - Pro - D - Phe
```

$\delta^{15}NH_2$ (Orn) = -344.8

in 0.25 M DMSO-methanol (4:1) solution (at 9.12 MHz and 308 K). The shifts were originally referenced to 5M NH_4NO_3 in 2M HNO_3 (conversion factor: -359.0 ppm).

Viomycin sulfate (H 14) is a cyclopeptide containing six peptide signals and seven other ^{15}N resonances (0.3 M solutions in H_2O - D_2O 9/1, pH = 2.8, 36.48 MHz and room temperature; the shifts were originally referenced to 5M NH_4NO_3 in 2M HNO_3, conversion factor: -359.0 ppm) (a, b) the assignment may be interchanged.

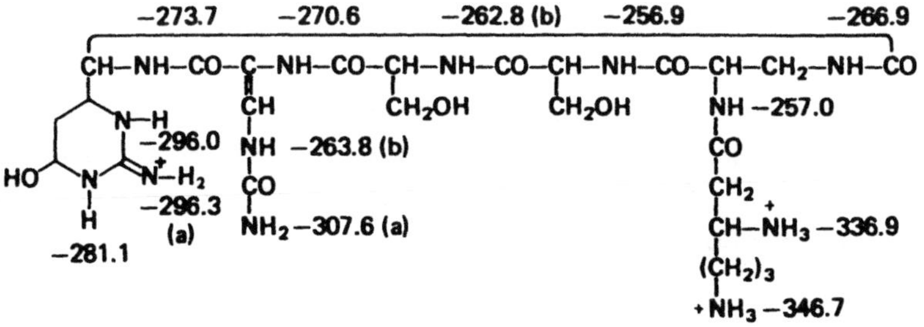

Oxytocin (L 22) is a peptide hormone which was studied in aqueous solution (100 $mg.ml^{-1}$) at pH = 4.0 (18.24 and 9.12 MHz) and room temperature; (shifts originally referenced to 5M NH_4NO_3 in 2M HNO_3, conversion factor: -359.0 ppm).

Cys 1	Gly 9	Gly 9	Asn 5		Gln 4	Cys 6	Ile 3	Leu 8	Tyr 2	Pro 7
NH_3^+	NH_2	NH	NH_2	NH	NH	NH	NH	NH	NH	N
-342	-273.2	-269.7	-267.9	-263.7	-260.3	-260.3	-260.3	-257.8	-256.4	-242.9

Nebramycin and related aminoglycoside derivatives (D 24). These amino sugar derivatives are classified in three series which are characteriszed by typical structural units: I (apramycin, oxyapramycin), II (kanamycin B, tobramycin) and III (nebramine, lividamine, neamine). At 10.09 MHz and for 0.5 to 1.0 M solution in D_2O (room temperature), three kinds of ^{15}N resonance are observed (reference of shifts: 2.9 M NH_4Cl in 1M HCl, conversion factor = -355.3 ppm).

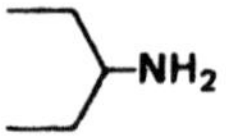	- 338.1 (pH 3) to - 356.7 (pH 11)	I, II, III
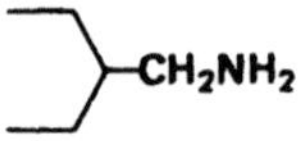	- 350.8 (pH 5) to - 363.8 (pH 10.4)	II
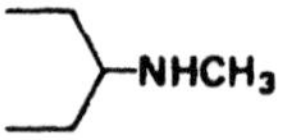	- 343.9 (pH 3) to - 355.3 (pH 11)	I

Thiostrepton and Siomycin-A (O 9) show nineteen ^{15}N signals at 298 K and 36.48 MHz (1 g in 4 ml $CDCl_3$ - MeOH solutions) which are assigned to the azine and peptide bond nitrogen atoms. (Shifts originally referenced to NO_3^-, H_2O; conversion factor: -4.6)

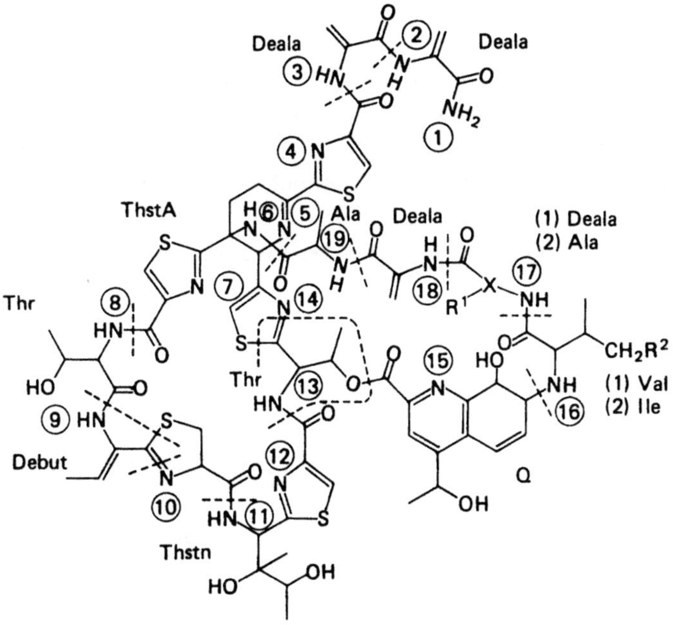

(1) X = C, R^1 = CH_2, R^2 = H Siomycin A

(2) X = CH, R^1 = R^2 = Me Thiostrepton

N_i	$\delta^{15}N_{(1)}$	$\delta^{15}N_{(2)}$
4, 5, 7, 10	-59.8/-68.2/-68.7/-71.2	-59.5/-68.1/-68.7/-70.7
12, 14, 15	-76.6/-89.9/-92.1	-76.8/-89.4/-92.0
17	-254.7	-
6	-257.7	-257.8
2,3	-258.4/-258.4	-258.5/-258.5
18	-	-259.4
17	-	-259.9
19	-261.4	-261.4
9	-263.8	-263.9
18	-264.1	-
8,13	-267.5/269.1	-269.3/-267.7
11	-270.2	-270.0
1	-285.0	-285.5
16	-341.1	-336.0

The ^{15}N spectrum of *nosiheptide*, an antibiotic structurally related to thiostrepton (D 28) and that of alumichrome (L 16, L 58) were also studied.

6.1.3.2 Proteins

Various natural proteins were studied using wide-bore superconducting solenoids. Lysozyme (from hen egg-white) studied at 18.25 MHz (G 17) displays a number of ^{15}N resonances in the amide range (-251.2 to -272.6 ppm) along with other signals corresponding to aminoacid residues (arginine, lysine) at a lower frequency (-297.8 to -351 ppm). Protamines and proteins of hemoglobin give relatively well resolved ^{15}N signals.

6.1.3.3 Nucleotides and Polynucleotides

Nicotinamide adenine dinucleotides have been studied at 10.14 MHz in D_2O solutions (pH 7) at room temperature; the ^{15}N chemical shifts of the pyridine fragment exhibit great changes during redox reactions which lead to rehybridization of the azine nitrogen (the shifts were originally referenced to 1M ND_4Cl in D_2O; conversion factor: -352 ppm) (O 10).

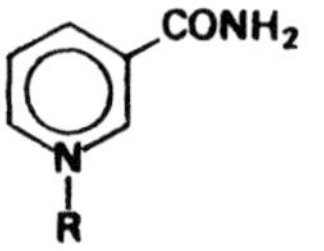

−170.3 −154.0
R = CH_3, ADP–ribose

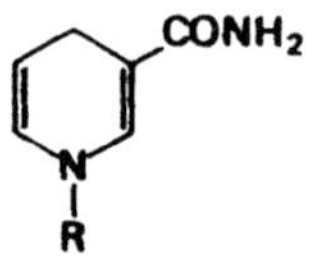

−278.9 −263.2 ppm
R = CH_3 ADP–ribose

As described in section 3, the ^{15}N-NMR spectrum of transfer RNA of yeast (M≃30.000) was studied in ~10^{-2}M aqueous solutions (G 17) and displayed signals characteristic of the structural units: guanosine, adenosine, and uridine. G(Guanosine-5'MP) - 309.4; A(Adenosine-5'MP) -304.1; C(Cystine-5'MP) -288.2; U (Uridine 5'MP) -224.5; G,U.: -235.2; C: -182.2/-186.2; G,A: -143.2/-159.2 ppm.

6.1.3.4 Miscellaneous Bioorganic Molecules

Vitamin B1 (C 20), vitamin B 12 (G 17) and some alkaloids (F 13) belonging to the series of yohimbine and reserpine were also studied:

Thiamine, HCl (Vitamine B1) (C20) (18.24 MHz and 300 K)

$HOCH_2$-CH_2OH		$HOCH_2$-CH_2OH + 1 equiv. NaOH
-136.7	(thiazole)	-140.2
-170.9	(N_3)	-168.0
-215.0	(N_1)	~ -141.2 (deprotonation)
-273.0	(NH_2)	-294.2

The Yohimbine-like alkaloids have a indoloquinolizidine skeleton

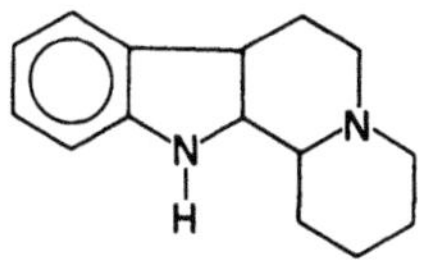

and the indole nitrogen are found between -265 and -255 ppm whereas the quinolizidine nitrogen atom resonates at -348 to -322 ppm (F 13). Similarly three nitrogen atoms are identified in the spectra of physostigmine (S 39). We shall also mention the use of ^{15}N NMR to corroborate the structure of an ^{15}N labelled antibiotic, chetomin (M 48).

The Ferric low-spin cyanide complexes of iron-bound haemin and haemoproteins are characterized by a great paramagnetic shift of the ^{15}N enriched cyano group (up to ≃1200 ppm). (M 29, M 30, M 49, M 50, M 51).

Finally, as already mentioned in Ch 3, ^{15}N spectra of whole cells have been investigated (I 3, L 40, L 59).

6.2 Correlation Analysis of ^{15}N Shifts

The large amount of data concerning ^{15}N chemical shifts offers well-suited bases for correlation analysis and a number of simple or multiple regressions have been computed involving $\delta^{15}N$ and various physical, topological or semi-empirical parameters. Since we are primarily concerned here with ^{15}N chemical shifts, we shall consider $\delta^{15}N$ as the dependent variable and the others parameters (P) as the independent variables. To conform to this definition, some linear correlations published in the alternate form, i.e., (P) = f($\delta^{15}N$) have been recalculated as follows: $\delta^{15}N = a + b(P)$. (a: intercept, b: slope). Moreover, since the collected $\delta^{15}N$ data are referenced with respect to nitromethane, using the recommended sign convention (I 8) (i.e. an increase in $\delta^{15}N$ corresponds to a high frequency-paramagnetic-shift and to a decrease in ^{15}N screening), the coefficients of the various regressions have been converted to the CH_3NO_2 scale. Thus in linear correlations the intercept a, generally appears as a negative number. The degree of significance of the simple or multiple correlations presented now can be appreciated through the correlation coefficient R, defined as $\frac{b\sigma(\delta^{15}N)}{\sigma(P)}$ (where σ_i is the standard deviation corresponding to the experimental value either of the dependent variable ($\delta^{15}N$) or of the independent variable (P)) and through the standard deviation σ or the mean standard deviation $\bar{\sigma}$ on the estimate of the dependent variable. The number of pairs (or multiple) of values, N, which have been used to define the degree of freedom of the correlation, are also given. We shall discuss, in turn, correlation analyses between $\delta^{15}N$ and:

- qualitative topological parameters
- chemical shifts of other nuclei
- electronic density
- Hammett-like substituent parameters
- free energy of activation of hindered rotation processes about N-X bonds
- various physical parameters

6.2.1 Correlation between $\delta^{15}N$ and Topological Qualitative Parameters

A multiple regression analysis is performed on a matrix constructed from a series of chemical shifts values associated with parameters which have the properties of a delta function, i.e. parameters which are equal to 1 or 0 according as whether a substituent is present or not in the considered position. This treatment, which has been very successful in ^{13}C spectroscopy, leads to additive substituent increments appropriate to different kinds of nitrogen-containing compounds. Such increments often present a less general character than their ^{13}C equivalents and are highly dependent on the solvent used (Sect. 5). Thus different series of increments corresponding to specific media are sometimes derived for the same series of compounds.

Amines and their hydrochlorides (D 12)

		δ_o	β	γ	Cor β	Cor γ	R	σ	N
RNH_2	C_6H_{12}	-378.8	22.6	-3.8	-4.9	-1.8	for C_6H_{12}		
	MeOH	-377.2	21.3	-3.4	-5.0	-0.6	0.997	1.4	34
	HCl, MeOH	-361.8	15.2	-2.2	-3.6	+0.4			
$(R)_2NH$	C_6H_{12}	-371.3	19.0	-3.8	-4.9	-1.8	for MeOH		
	MeOH	-369.8	18.3	-3.4	-5.0	-0.6	0.998	1.0	34
	HCl, MeOH	-356.6	13.5	-2.2	-3.6	+0.4			
$(R)_3N$	C_6H_{12}	-367.7	11.7	-2.2	-8.2	-1.8	for HCl, MeOH		
	MeOH	-362.1	10.5	-1.8	-5.7	-0.6	0.999	0.5	35
	HCl, MeOH	-349.2	9.1	-0.9	-2.2	+0.4			

The ^{15}N chemical shift of a given amine (or hydrochloride) may be computed from the equation

$$\delta^{15}N = \delta_o\ ^{15}N + n_1\beta + n_2\gamma + n_3(\text{Cor } \beta + \text{Cor } \gamma) \qquad \text{ppm}/CH_3NO_2$$

Cor β is the correction for branching at Cα and Cor γ the additional correction for branching at Cα if there is a γ carbon at the same residue

Example: 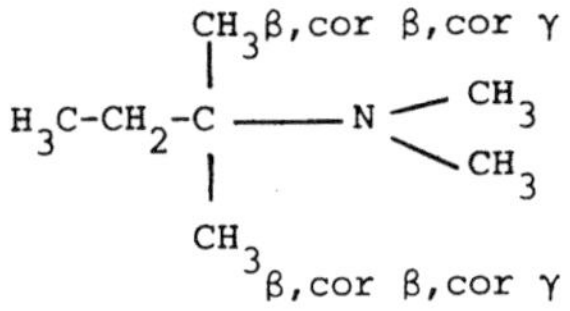

in CH_3OH, $\delta^{15}N$ obs = -344.9 ppm

$$\delta^{15}N \text{ calc} = -362.1 + 3 \times 10.5 - 1.8 - 2(5.7 + 0.6) = -345.0 \text{ ppm} .$$

Other tables of additive increments for amines, based on smaller sets of $\delta^{15}N$ data from the same laboratory, have also been published (L 25) (W 21).

According to the approximate Karplus-Pople theory of chemical shifts, at least a large part of the β parameter may be understood as an inductive effect acting mainly through the $\langle r_{2p}^{-3} \rangle$ contribution (K 35, K 36). The γ parameters depend on the conformational equilibria and are influenced by steric interactions between vicinal groups and nitrogen atoms.

Aminoacids (S 5)

$$\delta^{15}N = \delta_o\ {}^{15}N + n_1\beta + n_2\gamma + n_3\delta$$

$$\delta_o = -352.2\ ,\ \beta = 12.9\ ,\ \gamma = -3.5\ ,\ \delta = 1.2\ .$$

The corresponding ^{13}C shift parameters have roughly the same values (β = 9.4, γ = -2.5 and δ = +0.3), indicating a similarity in the factors affecting $\delta^{15}N$ and $\delta^{13}C$.

Methylanilines

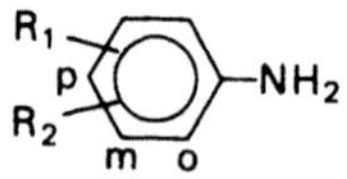

$R_1, R_2 = CH_3$ (L 26)

$$\delta^{15}N = -325.5 - n_o\ 2.37 - n_m\ 0.77 - n_p\ 2.55 - n_S\ 0.66 \quad (R = 0.997,\ \sigma = 0.13\text{ppm})$$

The subscript S corresponds to a steric effect operative only in 2,6- and 2,3- disubstituted compounds. Other substituent parameters are given:

$$C_o(Cl) = -0.92\ ,\ C_m(Cl) = +1.17\ ,\ C_o(Br) = +4.22\ ,$$

$$C_m(Br) = +1.95\ ,\ C_o(OCH_3) = -10.51\ .$$

Piperidines and decahydroquinolines and their hydrochlorides (D 13)(D 14)

A_i		δ_o	α		β		γ			$\alpha\beta$ inter	
			eq	ax	eq	ax	eq	ax	*gem*	$\alpha_{eq}\beta_{eq}$	$\alpha_{eq}\beta_{ax}$
Free bases (C_6H_{12})		-343.2	2.2	-20.2	17.8	9.0	-0.5	-10.3	3.7	-5.3	-13.5
Hydrochlorides	$CHCl_3$	-335.9	2.3	- 3.9	12.0	8.4	1.3	- 4.5	1.2		-4.3
	CH_3OH	-341.2	5.9	- 0.3	13.2	8.4	1.3	- 4.5	1.2		-4.3

The increment $\alpha\beta$ is added when two methyl groups are present in positions α and β ($\alpha_{ax}\beta_{ax}$ = 0)

$$\delta^{15}N = \delta_o + \sum_i n_i\ A_i \quad (R = 0.998,\ \sigma = 0.5\ \text{ppm},\ N = 56)$$

Example: *trans* 1,2-dimethyl piperidine hydrochloride (in $CHCl_3$): $\delta^{15}N_{exp} = -325.6$ ppm

$$\delta^{15}N_{calc} = -335.9 + 2.3 + 12.0 - 4.3 = -325.9$$

Protonation shifts of piperidines are also expressed in terms of additive parameters

$\delta_o = +3.1$, $\alpha_{eq} = 2.1$, $\beta_{eq} = -4.1$, $\beta_{ax} = -0.5$, $\gamma_{eq} = 0.8$, $\gamma_{ax} = 4.1$,

$\gamma_{gem} = -2.0$, $\alpha_{eq}\beta_{eq} = 2.1$, $\alpha_{eq}\beta_{ax} = 6.8$. (R = 0.989, σ = 0.7, N = 23)

Amides

C - C - CO - N - C - C (M 20d)

γ' β' α β

$$\delta^{15}N = -266.8 + \sum_i n_i S_i \qquad (R = 0.99_3,\ \sigma = 1.6)$$

Si	α	β	β'	γ'
	-5.9	14.6	-2.2	-2.2
	(±0.5)	(±0.5)	(±1.2)	(±0.7)

Example: CH_3-CH(CH_3)-CO-NH-CH_3 $\delta^{15}N_{exp} = -278.4$ ppm

$$\delta^{15}N_{calc} = -266.8 - 5.9 - 2.2 - 4.4 = -279.3 \text{ ppm}$$

Ureas and thioureas (M 9)

C - C - N - CA - N - C - C

δ' γ' α' α β

in H_2O/CH_3OH $\delta^{15}N = -303.3 + n'\alpha' + \sum_i n_i\, Si$ (R = 0.992, σ = 2.6, N = 29)

(18 shift values for ureas and 11 for thioureas)

(n' = 0 if A ≡ O and n' = 1 if A ≡ S)

Si =	α'	α	β	γ'	δ'
	30.6	-4.9	14.3	-2.4	0.8
	(±1)	(±0.6)	(±0.7)	(±0.6)	(±0.5)

Example: 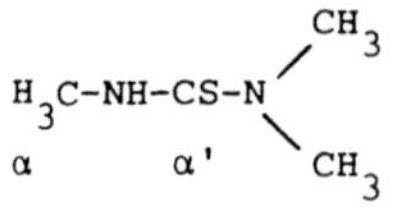 γ' $\delta^{15}N_{exp}$ = -281.4 ppm

$\delta^{15}N_{calc}$ = -303.3 + 30.6 - 4.9 - 2 x 2.4 = -282.4 ppm

Ureas

C - C - C - N - CO - N - C - C - C - C (S 16)

ε' δ' γ' α β γ δ

in $(CH_3)_2SO$ $\delta^{15}N = -307.6 + \sum_i n_i\ Si$ (R = 0.995, σ = 1.0, 28 shift values)

Si =	α	β	γ	δ	γ'	δ'	ε'
	-4.64	16.61	-1.81	-0.67	-0.99	0.51	-0.36
	(±0.4)	(±0.6)	(±0.5)	(±0.9)	(±0.4)	(±0.6)	(±0.9)

Example: H_3C-CH_2-NH-CO-NH-CH_2-CH_3 $\delta^{15}N_{exp}$ = -295.3 ppm

β α γ' δ'

$\delta^{15}N_{calc}$ = -307.6 - 4.64 + 16.61 - 0.99 + 0.51 = 296.1 ppm

Iminium and amidinium, guanidinium salts (R 13)

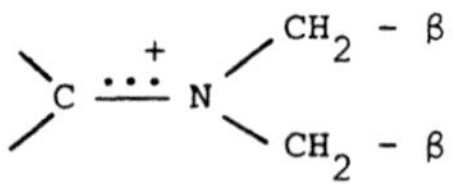

β = 11.4 ppm

Nitrosamines (G 14)

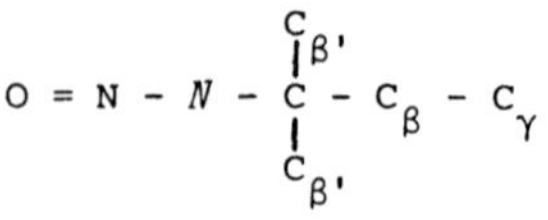

$\delta^{15}N = \delta_o + \sum_i n_i\ Si$

Si	δ_o	β	β'	γ	
	-147.0	+12.2	+7.9	-2.0	(R = 0.999, σ = 1.5)
		(±0.3)	(±0.2)	(±0.2)	

Aminophosphines

(G 10)

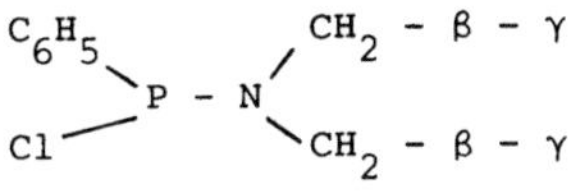

$\delta^{15}N = -333.2 + \sum_i n_i\ Si$ (R = 0.996, σ = 1.7, N = 8)

Si β = 11.6 (±0.5) , γ = -0.6 (±0.5)

Phosphoramidates

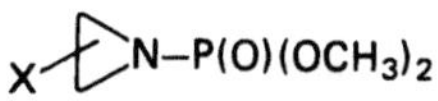

(G 6)

$\delta^{15}N = \delta_o + n_1\beta + n_2G + n_3V_c + n_4V_t$

δ_o is the value for the unsubstituted compound: -367.4 ppm

β = 15.3 G = -4.0 V_c = -1.2 V_t = -1.6 ppm

Example: 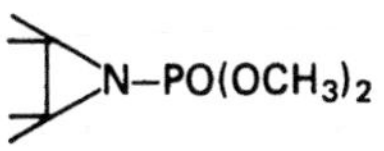

δ_{exp} = -320.6 ppm

δ_{calc} = -367.4 + 4 x 15.3 - 2 x 4.0 - 2 x 1.2 - 2 x 1.2 - 2 x 1.6 = -319.8 ppm

Miscellaneous values of the γ parameter

(N 4) (H 18)

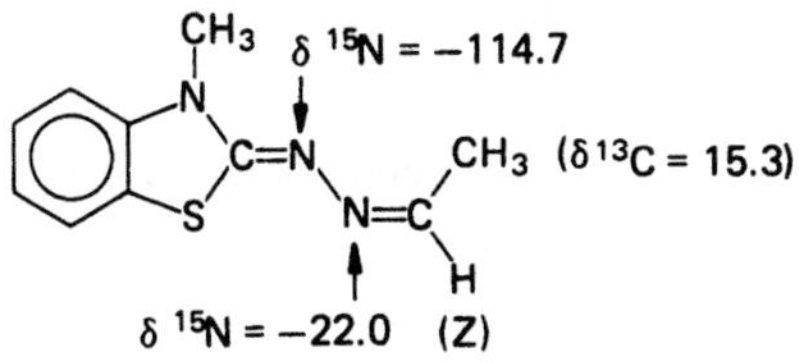

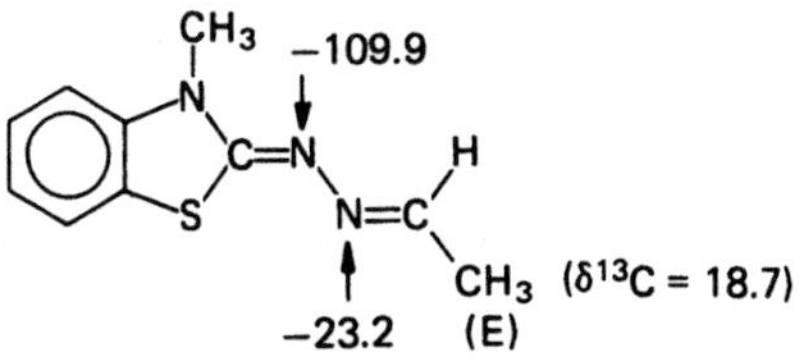

γ_{effect} = -4.8 ppm

(6.08 MHz 304 K 20% in $CDCl_3$ 50% ^{15}N enriched)

(H 18)

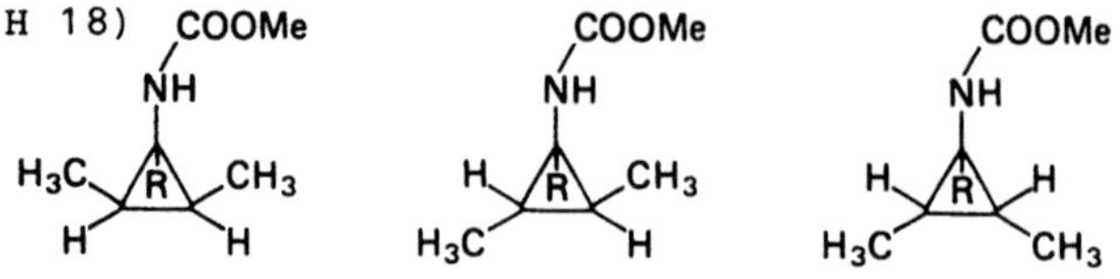

R = OEt

$\delta^{15}N$ -293.5 ppm -287.7 ppm -280.7 ppm (experimental conditions not specified)

$\bar{\gamma}_{effect.}$ = -6.2 ppm

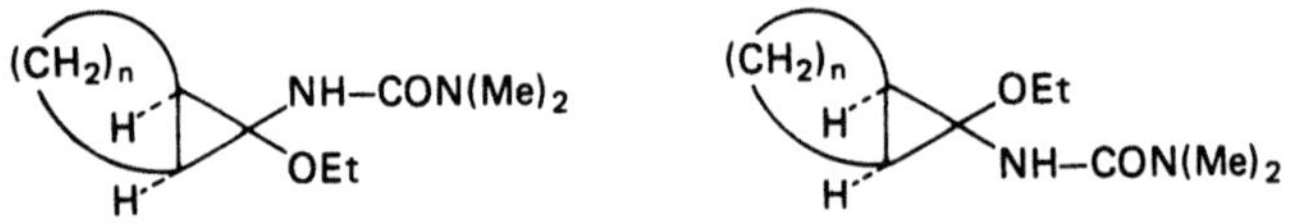

(n = 3, 4)

$\delta^{15}N$ -292.7 ppm -281.0 and -282.1 ppm

$\bar{\gamma}_{effect}$ = -5.6 ppm

Conclusion

Examination of these analyses shows that the value of the β effect ranges from +23 ppm to +8 ppm and lies frequently at about +15 ppm. Alkyl substitution therefore induces paramagnetic shifts. The γ shift, which is probably sterically induced, is diamagnetic and varies from -6 to -1 ppm.

6.2.2 Correlations between $\delta^{15}N$ and Chemical Shifts of Other Nuclei: $\delta^{15}N = a + b\delta(X)$

6.2.2.1 Correlations $\delta^{15}N$ vs. $\delta^{1}H$ (ppm/CH_3NO_2)

<u>Anilines</u> (A 4)

$$\delta^{15}N = -371.4 + 10.2\ \delta^{1}H\ (DMSO) \qquad (R = 0.97_1,\ \sigma = 1.5,\ N = 21)$$

$\delta^{1}H$ is the chemical shift of the amino proton. It appears that electron-withdrawing groups produce down-field shifts for both ^{1}H and ^{15}N.

6.2.2.2 Correlations $\delta^{15}N$ vs. $\delta^{13}C$

Amines (D 12)

$$\delta^{15}N = a + b\delta^{13}C$$

$\delta^{13}C$ is the shift value of the carbon atom which corresponds to the structurally analogous alkane and is situated in the same position as the nitrogen atom of the amine.

Example:

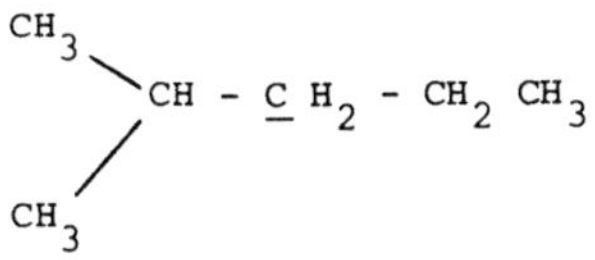

and

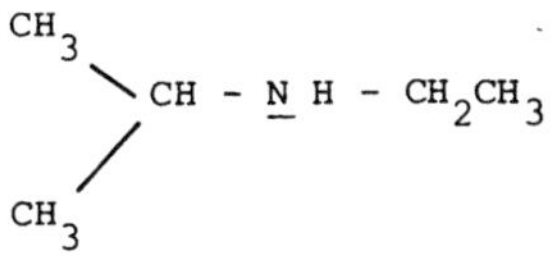

	RNH$_2$			(R)$_2$NH			(R)$_3$N		
	amine		HCl	amine		HCl	amine		HCl
	I	II	II	I	II	II	I	II	II
a	-387.5	-385.2	-367.3	-400	-396.3	-376.3	-397.6	-391.9	-378.5
b	2.06	1.95	1.40	1.96	1.86	1.38	1.39	1.38	1.36
R	0.998	0.999	0.997	0.994	0.992	0.989	0.906	0.953	0.952
N	14	15	15	7	7	7	17	16	16

I corresponds to cyclohexane solutions and II to methanol solutions.

Piperidines, decahydroquinolines and their hydrochlorides (D 13, D 14)

$$\delta^{15}N = a + b\delta^{13}C \qquad (\delta^{13}C \text{ of the hydrocarbon analogue})$$

	Secondary piperidines				Tertiary piperidines					
	Free bases		Hydrochlorides		Free bases				Hydrochlorides	
	C_6H_{12}	CH_3OH	$CHCl_3$	CH_3OH	C_6H_{12}		CH_3OH		$CHCl_3$	CH_3OH
					I	II	I	II		
a	-393.9	-392.0	-364.0	-376.0	-403.5	-404.2	-397.7	-400.4	-369.5	-380.6
b	1.90	1.80	1.11	1.36	1.82	1.97	1.73	1.95	1.11	1.39
R	0.997	0.999	0.976	0.993	0.997	0.999	0.999	0.995	0.987	0.992
N	14	11	9	17	10	5	8	3	24	22

I corresponds to tertiary piperidines with *N*-equatorial groups and II to the other piperidines.

The amplification factor of electronic effects for ^{15}N with respect to ^{13}C is about 2 for the free bases and 1.3 for the hydrochlorides. This behaviour is evidence of an important contribution of the nitrogen lone pair.

Enamines (S 10)

$$\Delta\delta^{15}N = -161.1 + 1.11\ \delta^{13}C_\beta \qquad (R = 0.96,\ \sigma = 0.5,\ N = 7)$$

where $\delta^{13}C_\beta$ is the chemical shift of the β carbon of the substituted anilinostyrenes I

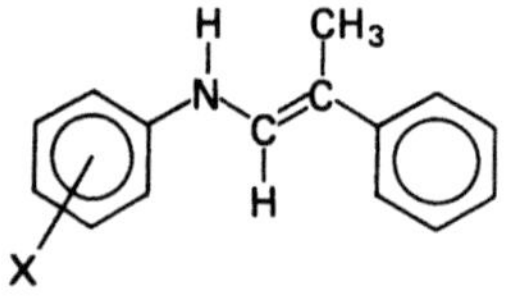

I

and $\Delta\delta^{15}N$ represents the ^{15}N shift of the anilinostyrene I minus the ^{15}N shift of the corresponding substituted aniline ($X = NO_2$, Cl, OCH_3, H, CH_3) (in DMSO d^6).

Aniline derivatives 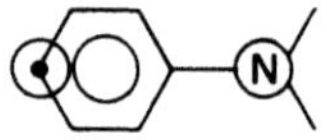 (L 26) (S 14)

N,N dimethylanilines

$$\Delta\delta^{15}N = \text{cst} - 2.3\ \Delta\delta^{13}C_p \qquad (R = 0.941,\ N = 9) \qquad (S\ 14)$$

C_p is the carbon atom situated in the *para* position with respect to the amine function and $\Delta\delta^{15}N = \delta^{15}N(NMe_2) - \delta^{15}N(NH_2)$, $\Delta\delta^{13}C_p = \delta^{13}C_p(NMe_2) - \delta^{13}C_p(NH_2)$.

The variations of both $\delta^{15}N$ and $\delta^{13}C_p$ are dominated by the torsion of the $N(Me)_2$ group under the influence of alkyl ring substitution. When torsion around the C-N bond reduces the electron delocalization, $\delta^{15}N$ decreases.

anilines

$$\delta^{15}N = -355.3 + 1.48\ \delta^{13}C \qquad (R = 0.997,\ \sigma = 0.45,\ N = 5) \qquad (L\ 26)$$

$\delta^{15}N$ is the chemical shift of ring substituted anilines and $\delta^{13}C$ is the methyl shift of the corresponding substituted toluene (substituents 2-Cl, 3-Cl, 2-Br, 3-Br, 2-OCH_3).

Methyl substituted anilinium ions (P 16)

The nitrogen resonances correlate moderately well with the methyl ^{13}C shifts of the corresponding methylbenzenes

$$\Delta\delta^{15}N = cst + 0.83\ \Delta\delta^{13}C \qquad (R = 0.972,\ N = 10)$$

The ^{15}N and ^{13}C shifts are referred to the resonances of the anilinium ion and of toluene, respectively.

Aminopyridines and aminopyrimidines (S 25)

aminopyridines:

$$\Delta\delta^{15}N = -4.62 + 3.73\ \Delta\delta^{13}C \qquad (DMSO)\ (R = 0.998,\ N = 7)$$

$\Delta\delta^{15}N$ and $\Delta\delta^{13}C$ concern the aminopyridine and the corresponding substituted aniline and are referred respectively to pyridine and benzene.

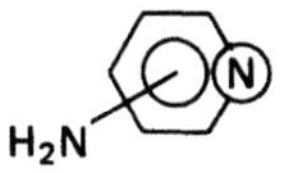

$$\Delta\delta^{15}N = \delta^{15}N(\text{aminopyridine}) - \delta^{15}N(\text{pyridine})$$

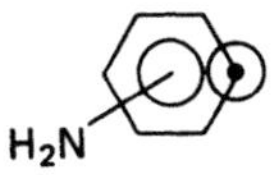

$$\Delta\delta^{13}C = \delta^{13}C(\text{aniline}) - \delta^{13}C(\text{benzene})$$

aminopyrimidines:

$$\Delta\delta^{15}N = -3.38 + 2.78\ \Delta\delta^{13}C \qquad (DMSO)\ (R = 0.990,\ N = 7)$$

$\Delta\delta^{15}N = \delta^{15}N(\text{aminopyrimidine I}) - \delta^{15}N(\text{pyrimidine})$ and

$\Delta\delta^{13}C = \delta^{13}C(\text{aminopyridine II}) - \delta^{13}C(\text{pyridine})$

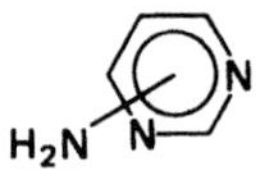

I

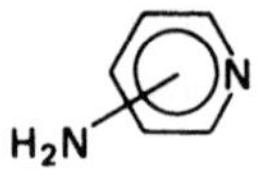

II

Amidinium salts (R 13)

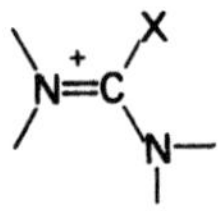

$\delta^{15}N = 76.5 - 2.25\ \delta^{13}C$ (R = 0.995, σ = 1.3, N = 8)

$\delta^{15}N$ increases as $\delta^{13}C$ decreases; (push-pull mechanism). This behaviour may be compared to that of aniline derivatives.

Imines (W 19)

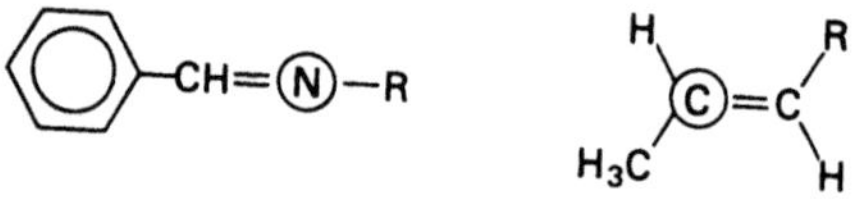

R = alkyl

$\delta^{15}N = -307.0 + 1.96\ \delta^{13}C$ (R = 0.998, N = 5)

Imines, amidines, hydrazones (N 5)

$\delta^{15}N = -405.0 + 2.75\ \delta^{13}C$ (R = 0.985, σ = 6.2, N = 17)

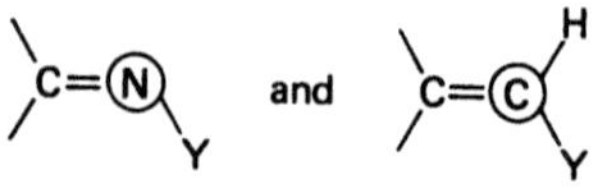

with Y = R, Ar, $N(R)_2$

Oximes (B 47)

$\delta^{15}N = -302.0 + 2.18\ \delta^{13}C$ (R = 0.986, N = 16)

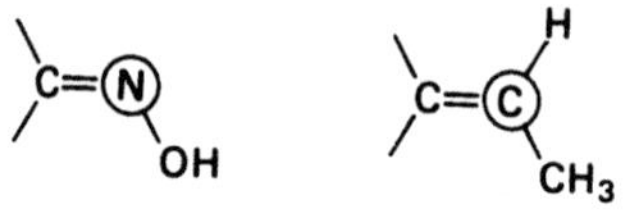

alkyl and aryl substituents

Z or *E* *Z* or *E*

6.2.2.3 *Correlations* $\delta^{15}N$ *vs.* $\delta^{15}N$

Aminophosphines

$X(Y)P\text{-}N(R)_2$ $\quad \delta^{15}N(R{=}CH_3) = 40.3 + 1.22\ \delta^{15}N(R{=}C_2H_5)$ $(R = 0.978,\ \sigma = 3,\ N = 8)$ (G 10)

$\delta^{15}N(R{=}C_2H_5) = 12.2 + 1.11\ \delta^{15}N(R{=}C_3H_7i)$ $(R{=}0.994,\ \sigma{=}1.3, N{=}8)$

Anilines

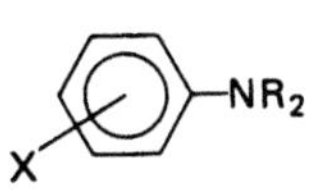

$\delta^{15}N(R{=}CH_3) = -26.6 + 0.96\ \delta^{15}N(R{=}H)$ (D 15)

$(R = 0.992,\ \sigma = 1.3,\ N = 8)$

Aminosilanes

$R_2N\text{-}Si(R',R'',R''')$ $\quad \delta^{15}N(R{=}CH_3) = -62.6 + 0.905\ \delta^{15}N(R = Et)$ (F 8)

$(R = 0.965,\ \sigma = 4.3,\ N = 7)$

6.2.2.4 *Correlations* $\delta^{15}N$ *vs.* $\delta^{19}F$

Anilines

methyl derivatives

$\delta^{15}N = -294.9 + 0.27\ \delta^{19}F$ $\quad (R = 0.999,\ \sigma = 0.1,\ N = 5)$ (L 26)

$\delta^{19}F$ is the chemical shift of the corresponding substituted fluorobenzene referred to $CFCl_3$.

electronegatively substituted derivatives (2 Cl, 3 Cl, 2 Br, 3 Br, 2 OCH_3)

$\delta^{15}N = -266.4 + 0.505\ \delta^{19}F$ $\quad (R = 0.996,\ \sigma = 0.6,\ N = 5)$

$\delta^{15}N$ (ppm/CH_3NO_2) and $\delta^{19}F$ (ppm/$CFCl_3$) are expressed using the same sign convention. As $\delta^{15}N$ increases (paramagnetic shift), $\delta^{19}F$ also increases but the fluorine atom is a better electronic sensor than nitrogen.

6.2.2.5 *Correlation $\delta^{15}N$ vs. $\delta^{29}Si$*

Silatranes

$H_3C\text{-}Si\text{-}(O\text{-}CH_2CH_2)_3N$. Solvent effects on $\delta^{15}N$, $\Delta\delta^{15}N$ and on $\delta^{29}Si$, $\Delta\delta^{29}Si$. (P 12)

$\Delta\delta^{15}N = -0.29 - 0.77_2\ \Delta\delta^{29}Si$ (R = 0.987, σ = 0.55, N = 12)

this correlation was computed from solvent effects.

6.2.2.6 *Correlation $\delta^{15}N$ vs. $\delta^{31}P$*

Aminophosphines

$\delta^{15}N = a + b\delta^{31}P$ (or $^{1}J_{P\text{-}N}$) (G 10)

		a	b	R	σ	N
$X(Y)P\text{-}N(Me)_2$	δP	-428.6	0.64	0.97	3	13
	$J_{P\text{-}N}$	-408.9	1.00	0.98	2.3	13
$X(Y)P\text{-}N(C_2H_5)_2$	δP	-393.6	0.59	0.97	3	10
	$J_{P\text{-}N}$	-383.1	0.96	0.98	2.8	10
$X(Y)P\text{-}N(C_3H_7i)_2$	δP	-340.1	0.38	0.97	3	8
$X(Y)P\text{-}N(R)_2$	δP	-409.7	0.53	0.92	4.7	31
	$J_{P\text{-}N}$	-402.3	0.89	0.91	5.2	31

The $\delta^{31}P$ values are referred to H_3PO_4. It appears that as $\delta^{15}N$ increases (paramagnetic shift), both $\delta^{31}P$ and $^{1}J_{P\text{-}N}$ increase.

Conclusion

The magnitude and the sign of the slope b of the above mentioned correlations are representative of the way in which the electronic effects are transmitted in a molecule.

The slope is mainly positive when the atoms under consideration are situated in equivalent molecular positions and that means that the electronic effects act in the same direction for both atoms. These cases represent the more simple situations, but interestingly, the magnitude of b can be considered as the sensor ability of a given atom towards the electronic effects in a molecule. A rough scale of amplification factors may be established from the limited series of results considered

^{1}H	^{13}C	^{15}N	^{19}F
1	5	10	25

In fact, these factors correspond to the relative ranges of chemical shifts of the different nuclei considered here. When the slope is negative (i.e. anilines, iminium salts) a push-pull mechanism, i.e. one atom donates electron density to the other atom, may often be involved.

6.2.3 Correlations between $\delta^{15}N$ and Computed Electronic Density Q

$$\delta^{15}N = a + b(Q)$$

Alkylpyridines

No evident correlation is found between the chemical shifts $\Delta\delta^{15}N$ and the π and σ electron densities calculated in the INDO-MO framework: Q_{π}, Q_{σ}, $Q_{\pi,\sigma}$.

Triazenes (aryl - benzothiazolinyl - triazenes) (D 16)

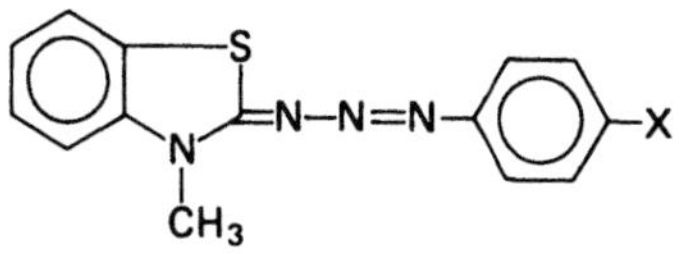

X = H, OCH_3 (*cis* and *trans* isomers)

$$\delta^{15}N = 845.9 - 723.3\ Q_{\pi} \qquad (R = 0.987,\ \sigma = 11,\ N = 15)$$

The π electron densities Q_π are calculated in the Pariser-Parr-Pople approximation. The ^{15}N chemical shifts of the three nitrogens increase as their π electron density decreases.

Methylanilines (P 16)

$$\delta^{15}N = a + b\Delta Q_N$$

$\delta^{15}N$ is the nitrogen chemical shift of the methylaniline referenced to CH_3NO_2 and ΔQ is the difference in π,σ or π-σ electron density at the nitrogen, between the considered methylaniline and aniline. A positive ΔQ value denotes a higher electron density than in aniline. These electron densities result from INDO-MO calculations.

ΔQ_π	a = -325.0	b = +1322 ppm·electron^{-1}	R=0.970,σ=0.4,N=9
ΔQ_{σ} (non-*para*)	a = -325.2	b = -753 ppm·electron^{-1}	R=0.997,σ=0.1,N=6
(all points)	-326.1	-649	R=0.870,σ=0.8,N=9
$\Delta Q_{\pi,\sigma}$ (non-*para*)	a = -326.0	b = -1580 ppm·electron^{-1}	R=0.978,σ=0.4,N=6

$\delta^{15}N$ moves towards high frequencies (paramagnetic shift) as Q increases and as Q_σ or $Q_{\pi,\sigma}$ decreases. These results correct errors discovered in previous publications (A 28) (L 26).

Substituted anilines (A 4)

$$\delta^{15}N = -139.15 - 106.4\ Q_\pi \qquad (R = 0.998,\ \sigma = 1.2,\ N = 8)\ (DMSO)$$

Q_π is the π electron density at the nitrogen calculated in the Hückel framework. Electronegative substituents increase the *p*-π interaction between the nitrogen lone pair and the aromatic ring and the concomitant decrease in π electron density at the nitrogen is associated with deshielding. The same behaviour was observed in ^1H spectroscopy (L 48) where $\delta_{NH_2} = 22.4 - 10.2\ Q_\pi$ ($R = 0.992$, $\sigma = 0.88$, $N = 14$).

N-substituted acetamides (K 37)

For *N*-alkylacetamides:

$$\delta^{15}N = -266.8 + 3300\ \Delta Q_\pi \qquad (R = 0.947,\ \sigma = 3.5,\ N = 10)$$

ΔQ_π is the π electron density calculated by the CNDO/2 method and referenced to that of the parent *N*-methylacetamide. A paramagnetic shift accompanies the increase in the π electron density on the nitrogen.

For *N*-arylacetamides, it is shown that $\delta^{15}N$ is much less sensitive to the electron effects of *para* substituents that $\delta^{15}N$ of the corresponding benzenamines (W 17) (A 4).

Conclusion

Correlations between $\delta^{15}N$ and theoretical parameters can provide useful information about both the transmission of the electronic effects within different types of molecules and the factors which govern the nitrogen shifts. However, it should be kept in mind that stringent checks must be applied to ensure that the correlations are meaningful. In fact, correlations between a physical parameter [E(PES), $\delta^{15}N$, $\delta^{13}C$ ...] and an ordered computed parameter (Q_π, Q_σ ...) are always characterized by correlation coefficients higher than 0.9 (H 22).

6.2.4 Correlation between $\delta^{15}N$ and Hammett-like Substituent Parameters

The Hammett σ, Taft, σ_I and σ_R, Swain-Lupton F and R parameters are those which have been the most frequently used to discuss the $\delta^{15}N$ changes in terms of electronic effects. The commonly adopted values of these parameters are given in Table 6.62.

Anilines, benzenesulfonamides, acetanilides (S 12)

$$\Delta\delta^{15}N = \rho_I\sigma_I + \rho_{R^-}\sigma_{R^-}$$

where $\Delta\delta^{15}N = \delta^{15}N\ (X{=}X) - \delta^{15}N\ (X{=}H)$ in pX-C_6H_4-NH_2 I

pX-C_6H_4-NH-$SO_2C_6H_5$ II

pX-C_6H_4-NH-CO-CH_3 III

σ_I, σ_{R^-} are the Taft substituent parameters which reflect the inductive and resonance effects of the substituent X.

Compounds	ρ_I	ρ_{R^-}	R	σ	N	Exper. Conditions
I	+12.11	+21.64	0.995	0.8	9	(DMSO solution 9 mol-%)
II	+ 4.42	+ 8.61	0.998	0.2	7	(DMSO solutions 1M)
III	+ 0.61	+ 5.27	0.997	0.15	5	(DMSO solutions 10-16 mol-%)

The transmission of electronic effects from X to ^{15}N decreases in the series I > II > III.

Enamines (S 10)

$\Delta\delta^{15}N = -39.6 + 5.27\ \sigma$ (R = 0.98_1, σ = 0.35, N = 7)

where $\Delta\delta^{15}N$ represents the ^{15}N chemical shift of the anilinostyrene I referred to the ^{15}N shift of the corresponding substituted aniline

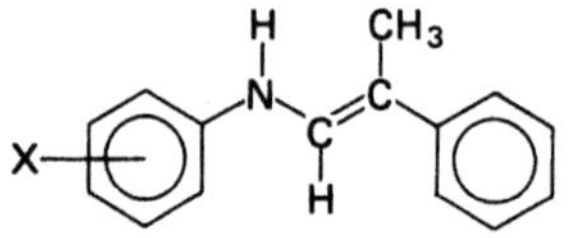

(I)

Benzamides $X-C_6H_4CONH_2$ (P 14) (J 10)

$\delta^{15}N = -279.3 + 3.57\ F + 5.10\ R$ (R = 0.998, σ = 0.1, N = 8)

(F and R are the Swain-Lupton parameters of X)

From the experimental values of $\delta^{15}N$ given for benzamides in (P 14), the above correlation was computed (Table 6.62) and applied to the discussion of electronic delocalization in the N-C bond (see Sect.6.2.5).

Imines, oximes, hydrazones X - Ar - CH = N - Y

$\delta^{15}N = a + b\sigma$ σ is the Hammett constant (see Sect. 5.2.2)

The slope is positive, denoting a paramagnetic shift when an electron withdrawing group X is substituted in the *para* position of the aromatic ring.

<u>Porphyrins</u> (Zn tetraphenylporphyrin - substituted pyridine complexes) (G 15)

The ^{15}N shift of porphyrin (95% ^{15}N enriched) is correlated with the Hammett parameter, σ, of the group X substituted in the *para* or *meta* position of the pyridine ligand and with the stability constant K

$$\Delta\delta^{15}N = -1.53\ \sigma \qquad (R = 0.955,\ \sigma = 0.2_5,\ N = 11)$$

$$\Delta\delta^{15}N = +1.16\ \log K - 4.48 \qquad (R = 0.986,\ \sigma = 0.1_5,\ N = 7)$$

$\Delta\delta^{15}N$ is defined as $\delta^{15}N$ (X-pyridine) - $\delta^{15}N$ (pyridine) and is found in a 2.25 ppm range. The nitrogen shielding decreases with complex formation and with the electron donating properties of X.

<u>Aminophosphines</u> (G 10)

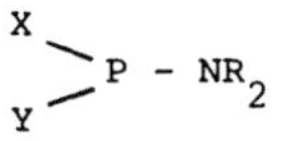

$$\delta^{15}N = -328.5 + 69.5\ \sigma_I\ (X,Y) \qquad (R = 0.90,\ \sigma = 6,\ N = 13)$$

Variations towards high frequencies are roughly observed when the inductive electron withdrawing properties of the substituents increase. No evident correlation is found between $\delta^{15}N$ and σ_R.

<u>Silatranes</u> $R\text{-}Si(OCH_2CH_2)_3N$ (L 42) (P 20)

$$\delta^{15}N = -357.3 + 3.33\ \sigma^* \qquad (R = 0.97,\ \sigma = 1.28,\ N = 6) \qquad (L\ 42)$$

$$-356.8 + 3.54\ \sigma^* \qquad (P\ 20)$$

The increase of $\delta^{15}N$ with the aliphatic substitution constant σ^* indicates an increase of the N ⟶ Si bond polarity. $\delta^{15}N$ values appear also to be correlated with the electric moment of the N-Si bond.

6.2.5 Correlations between $\delta^{15}N$ and Free Energies of Activation of Hindered Rotation Processes about N-X Bonds

The problem of delocalization of the electron lone pair at nitrogen in conjugated systems is very important and is frequently discussed on the basis of activation parameters associated with rotation around

N-X bonds. However a number of limitations are inherent in dynamic NMR and it is interesting to examine if, and to what extent, the nitrogen chemical shift may reflect the degree of electron delocalization. In fact, significant correlations between $\delta^{15}N$ and ΔH^* or ΔG^* parameters are found in certain series of compounds (M 22) and may be a source of information about the electron distribution. Moreover, anomalies with respect to such correlations allow special properties either of the nitrogen shift or of the rotational barrier to be detected and explained.

Anilines (D 15)

m, p-substituted $\delta^{15}N = -367.2 + 1.78_5\ \Delta G^*\ (kJ \cdot mol^{-1})$

o -substituted $\delta^{15}N = -378.5 + 1.78_5\ \Delta G^*\ (kJ \cdot mol^{-1})$

The correlations were computed from the equation: $\delta^{15}N\ (NH_2)$ vs. $\delta^{15}N\ (NMe_2)$ in Sect. 6.2.2 and the following relation, established for *N,N*-dimethylanilines.

N,N-Dimethylanilines and -pyridines (D 15)

$\delta^{15}N = -370.0 + 1.69_5\ \Delta G^*\ (kJ \cdot mol^{-1})$ $(R = 0.99,\ \sigma = 1.2,\ N = 11)$

Enamines (enamino ketones) (D 15)

$\Delta\delta^{15}N = cst + 1.45\ \Delta(\Delta G^*)\ (kJ \cdot mol^{-1})$ $(R = 0.99,\ \sigma = 1.4,\ N = 10)$

a) $\Delta\delta^{15}N = \delta^{15}N$ (enamine) $-$ $\delta^{15}N$ (amine) and $\Delta(\Delta G^*) = \Delta G^*$(enamine) $-$ ΔG^*(amine)

Example: $(CH_3)_2N\text{-}CH = CHCH_3$ and $(CH_3)_2N\text{-}CH_2CH_2CH_3$

b) $\Delta\delta^{15}N = \delta^{15}N$ (enaminocetone) $-$ $\delta^{15}N$ (enamine)

Example: $(CH_3)_2N\text{-}CH = CH\text{-}CHO$ and $(CH_3)_2N\text{-}CH = CH\text{-}CH_3$

A series of merocyanines has also been studied recently (R 23). The slope of the correlation between $\delta^{15}N$ and ΔG^* is found to be about 0.85.

Enamines (S 10)

$\Delta\delta^{15}N = 11.1 - 1.20\ \Delta G^*\ (kJ \cdot mol^{-1})$ $(R = 0.97,\ \sigma = 3.5,\ N = 7)$

$\Delta\delta^{15}N - \delta N$ (amine) $-$ δN (enamine)

Example: 

The ΔG^* values vary between about 34 and 66 $kJ \cdot mol^{-1}$ and the $\Delta\delta^{15}N$ values between -29 and -66 ppm.

Benzamides

On the basis of the correlations observed between ΔG^* and $\delta^{15}N$ on one hand, and substituent parameters on the other hand (K 34) (P 14), it has been suggested that $\delta^{15}N$ and ΔG^* should be closely related (J 10). Thus for *p*-substituted *N,N*-dimethylbenzamides $X\text{-}C_6H_4\text{-}N(CH_3)_2$

$$\Delta G^*_{298} = +64.4 + \underset{(\pm 0.4)}{2.68}\,F + \underset{(\pm 0.7)}{8.32}\,R \qquad (R = 0.988,\ \sigma = 0.4,\ N = 9)$$

where *F* and *R* are the Swain-Lupton parameters of X.
ΔG^* can also be expressed in terms of $\delta^{15}N$ but the correlation, although significant, is characterized with a relatively small value of the coefficient R.

Ureas (and thioureas) (M 9)

ureas $\delta^{15}N = -356.2 + 1.07\ \Delta G^*_T + 7.6\ I_S$ $(R = 0.98_7,\ \sigma = 2.9,\ N = 6)$

thioureas $\delta^{15}N = -335.4 + 0.59\ \Delta G^*_T + 2.7\ I_S$ $(R = 0.98,\ \sigma = 5.4,\ N = 6)$

I_S is a parameter which takes into account the steric interaction between the number n or n' of alkyl substituents on each nitrogen atom $I_S = n + n' - 1$.

Amides (and related compounds) (M 22)

$$\delta^{15}N = cst + 1\ \Delta G^*_T\ (Ea) \qquad (R = 0.98,\ \sigma = 1.8,\ N = 9)$$

Thioamides (and related compounds) (M 22)

$$\delta^{15}N = cst + 0.8\ \Delta G^*_T\ (Ea) \qquad (R = 0.97_5,\ \sigma = 2,\ N = 5)$$

<u>Nitrosamines and related derivatives</u> (G 14)

$(CH_3)_2N\text{-}N = X$

$\delta^{15}N = -\ 341.9 + 2.0\ \Delta G^*\ (kJ \cdot mol^{-1})$ (R = 0.99, σ = 1.6, N = 8)

This trend is confirmed by the satisfactory correlations between $\delta^{15}N$ and the N-N bond lengths which are expected to also reflect electron delocalization (see following section). ΔG^* is assumed to be free of entropy contribution due to steric or solvation effects.

Conclusion

It can be considered that a certain degree of correlation exists between $\delta^{15}N$ and $E_a(\Delta H^*)$ parameters. However, some discrepancies may arise, in principle, when ΔG^* values are considered, since entropy effects may cancell a part of the effect of electron delocalization, assumed to be expressed in Ea(ΔH^*) terms. Such correlations are of practical use in the rough estimation either of $\delta^{15}N$ or of ΔG^*, as far as structurally comparable compounds (D 15) (A 29) (S 16) (R 13) are concerned. Moreover, as already emphasized, such correlations can be very useful in discussing the influence of factors such as steric hindrance (M 9) or solvent effects (M 36). Indeed, these are liable to induce specific changes in $\delta^{15}N$ or ΔG^*_T, which are not necessarily intercorrelated. Interestingly, special behaviour either of $\delta^{15}N$ or of ΔG^*_T can then be detected. Thus, in compounds such as the guanidinium salts, no correlation can be found and this is explained on the basis of a chiral propeller-like structure. In this case, strong steric effects occur in the rotational transition state and ΔG^* may in no way reflect the electron delocalization in the ground state. $\delta^{15}N$ is then a better characteristic of delocalization (R 13).

6.2.6 Correlations between $\delta^{15}N$ and Various Parameters (λ, IP, d_{N-N}, ν_{IR}, pH)

<u>*N,N*-Dimethyl anilines</u> (ionization potential IP) (S 14)

$\delta^{15}N = cst + 20.6\ \Delta(IP)$ (R = 0.97, N = 7)

Δ(IP) is deduced from vertical ionization potentials obtained from photoelectron spectroscopy. $\Delta(IP) = \pi_4 - \pi_2$ represents the difference between the first π_4 and the third π_2 ionization bands and is expected to reflect the extent of lone-pair delocalization. $\delta^{15}N$ also correlates moderately well with the π_2 ionization potential and the slope is now 20.2 ppm/eV.

$$\delta^{15}N = cst + 20.2\ \Delta(IP)\ (\pi_2) \qquad (R = 0.95,\ N = 7)$$

Since the ionization potentials have been related to the cosines of the torsional angles, the chemical shifts can also be used to estimate these angles.

Amides (ionization potential IP) (M 20d, S 16)

As noticed in Ref. (S 16), the nitrogen chemical shifts of methyl formamides and acetamides (D 18) correlate with the ionization potential of the highest occupied π orbitals measured using photoelectron spectroscopy. Thus,

formamides:	$\delta^{15}N = -329.9 + 6.0\ IP\ (\pi_2)$	$(R = 0.98,\ \sigma = 1.1,\ N = 3)$
acetamides:	$\delta^{15}N = -365.0 + 9.3\ IP\ (\pi_2)$	$(R = 0.98_5,\ \sigma = 1.4,\ N = 3)$
formamides and acetamides :	$\delta^{15}N = -351.0 + 8.0\ IP\ (\pi_2)$	$(R = 0.93,\ \sigma = 2,\ N = 6)$

Ureas (ionization potential IP) (S 16)

The ^{15}N shifts are correlated with ionization potentials, π_1, and π_2 obtained in photoelectron spectroscopy. It is assumed that, for unsymmetrically substituted ureas, π_1 and π_2 may be assigned to the more and to the less highly substituted nitrogens respectively. Δ(IP) corresponds to the difference between each of these values and the corresponding values for urea.

$$\Delta\delta^{15}N\ \alpha + 0.6\ \Delta(IP)$$

$\Delta\delta^{15}N$ is the chemical shift difference with respect to urea. Thus, a diamagnetic shift corresponds to a decrease in IP. (The slope is computed from the values given in Table 1 of Ref. S 16 rather than from Fig. 1 of S 16 since some points do not correspond.) This trend further corroborates the relation between $\delta^{15}N$ and lone-pair delocalization discussed in the previous section.

Nitrosamines (bond length)

$$\delta^{15}N = 1941 - 1556\, d(N\text{-}N) \quad (d_{N\text{-}N} \text{ is in } \mathring{A}) \quad (R = 0.998,\ N = 4,\ \sigma = 1.5)$$

$$(CH_3)_2N\text{-}N = X \quad (X \equiv O,\ O_2,\ NAr,\ CHR,\ H_2)$$

A paramagnetic shift accompanies the decrease in $d_{(N-N)}$.

Sulfamates $nHN\text{-}(SO_3\text{-}NH_4)_{3-n}$ (IR stretching frequency ν_{N-S} (cm^{-1})) (N 6)

$$\delta^{15}N \ \alpha + 1.4\ \nu_{N-S}$$

$\delta^{15}N$ increases with the increase of the force constant of the N-S bond.

Diazo derivatives $\bar{X} - \overset{+}{N} \equiv N$ (λ_m, UV) (M 37)

The shifts of the terminal nitrogen of diazoalkanes and related compounds are correlated with the lowest energy transitions measured in UV spectroscopy.

X	Fluorene	Camphor	CH_2	O	CH_3	$\bar{N}$=C=O	$\bar{N}$=C=S	$\bar{N}$	
$\delta^{15}N(I)$	+67.5	+30.5	+23.5	-68.5	-168	-286.5	-163.5	-275.5	ppm/CH_3NO_2
λ_{max}	493	403	395	290	287	300	300	225	nm

Using the 8 pairs of the above values, the following relationship may be calculated:

$$\delta^{15}N = -575.4 + 1.40\ \lambda_{max} \qquad (R = 0.87,\ \sigma = 75,\ N = 8)$$

The correlation coefficient is slightly improved if the isocyanate ion is not considered (conversion factor: -353.5 ppm, shifts originally referenced to NH_4Cl).

Correlations: $\delta^{15}N$ vs. pH

The variations of the ^{15}N chemical shift may be used in the same way as $\delta^{13}C$ variations to obtain pK_a values from the curves $\delta^{15}N = f(pH)$. Thus, the pK_a values of aminoacids, which are well known parameters, have been computed from ^{15}N data and shown to be consistent with the accepted values (L 49) (K 41) (I 4). In this respect, the following values of the pK_a of the *Z* and *E* isomers of alanyl-proline dipeptide are derived: 3.23 (*Z*) and 2.75 (*E*) for the carboxylate group and 8.7 (*Z* and *E*) for the amino group (B 20). Similarly, titration curves have been used to obtain pK_a values for pyrazoles (S 21) and for the individual nitrogens of nebramycin or apramycin aminoglycosides (D 24) (P 13). However, it should be remembered when considering $\delta^{15}N$ variations that $\delta^{15}N$ is much more sensitive than $\delta^{13}C$ to solvent effects (see Sect. 5). The important effect of pH on the $\delta^{15}N$ line shapes is discussed in Sect. 8.

Table 6.1. ^{15}N chemical shifts of acyclic amines and hydrochlorides in methanol with respect to 1M HNO_3 in D_2O. (Ref. D 12 at 18.25 MHz and 293/303K) Solvent effect in amines $\Delta\delta = \delta_{CH_3OH} - \delta_{C_6H_{12}}$

RNH_2 + 2.5 R_2NH + 0.7 R_3N + 5.2 ppm. (conversion factor: -6.2 ppm)

R(R')N — R'' R	R'	R''	$\delta^{15}N(ppm/CH_3NO_2)$ (Amine)	$\delta^{15}N(ppm/CH_3NO_2)$ (Hydrochloride)
CH_3	H	H	-377.2 (a)	-361.4
C_2H_5	H	H	-355.4 (a)	-346.7
nC_3H_7	H	H	-359.6	-349.0
$(CH_3)_2CH$	H	H	-338.1	-334.0
nC_4H_9	H	H	-359.4	-348.8
CH_3CH_2-$CH(CH_3)$	H	H	-342.2	-335.9
$(CH_3)_2CH$-CH_2	H	H	-362.7	-351.1
$(CH_3)_3C$	H	H	-324.3	-323.7
$(CH_3)_2CH$-CH_2-CH_2	H	H	-359.3	-348.7
$(CH_3)_3C$-CH_2	H	H	-367.6	-354.0
CH_3CH_2-$C(CH_3)_2$	H	H	-328.6	-325.2
$CH_2CH{=}CH_2$	H	H	-361.7 (b)	/
CH_3	CH_3	H	-369.5 (a)	-356.6
C_2H_5	CH_3	H	-352.0	-343.7
C_2H_5	C_2H_5	H	-333.7	-330.1
nC_3H_7	nC_3H_7	H	-340.5	-333.8
$(CH_3)_2CH$	$(CH_3)_2CH$	H	-306.5	-310.0
nC_4H_9	nC_4H_9	H	-340.0	-333.5
$(CH_3)_2CHCH_2$	$(CH_3)_2CHCH_2$	H	-345.3	-337.8
CH_3	CH_3	CH_3	-363.1	-349.4
C_2H_5	CH_3	CH_3	-351.3	-340.0
C_2H_5	C_2H_5	CH_3	-340.5	-331.1
C_2H_5	C_2H_5	C_2H_5	-332.0	-322.8
nC_3H_7	CH_3	CH_3	-352.8	-340.9
$(CH_3)_2CH$	CH_3	CH_3	-347.5	-333.7
nC_4H_9	CH_3	CH_3	-352.8	-340.9
CH_3CH_2-$CH(CH_3)$	CH_3	CH_3	-349.6	-333.6
$(CH_3)_2CH$-CH_2	CH_3	CH_3	-354.8	-342.3
$(CH_3)_3C$	CH_3	CH_3	-339.8	-326.5

Table 6.1 (continued)

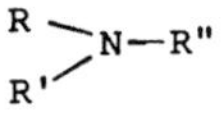			$\delta^{15}N(ppm/CH_3NO_2)$	$\delta^{15}N(ppm/CH_3NO_2)$
R	R'	R"	(Amine)	(Hydrochloride)
$(CH_3)_2CH-CH_2-CH_2$	CH_3	CH_3	-352.9	-340.8
$(CH_3)_3C-CH_2$	CH_3	CH_3	/	-343.5
$CH_3CH_2-C(CH_3)_2$	CH_3	CH_3	-344.9	-326.7
nC_3H_7	nC_3H_7	CH_3	-343.8	-332.3
$(CH_3)_2CH$	$(CH_3)_2CH$	CH_3	-330.3	-317.0
nC_4H_9	nC_4H_9	CH_3	-343.6	-332.4
$(CH_3)_2CHCH_2$	$(CH_3)_2CHCH_2$	CH_3	-349.4	-335.1

(a) At 273/278K (b) From Ref. L25

Table 6.2. ^{15}N chemical shifts of bifunctional amines

Amine	^{15}N (ppm/CH_3NO_2)	Conditions	Reference of Shifts		Ref.
$H_2N-(CH_2)_nNH_2$					
n					
2	-364.8	neat	(a) NO_3H	10M	L25
	-351.6	TFA	(b) NO_3Na		K18
3	-360.3	neat	(a) NO_3H	10M	L25
	-350.0	TFA	(b) NO_3Na		K18
4	-360.0	neat	(a) NO_3H	10M	L25
	-349.5	TFA	(b) NO_3Na		K18
5	-360.0	neat	(a) NO_3H	10M	L25
6	-359.7	neat	(a) NO_3H	10M	L25
	-349.4	TFA	(b) NO_3Na		K18
8	-349.4	TFA	(b) NO_3Na		K18
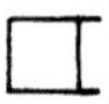 NH_2 trans	-342.0	neat	(a) NO_3H	10M	L25
NH_2 cis	-355.5	neat	(a) NO_3H	10M	L25
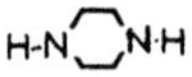	-346.4	dioxane sat.	(a) NO_3H	10M	L25
$(Me)_2N-(CH_2)_2N(Me)_2$	-359.6	neat	(a) NO_3H	10M	L25
$HO-(CH_2)_2NH_2$	-365.7	neat	(a) NO_3H	10M	L25
	-361.7	neat	(c) Me_4NCl		L21
$HO-(CH_2)_3NH_2$	-357.7	neat	(c) Me_4NCl		L21
$HO-CH_2CH(Me)NH_2$	-346.0	neat	(a) NO_3H	10M	L25
$HOCH_2CH(Et)NH_2$	-351.0	neat	(a) NO_3H	10M	L25
$HOCH_2C(Me)_2NH_2$	-332.9	neat	(a) NO_3H	10M	L25
$(HO-CH_2-CH_2)_3N$	-359.3	neat	(d) NO_3Na		M20b
$CH_3O(CH_2)_2NH_2$	-367.8	neat	(a) NO_3H	10M	L25
$NC-(CH_2)_2NH_2$	-361.7	neat	(a) NO_3H	10M	L25
H 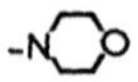	-349.8	neat	(a) NO_3H	10M	L25
	-351.6	neat	(e) NO_3NH_4		S10
CH_3	-349.2	neat	(e) NO_3NH_4		S10
nC_3H_7	-339.0	neat	(e) NO_3NH_4		S10
$secC_4H_9$	-340.2	neat	(e) NO_3NH_4		S10
$(Ar)_2CHCH_2$	-340.4	neat	(e) NO_3NH_4		S10
Methylthiazolidines					
2(g) 4(g) 5(g)					
H H H H H H	-320.7	40% $CHCl_3$	(f) NO_3Na		M20a
CH_3 H H H H H	-303.1	neat	(f) NO_3Na		M20a
H H H H CH_3 H	-306.0	40% $CHCl_3$	(f) NO_3Na		M20a
CH_3 H H H CH_3 CH_3	-277.1	neat	(f) NO_3Na		M20a
CH_3 H CH_3 CH_3 H H	-304.6	40% $CHCl_3$	(f) NO_3Na		M20a
CH_3 CH_3 H H CH_3 H	-276.3	neat	(f) NO_3Na		M20a

Table 6.2 (continued)

Amine		^{15}N (ppm/CH_3NO_2)	Conditions	Reference of Shifts	Ref.
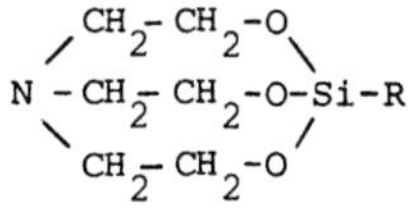	R ≡ CH_3	-357.3	$CDCl_3$	(h) HNO_3	L42
		-357.2	CD_3COCD_3	(h) HNO_3	L42
	CH_2=CH-	-356.0	CD_3COCD_3	(h) HNO_3	L42
	C_6H_5	-355.5	CD_3COCD_3	(h) HNO_3	L42
	CH_2Cl	-352.5	$CDCl_3$	(h) HNO_3	L42
		-353.0	CD_3COCD_3	(h) HNO_3	L42
	CH_3O	-351.0	CD_3COCD_3	(h) HNO_3	L42
	C_2H_5O	-352.4	CD_3COCD_3	(h) HNO_3	L42
	$(CH_2)_3CN$	-357.1	$CDCl_3$ 2M	(i) NO_3Na	M20f
	$(CH_2)_3SH$	-368.9	$CDCl_3$ 2M	(i) NO_3Na	M20f

(a) At 6.07 MHz and 323K (conversion factor = -6.1 ppm)

(b) At 9.12 MHz and room temperature (conversion factor = -4.6 ppm)

(c) At 27.4 MHz and room temperature (conversion factor = -336.7 ppm)

(d) At 9.12 MHz and 301K (conversion factor = -1.5 ppm)

(e) From Ref. S10, ext NH_4NO_3, 31.3 mg/ml D_2O at 10.1 MHz and room temperature (conversion factor = -5.2 ppm)

(f) From Ref. M20a at 9.12 MHz and 301K (conversion factor = -1.5 ppm)

(g) Methylthiazolidines NH-C-S-C-C (ring; positions 2, 4, 5)

(h) At 9.12 MHz and room temperature; ^{15}N spectra were obtained from ^{15}N enriched samples, referenced to HNO_3 and corrected to the CH_3NO_2 scale by the authors

(i) At 9.12 MHz and 301 K (natural abundance of the ^{15}N isotope) (conversion factor: -3.1 ppm)

^{15}N spectra of stannatranes have also been studied:

$N(CH_2CH_2O)_3Sn-CH_3$ $\delta^{15}N$ = -358.1 and -364.7 ppm $CDCl_3$ (9.1 MHz 303 K and Ref. NO_3^-).

The two signals arise from molecular association (J6)

Table 6.3. ^{15}N chemical shifts of amino acids and methyl esters

Amino Acids $H_2N-CH(R)-COOR'$			$\delta^{15}N$ (ppm/CH_3NO_2)	Conditions	Reference of Shifts	Ref.
R	R'	Symbol				
H	H	Gly	-351.0	H_2O pH 2.8	(a) NO_3H	C23
			-350.2	H_2O pH 6.3	(a) NO_3H	
			-362.6	H_2O pH 13.6	(a) NO_3H	
			-346.0	H_2O pH 2.3	(b) NH_4Cl	L36
			-344.1	H_2O pH 5.0	(b) NH_4Cl	
			-348.9	H_2O pH 9.1	(b) NH_4Cl	
			-347.3	cation	(c) Glycine H^+	B39
			-345.4	zwitterion	(c) Glycine H^+	
			-354.7	anion	(c) Glycine H^+	
			-344.4	6N HCl	(d) Glycine	S5
			-344.1	H_2O	(d) Glycine	
			-348.7	6N NaOH	(d) Glycine	
			-354.5	TFA	(e) $\underline{NO_3}$ NH_4	K26
H	CH_3	ac Gly	-349.0	H_2O pH < 2	(f) $\underline{NH_4}$ NO_3	P8
CH_3	H	Ala	-331.3	H_2O	(d) Glycine	S5
			-331.1	6N HCl	(d) Glycine	S5
			-341.2	TFA	(e) $\underline{NO_3}$ NH_4	K26
CH_3	CH_3	ac Ala	-335.8	H_2O pH < 2	(f) $\underline{NH_4}$ NO_3	P8
$(CH_3)_2CH$	H	Val	-338.1	H_2O	(d) Glycine	S5
			-338.0	6N HCl	(d) Glycine	S5
			-347.2	TFA	(e) $\underline{NO_3}$ NH_4	K26
			-342.1	pH 0.3	(g) NO_3H	S19
$(CH_3)_2CH$	CH_3	ac Val	-340.7	H_2O pH < 2	(f) $\underline{NH_4}$ NO_3	P8
$(CH_3)_2CHCH_2$	H	Leu	-332.7	6N HCl	(d) Glycine	S5
			-343.3	TFA	(e) $\underline{NO_3}$ NH_4	K26
$(CH_3)_2CHCH_2$	CH_3	ac Leu	-337.0	H_2O pH < 2	(f) $\underline{NH_4}$ NO_3	P8
$CH_3CH_2CH(CH_3)$	CH_3	ac Ile	-340.1	H_2O pH < 2	(f) $\underline{NH_4}$ NO_3	P8
C_6H_5	H	Phg	-337.7	TFA	(e) $\underline{NO_3}$ NH_4	K26
$C_6H_5CH_2$	H	Phe	-335.0	H_2O	(d) Glycine	S5
			-345.1	TFA	(e) $\underline{NO_3}$ NH_4	K26
$C_6H_5CH_2$	CH_3	ac Phe	-338.6	H_2O pH < 2	(f) $\underline{NH_4}$ NO_3	P8
p-HO-$C_6H_4CH_2$	H	Tyr	-334.5	6N HCl	(d) Glycine	S5
p-HO-$C_6H_4CH_2$	CH_3	ac Tyr	-339.6	H_2O pH < 2	(f) $\underline{NH_4}$ NO_3	P8
HO-CH_2	H	Ser	-342.1	pH 2.4	(g) NO_3H	S19
HS-CH_2	H	Cys	-336.2	pH 0	(g) NO_3H	S19
			-338.3	pH 1	(h) NO_3Na	M20g
$H_2N-(CH_2)_3$	H	Orn	-333.1	H_2O	(d) Glycine	S5
			-332.9	6N HCl	(d) Glycine	S5
$H_2N-(CH_2)_4$	H	Lys (j)	-337.0	pH 2.6	(g) NO_3H	S19
			-337.8	pH 7	(h) NO_3Na	M20g
			-337.9	cation	(c) Glycine H^+	B39
			-336.1	zwitterion	(c) Glycine H^+	B39
			-342.8	anion	(c) Glycine H^+	B39

Table 6.3 (continued)

Amino Acids $H_2N-CH(R)-COOR'$	R'		$\delta^{15}N$ (ppm/CH_3NO_2)	Conditions	Reference of Shifts	Ref.
$H_2N-(CH_2)_4$	CH_3	ac Lys (j)	-337.5	H_2O pH < 2	(f) $\underline{N}H_4$ NO_3	P8
$HN{=}C(NH_2)-NH(CH_2)_3$	H	Arg (j)	-336.8	pH 2.1	(g) NO_3H	S19
			-338.2	pH 7 2M	(h) NO_3Na	M20g
			-348.9	pH 11.5 1M	(i) NO_3H	K41
$HN{=}C(NH_2)-NH(CH_2)_3$	CH_3	ac Arg	-338.0	H_2O pH < 2	(f) $\underline{N}H_4$ NO_3	P8
$N{=}CH-NH-CH{=}C-CH_2$ (ring)	CH_3	ac His	-337.5	H_2O pH < 2	(f) $\underline{N}H_4$ NO_3	P8
$HOCO-CH_2$	H	Asp	-334.3	6N HCl	(d) Glycine	S5
$HOCO-(CH_2)_2$	H	Glu	-332.9	6N HCl	(d) Glycine	S5
$H_2NCO-CH_2$	H	Asn (j)	-336.2	pH 0.8	(g) NO_3H	S19
$H_2NCO-(CH_2)_2$	H	Gln (j)	-337.8	pH 1.0	(g) NO_3H	S19
$H_2NCO-NH(CH_2)_3$	H	Cit	-339.0	pH 0.1	(g) NO_3H	S19
$HN-CH_2-CH_2-CH_2-CH-COOH$ (ring)		Pro	-323.2	cation	(c) Glycine H^+	B39
			-321.1	zwitterion	(c) Glycine H^+	B39
			-328.7	TFA	(e) $\underline{N}O_3$ NH_4	K26

(a) At 6.07 MHz and room temperature (conversion factor: -6.1 ppm)

(b) At 9.12 MHz and room temperature (conversion factor: -352.9 ppm)

(c) At 9.12 MHz and 308 K; samples shaken with Chelex 100 to remove paramagnetic impurities. The shifts were measured with respect to 200 mg 95% ^{15}N glycine in 1 ml of 6N HCl (conversion factor: -346.0 ppm).

(d) At 9.12 MHz and 300 K. The shifts were referenced to aqueous glycine at iso-electric point (conversion factor: -344.1 ppm)

(e) At 9.12 MHz and 305 K, the shifts were referenced to NO_3NH_4 in D_2O (conversion factor: -4.6 ppm).

(f) At 9.12 MHz and room temperature; 5 to 9 molar aqueous solutions, pH = 0.5 to 2.0. The shifts were referenced to 5 M NH_4NO_3 in 2M HNO_3 (conversion factor: -359.0 ppm).

(g) At 10 MHz and room temperature; 20 to 40 w/v % aqueous solutions (conversion factor: -6.2 ppm).

(h) At 9.118 MHz and 301 K; (conversion factor: -3.1 ppm).

(i) At 18.25 MHz and 298 K; (conversion factor: -6.2 ppm).

(j) The ^{15}N resonances of other nitrogen atoms were also detected

Lys $\delta NH_2(\omega)$ - 344.0 (S 19) (g) - 345.4 (M 20 g) (h)

Arg δNH_2 - 305.5 (S 19) (g) - 307.0 (M 20g) (h) - 310.6 (K 41) - 304.8 (B 39) (c)

δNH_2 - 293.0 (S 19) (g) - 294.5 (M 20g) (h) - 296.5 (K 41)

Asn $\delta NH_2(\omega)$ - 266.2 (S 19) (g)

Gln $\delta NH_2(\omega)$ - 266.0 (S 19) (g) - 264.5 (B 39) (c)

Table 6.4. ^{15}N chemical shifts of amines containing a cycloalkyl substituent

R(R')N — R"			$\delta^{15}N$ (ppm / CH_3NO_2)		Conditions		Reference of Shifts	Ref.
R	R'	R"	amine	hydrochloride				
	H	H	-352.6		neat	(b)	NO_3H 10M	L25
	H	CH_3	-358.9		neat	(b)	NO_3H 10M	L25
	H	H	-337.7		neat	(b)	NO_3H 10M	L25
	H	H	-343.0		neat	(b)	NO_3H 10M	L25
	C_2H_5	C_2H_5	-327.3		20% C_6H_{12}	(c)	NO_3D 1M	W16
	H	H	-340.4	-335.3	2M CH_3OH	(c)	NO_3D 1M	D12
tC_4H_9 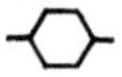	H	H						
trans			-341.0	-335.7	2M CH_3OH	(c)	NO_3D 1M	D12
cis			-349.7	-341.6	2M CH_3OH	(c)	NO_3D 1M	D12
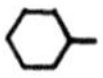	CH_3	CH_3	-347.7	-333.6	2M CH_3OH	(c)	NO_3D 1M	D12
	C_2H_5	C_2H_5	-327.2		20% C_6H_{12}	(c)	NO_3D 1M	W16
tC_4H_9 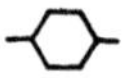	CH_3	CH_3						
trans			-353.7 (a)	-334.0	2M CH_3OH	(c)	NO_3D 1M	D12
1-adamantyl	H	H	-323.7	-323.3	2M CH_3OH	(c)	NO_3D 1M	D12
2-adamantyl	H	H	-346.5	-339.5	2M CH_3OH	(c)	NO_3D 1M	D12
1-adamantyl	CH_3	CH_3	-348.9 (a)	-326.0	2M CH_3OH	(c)	NO_3D 1M	D12

(a) In C_6H_{12}

(b) At 6.07 MHz and 323K (conversion factor = -6.1 ppm)

(c) At 18.25 MHz ans 310K (conversion factor = -6.2 ppm)

Table 6.5. ^{15}N chemical shifts of saturated heterocyclic amines in 2M CH_3OH solutions unless otherwise indicated. (The hydrochlorides were dissolved in CH_3OH, 4 to 5 mol-%). Spectra recorded at 18.25 MHz and 310K and chemical shifts referenced to 1M HNO_3 in D_2O (conversion factor = -6.2 ppm)

 N - R			$\delta^{15}N$ (ppm/CH_3NO_2)		
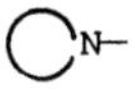	R		Amine (D13)		Hydrochloride (D14)
	H		-393.3		
	H		-342.1 -343.4 (a) -343.5 (b)		-333.4
	CH_3		-336.4 -339.5 (b)		-325.7
	nC_3H_7		-328.9 (b)		
	H		-343.2 -342.6 (a) -343.2 (b)		-351.0
	CH_3		-340.6 -343.5 (b)		-335.3
	C_2H_5		-328.2		-326.0
	nC_3H_7		-333.2 (b)		
	iC_3H_7		-324.2		-319.2
	iC_4H_9		-334.1 (b)		
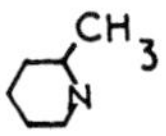	H		-326.8		-328.3
	CH_3		-329.3	cis trans	-332.1 -325.9
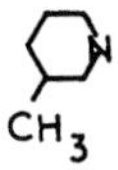	H		-343.3		-340.0
	CH_3		-340.6	cis trans	-332.7 (e) -338.1 (e)
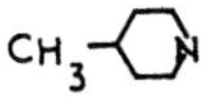	H		-344.2		-341.3
	CH_3		-340.8	cis trans	-336.8 (e) -334.0 (e)
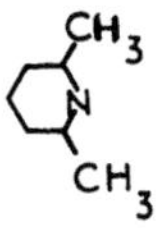	H	cis trans	-309.7 -317.7		-314.6 -319.2
	CH_3	cis	-317.2	eq ax	-316.6 -324.1
		trans	-331.7		-322.2
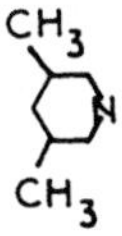	H	cis	-342.6		-338.3
		trans	-353.6		-344.9
	CH_3	cis	-340.8	eq ax	-331.9 (e) -336.6 (e)
		trans	-347.9		-339.0
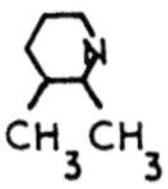	H	cis trans	-336.1 -324.8		-331.8 -326.7

Table 6.5 (continued)

N - R			$\delta^{15}N$ (ppm/CH_3NO_2)		
N	R		Amine (D13)	Hydrochloride (D14)	
	CH_3	cis		cis-cis	-328.1
				trans-cis	-330.9
		trans	-333.1 (c)	eq	-325.6
				ax	-332.1
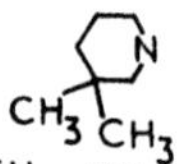	H		-350.8		-343.8
	CH_3		-344.9		-337.6
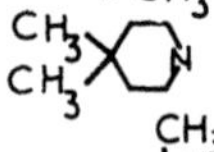	H		-343.9 (c)		-340.5
	CH_3		-342.9 (c)		-334.9
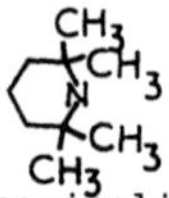	H		-298.8		-302.3
	CH_3		-324.7		-309.1
decahydroquinoline	H	cis	-337.8 (c)		-334.7
		trans	-327.3 (c)		-328.7
	CH_3	cis	-345.7 (c)	(i)	-334.0
				(o)	-328.4
		trans	-333.6 (c)	eq	-326.6
				ax	-333.4
8(eq)Me-decahydroquinoline	H	trans	-332.8 (c)		-332.2
	CH_3	trans	-356.4 (c)		-336.8
3(eq)Me-quinolizidine	H		-318.4 (c)		-319.9
3(ax)Me-quinolizidine	H		-327.3 (c)		-326.2
2-azaadamantane	H		-323.5		-325.9
	CH_3		-337.5		-326.2
quinuclidine			-362.2 (d)		-348.5
atropine			-321.1 (e)		
scopolamine			-342.6 (e)		
nicotine (2-pyridine)	CH_3		-330.1 (f)		
(3-pyridine)			-327.6 (f)		
(4-pyridine)			-329.4 (f)		

(a) From Ref. L25 10M HNO_3 at 6.07 MHz and 323K (conversion factor = -6.1 ppm)

(b) From Ref. S10 NH_4NO_3 31.3 mg/ml D_2O; 10.1 MHz and room temperature (conversion factor = 3.7 ppm)

(c) In C_6H_{12} (d) In H_2O (e) In $CHCl_3$

(f) From Ref. W15 neat CH_3NO_2 at 10.1 MHz and room temperature

(i) inside (o) outside (eq.) equatorial (ax.) axial

Table 6.6. ^{15}N chemical shifts of saturated heterocyclic cycloalkylamines

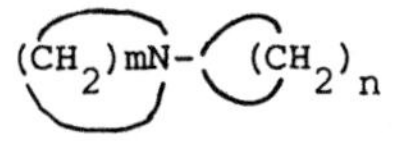

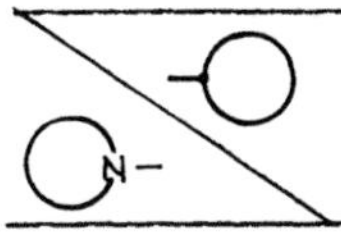		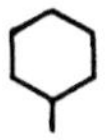	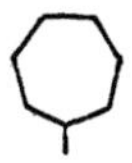	
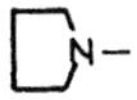	-315.9 (a) -314.2 (b)	-318.6 (a) -319.4 (b)	-319.5 (b)	-317.7 (b)
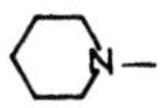	-320.9 (a) -320.5 (b)	-327.1 (a) -327.0 (b)	-324.1 (b)	
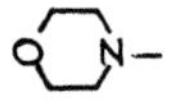	-324.7 (b)	-330.2 (a) -330.1 (b)	-328.7 (b)	
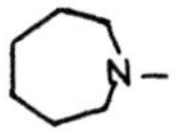	-327.1 (b)	-331.5 (b)		

(a) From Ref. S10, ext. NH_4NO_3 31.3 mg/ml D_2O at 10.1 MHz and room temperature in 80% v/vC_6D_6 solution (conversion factor = 3.7 ppm)

(b) From Ref. W16; solution 20 mol-% in C_6H_{12}; ext. 1.0 M DNO_3 at 18.25 MHz and room temperature (conversion factor = -6.2 ppm). Nicotines are listed in Table 6.15

Table 6.7. ^{15}N chemical shifts of amino sugars

Sugar	$\delta^{15}N$ (ppm/CH_3NO_2) α	β	Conditions	Reference of Shifts	Ref.
D-glucose					
2-amino-2-deoxy, HCl	-345.1	-346.5	(a)	NO_3NH_4	C 19
	-346.5	-348.1	25% in H_2O (b)	NO_3H	B 45
2-acetamido-2-deoxy	-256.8	-257.6	(a)	NO_3NH_4	C 19
	-258.6	-259.3	25% in H_2O (b)	NO_3H	B 45
2-acetamido-2-deoxy 1,3,4,6-tetra-o-acetyl	-268.1		(c)	NO_3NH_4	C 19
D-mannose					
2-amino-2-deoxy,HCl	-348.5	-356.0	(a)	NO_3NH_4	C 19
	-350.4	-357.8	25% in H_2O (b)	NO_3H	B 45
2-acetamido-2-deoxy	-260.5	-267.3	(a)	NO_3NH_4	C 19
	-261.8	-268.5	25% in H_2O (b)	NO_3H	B 45
D-galactose					
2-amino-2-deoxy,HCl	-346.1	-347.9	(a)	NO_3NH_4	C 19
	-347.8	-349.2	25% in H_2O (b)	NO_3H	B 45
2-acetamido-2-deoxy	-257.1	-257.9	(a)	NO_3NH_4	C 19
	-258.8	-259.6	25% in H_2O (b)	NO_3H	B 45
6-amino-6-deoxy 1,2:3,4-di-o isopropylidene	-365.2		(d)	NO_3NH_4	C 19
6-phthalimido-6-deoxy 1,2:3,4-di-o-isopropylidene	-224.3		(d)	NO_3NH_4	C 19

(a) Solution in 9/1 v/v H_2O DMSO-d_6 at 9.12 MHz and room temperature (conversion factor = -4.0 ppm)

(b) At 18.25 MHz and 308K (conversion factor = -6.2 ppm)

(c) $CDCl_3$ solutions; 0.05 M $Cr(acac)_3$ added

(d) Solutions in 9/1 v/v $CDCl_3/C_6F_6$

(e) Other phthalimido derivatives have been studied but their ^{15}N shifts are nearly constant (Ref. C 19, C 16 and C 30)

Table 6.8. ^{15}N chemical shifts of N,N-dimethylenamines

$(CH_3)_2N-C(R)=C(R')-R''$

R	R'	R''	E/Z	$\delta^{15}N$ (ppm/CH_3NO_2)	Conditions	Ref.
H	H	CH_3	E	-350.8	neat (a)	D 15
H	H	C_6H_5	E	-326.8 -324.0	80% C_6D_6 (b) neat (a)	S 10 D 15
H	H	CHO	E	-289.0	neat (a)	D 15
H	H	$COCH_3$	E	-298.5	neat (a)	D 15
H	H	$CO-C_6H_5$	E	-291.5	$CHCl_3$ (a)	D 15
H	H	$CH=CH_2$	E	-324.5	neat (a)	D 15
H	H	CH=CH-CHO	E	-295.5	$CHCl_3$ (a)	D 15
H	C_6H_5	C_6H_5	-	-329.3	80% C_6D_6 (b)	S 10
CH_2COOH	H	COOEt	Z	-274.6	80% C_6D_6 (b)	S 10
CH_3	H	$COCH_3$	E	-297.9	80% C_6D_6 (b)	S 10
H	CH_3	CH_3	-	-351.2	80% C_6D_6 (b)	S 10
	cyclopent-1-enyl		-	-335.1	80% C_6D_6 (b)	S 10
	cyclohex-1-enyl		-	-333.8	80% C_6D_6 (b)	S 10
	cyclohex-1-enyl-3-one		-	-299.0	80% C_6D_6 (b)	S 10
	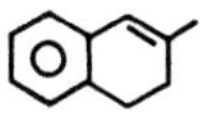		-	-326.9	80% C_6D_6 (b)	S 10
	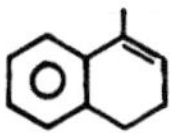		-	-313.0	80% C_6D_6 (b)	S 10

(a) From Ref. D15, ext. NO_3 Na in D_2O, H^+ (conversion factor = -1.5 ppm) at 9.12 MHz and 301 K

(b) From Ref. S22, ext. NH_4NO_3 31.3 mg/ml D_2O at 10.1 MHz and room temperature (conversion factor = -3.7 ppm)

Table 6.9. ^{15}N chemical shifts of cyclic enamines

 N– / – C = C –				
cyclopent-1-enyl	-310.3 (a) -310.4 (b)	-310.2 (a) -310.9 (b)	-315.0 (a) -315.3 (b)	-312.7 (a)
cyclohex-1-enyl	-306.9 (a) -307.7 (b)	-307.3 (a) -307.3 (b)	-311.6 (a) -312.1 (b)	-311.9 (a)
cyclohex-1-enyl methyl-2	-323.8 (a)	-326.1 (a)		
cyclohex-1-enyl methyl-6	-309.1 (a)	-313.6 (a)		
cyclohept-1-enyl	-305.5 (a)	-304.8 (a)	-309.4 (a)	
cycloct-1-enyl	-307.4 (a)			
$(CH_3)_2C = CH -$	-322.2 (b)	-325.6 (b)	-330.1 (b)	
$C_6H_5CH = CH -$ (E)	-301.1 (b)	-303.6 (b)	-309.8 (b)	
$(C_6H_5)_2C = CH -$	-301.9 (b)	-305.0 (b)	-312.9 (b)	
$NO_2-CH = CH -$ (E)		-266.7 (b)		
cyclohex-1-enyl-3-one	-274.6 (b)	-280.4 (b)	-291.7 (b)	
$CH_3CO - CH = C(CH_3)-$	-272.9 (b)	-276.8 (b)	-284.8 (b)	
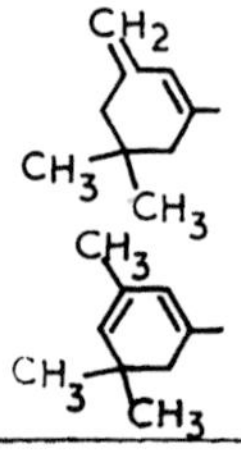 63% / 37%	-299.7 (c)			

(a) From Ref. W18; 20% mol in C_6H_{12}; 1.0 M DNO_3 at 18.25 MHz and room temperature (conversion factor = -6.2 ppm).

(b) From Ref. S10; ext. NH_4NO_3 31.3 mg/ml D_2O at 10.1 MHz and room temperature; solution 80% v/v in C_6D_6 (conversion factor = -3.7 ppm)

(c) From Ref. G12; neat; $NaNO_3$, D_2O at 6.07 MHz (conversion factor = -1.5 ppm)

Table 6.10. ^{15}N chemical shifts of alkyl substituted anilines

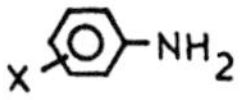

Ortho (2)	Ortho (6)	Meta (3)	Meta (5)	Para (4)	δNH_2 (ppm/CH_3NO_2)	Conditions		Reference of Shifts	Ref.
H	H	H	H	H	-320.3	25% DMSO	(a)	NO_3Na	D 15
					-323.7	neat	(a)	NO_3Na	D 15
					-325.5	neat	(b)	NO_3H	L 26
					-320.7	DMSO	(c)	Me_4NI	A 4
CH_3	H	H	H	H	-325.8	neat	(a)	NO_3Na	D 15
					-327.8	neat	(b)	NO_3H	L 26
H	H	CH_3	H	H	-324.5	neat	(a)	NO_3Na	D 15
					-326.3	neat	(b)	NO_3H	L 26
					-321.5	DMSO	(c)	Me_4NI	A 4
H	H	H	H	CH_3	-326.5	neat	(a)	NO_3Na	D 15
					-328.0	neat	(b)	NO_3H	L 26
					-323.3	DMSO	(c)	Me_4NI	A 4
CH_3	H	CH_3	H	H	-329.2	neat	(b)	NO_3H	L 26
CH_3	H	H	H	CH_3	-330.3	neat	(b)	NO_3H	L 26
CH_3	H	H	CH_3	H	-328.6	neat	(b)	NO_3H	L 26
CH_3	CH_3	H	H	H	-331.0	neat	(b)	NO_3H	L 26
H	H	CH_3	CH_3	H	-327.0	neat	(b)	NO_3H	L 26
					-321.9	DMSO	(c)	Me_4NI	A 4
H	H	CH_3	H	CH_3	-329.1	neat	(b)	NO_3H	L 26

(a) At 9.118 MHz and 301K (conversion factor = -1.5 ppm)

(b) At 6.07 MHz and room temperature (conversion factor = -6.2 ppm)

(c) From INDOR $1H\{^{15}N\}$ spectra; $\delta^{15}N(C_6H_5NH_2)$ = 6.080112 MHz (TMS: 60.0 MHz); (conversion factor = -337.3 ppm)

Table 6.11. ^{15}N chemical shifts of substituted anilines

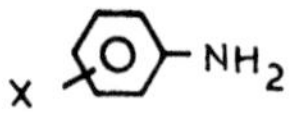

Ortho 2	6	Meta 3	5	Para 4	δNH_2 (ppm/CH_3NO_2)	Conditions	Reference of Shifts	Ref.
H	H	H	H	H	-320.3	DMSO 25% (a)	NO_3Na	D 15
CF_3	H	H	H	H	-318.3	acetone 1M(b)	aniline	A 21
CN	H	H	H	H	-311.9	DMSO (a)	NO_3Na	D 15
I	H	H	H	H	-304.4	acetone 1M(b)	aniline	A 21
$COCH_3$	H	H	H	H	-311.6	neat (a)	NO_3Na	D 15
COC_6H_5	H	H	H	H	-305.4	acetone 1M(b)	aniline	A 21
NO_2	H	H	H	H	-302.7	acetone 1M(b)	aniline	A 21
NH_2	H	H	H	H	-328.3	DMSO 25% (a)	NO_3Na	D 15
F	H	H	H	H	-328.1	acetone 1M(b)	aniline	A 21
Cl	H	H	H	H	-326.3	neat (c)	NO_3H	L 26
					-316.9	acetone (b)	aniline	A 21
Br	H	H	H	H	-321.0	neat (c)	NO_3H	L 26
					-312.1	acetone 1M(b)	aniline	A 21
OCH_3	H	H	H	H	-335.9	neat (c)	NO_3H	L 26
					-326.7	acetone 1M(b)	aniline	A 21
H	H	NH_2	H	H	-320.6	DMSO 25% (a)	NO_3Na	D 15
H	H	NO_2	H	H	-315.8	DMSO (d)	NMe_4I	A 4
H	H	CF_3	H	H	-318.0	DMSO (d)	NMe_4I	A 4
H	H	Cl	H	H	-318.2	DMSO (d)	NMe_4I	A 4
					-324.2	neat (c)	NO_3H	L 26
H	H	Br	H	H	-318.4	DMSO (d)	NMe_4I	A 4
					-323.4	neat (c)	NO_3H	L 26
H	H	I	H	H	-319.0	DMSO (d)	NMe_4I	A 4
H	H	CH_3O	H	H	-320.0	DMSO (d)	NMe_4I	A 4
H	H	H	H	NH_2	-327.3	DMSO 25% (a)	NO_3Na	D 15
H	H	H	H	OCH_3	-327.0	neat(333K)(a)	NO_3Na	D 15
					-326.5	DMSO (d)	NMe_4I	A 4
H	H	H	H	OH	-327.0	DMSO 25% (a)	NO_3Na	D 15
H	H	H	H	Br	-319.7	DMSO (d)	NMe_4I	A 4
H	H	H	H	F	-324.0	DMSO (d)	NMe_4I	A 4
					-319.7	acetone 1M(b)	aniline	A 21
H	H	H	H	$COCH_3$	-309.0	DMSO 25% (a)	NO_3Na	D 15
H	H	H	H	NO_2	-301.4	acetone 1M(b)	aniline	A 21
					-300.0	DMSO 25% (a)	NO_3Na	D 15
					-301.0	DMSO (d)	NMe_4I	A 4

Table 6.11 (continued)

Ortho 2	6	Meta 3	5	Para 4	δNH_2 (ppm/CH_3NO_2)	Conditions	Reference of Shifts	Ref.
H	H	H	H	I	-319.3 -315.6	DMSO (d) acetone 1M(b)	NMe_4I aniline	A 4 A 21
H	H	H	H	Cl	-320.4 -316.8	DMSO (d) acetone 1M(b)	NMe_4I aniline	A 4 A 21
Br	H	H	H	CH_3	-323.6	neat (c)	NO_3H	L 26
Cl	CH_3	H	H	H	-328.4	neat (c)	NO_3H	L 26
NO_2	H	H	H	NO_2	-288.0	DMSO (d)	NMe_4I	A 4
Br	H	H	H	Br	-314.9	DMSO (d)	NMe_4I	A 4
H	H	CH_3O	CH_3O	H	-320.4	DMSO (d)	NMe_4I	A 4
Br	Br	H	H	Br	-309.2	DMSO (d)	NMe_4I	A 4
H	H	CH_3	CH_3	NO_2	-311.3	DMSO (d)	NMe_4I	A 4
Cl	H	H	H	NO_2	-301.1	acetone 1M(b)	aniline	A 21
NO_2	H	H	H	Cl	-299.4	acetone 1M(b)	aniline	A 21
Br	Br	H	H	Br	-304.7	acetone 1M(b)	aniline	A 21

(a) At 9.118 MHz and 301K (conversion factor = -1.5 ppm)

(b) From INDOR $^{1}H\{^{15}N\}$spectra at room temperature (conversion factor = 316.3 ppm)

(c) At 6.07 MHz and room temperature (conversion factor = -6.1 ppm)

(d) From INDOR $^{1}H\{^{15}N\}$ spectra ($\delta^{15}N$ aniline = 6.080112 MHz) (TMS = 60.0 MHz) at room temperature (conversion factor = -320.7 ppm). This factor corresponds to the $\delta^{15}N$ value of aniline in DMSO (T 6.10)

Table 6.12. ^{15}N chemical shifts of substituted anilinium salts dissolved in 1M $HFSO_4$

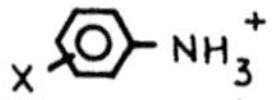

Ortho 2	Ortho 6	Meta 3	Meta 5	Para 4	$\delta\ NH_3^+$ (ppm/CH_3NO_2) (a)
H	H	H	H	H	-333.0 (b)
F	H	H	H	H	-342.7
NO_2	H	H	H	H	-335.6
Cl	H	H	H	H	-335.1
Br	H	H	H	H	-330.2
I	H	H	H	H	-322.6
H	H	NO_2	H	H	-332.9
H	H	Br	H	H	-333.6
H	H	H	H	F	-334.6
H	H	H	H	Cl	-334.2
H	H	H	H	Br	-333.9
H	H	H	H	I	-333.6
H	H	H	H	NO_2	-331.9
NO_2	H	H	H	Cl	-336.4
Cl	H	H	H	NO_2	-333.5
Br	Br	H	H	Br	-327.3

(a) The shifts are referenced to $C_6H_5NH_3^+$, CF_3COO^- unless otherwise indicated and were obtained from INDOR $^1H\ \{^{15}N\}$ spectra (A21) (conversion factor = -333.0 ppm)

(b) From $NaNO_3$ in D_2O, H^+ at 9.118 MHz and 301 K (M20b) solution 1/1 in CF_3COOH (conversion factor = -1.5 ppm)

Table 6.13. ^{15}N chemical shifts of N,N-dimethylanilines and anilinium trifluoroacetates

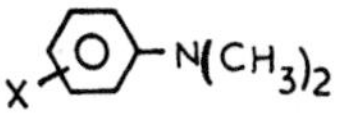

Ortho 2	Ortho 6	Meta 3	Meta 5	Para 4	Anilines $\delta^{15}N$ (ppm/CH_3NO_2)	Anilines Conditions	Anilinium Salts $\delta^{15}N$ (c) (ppm/CH_3NO_2)	Reference of Shifts	Ref.
H	H	H	H	H	-337.6	neat (a)		NO_3Na	D 15
					-335.0	DMSO 25%(a)		NO_3Na	D 15
					-332.2	neat (b)	-329.2	NH_4Cl	S 14
CH_3	H	H	H	H	-342.9	neat (b)	-330.7	NH_4Cl	S 14
H	H	CH_3	H	H	-333.1	neat (b)	-329.2	NH_4Cl	S 14
H	H	H	H	CH_3	-339.9	neat (a)		NO_3Na	D 15
					-334.8	neat (b)	-330.1	NH_4Cl	S 14
H	H	H	H	OCH_3	-341.0	neat (a)		NO_3Na	D 15
H	H	H	H	Br	-335.04	neat (a) 338 K		NO_3Na	D 15
H	H	H	H	COC_6H_5	-324.4	neat (a) 373 K		NO_3Na	D 15
H	H	H	H	CN	-323.2	neat (a) 373 K		NO_3Na	D 15
H	H	H	H	CHO	-320.9	neat (a) 363 K		NO_3Na	D 15
H	H	H	H	NO_2	-314.3	DMSO 10%(a)		NO_3Na	D 15
H	H	H	H	NO	-303.3	neat (a) 393 K		NO_3Na	D 15
CH_3	H	CH_3	H	H	-343.4	neat (b)	-331.0	NH_4Cl	S 14
CH_3	CH_3	H	H	H	-359.7	neat (b)	-330.4	NH_4Cl	S 14
CH_3	CH_3	H	H	CH_3	-361.4	neat (b)	-331.1	NH_4Cl	S 14
C_2H_5	C_2H_5	H	H	H	-363.7	neat (b)	-331.0	NH_4Cl	S 14
iC_3H_7	iC_3H_7	H	H	H	-366.4	neat (b)	-332.6	NH_4Cl	S 14

(a) At 9.12 MHz and 301K unless otherwise indicated (conversion factor = -1.5 ppm)

(b) At 10.1 MHz and room temperature (conversion factor = -351.8_5 ppm)

(c) 1 to 2 M solutions in C_6D_6 at 10.1 MHz and room temperature (conversion factor = -351.8_5 ppm)

Table 6.14. ^{15}N chemical shifts of N-substituted anilines

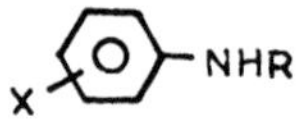 R	X	$\delta^{15}N$ (ppm/CH_3NO_2)	Conditions	Reference of Shifts	Ref.
H	H	-320.3	DMSO 25% (a)	NO_3Na	M 20c
		-320.3	neat (b)	CH_3NO_2	S 14
CH_3	H	-324.0	neat (b)	CH_3NO_2	S 14
		-337.6	neat (a)	NO_3Na	M 20c
CH_3	$pOCH_3$	-331.8	neat (a)	NO_3Na	M 20c
C_6H_5-$C(CH_3)$=CH	H	-285.3	DMSO d_6 (c)	NO_3NH_4	S 10
	mNO_2	-283.4	DMSO d_6 (c)	NO_3NH_4	S 10
	mCl	-284.7	DMSO d_6 (c)	NO_3NH_4	S 10
	$mOCH_3$	-284.7	DMSO d_6 (c)	NO_3NH_4	S 10
	pNO_2	-274.6	DMSO d_6 (c)	NO_3NH_4	S 10
	pCl	-285.8	DMSO d_6 (c)	NO_3NH_4	S 10
	$pOCH_3$	-289.4	DMSO d_6 (c)	NO_3NH_4	S 10
	pCH_3	-286.6	DMSO d_6 (c)	NO_3NH_4	S 10
C_6H_5	H	-288.8	DMSO 25% (a)	NO_3Na	M 20c

(a) At 9.12 MHz and 301K (conversion factor = -1.5 ppm)

(b) At 10.1 MHz and room temperature in C_6D_6 (conversion factor = -351.8_5 ppm)

(c) At 10.1 MHz and room temperature, NH_4NO_3 31.3 mg/ml D_2O, 80% solution v/v in C_6D_6 (conversion factor = -5.2 ppm)

Footnote to Table 6.15

(a) Neat compounds; at 10.1 MHz and room temperature unless otherwise indicated ($Cr(acac)_3$ added). CH_3NO_2 20% (internal), or $CH_3{}^{15}NO_2$, 20% enriched, in C_6D_6, (external)

(b) At 6.07 MHz and room temperature (CW spectra)

(c) At 18.25 MHz and room temperature 1M HNO_3 (conversion factor = -6.2 ppm)

(d) In C_6D_6

(e) In DMSO, δ_{NH_2} = -301.6 ppm/HNO_3; at 18.25 MHz and 333K

(f) In DMSO, δ_{NH_2} = -306.6 ppm/HNO_3; at 18.25 MHz and 333K

(g) 1/2 solution in CF_3COOH at 10.1 MHz and room temperature (conversion factor = -351.8_5 ppm)

(h) MP = N-methylpyrrolidine (nicotine derivatives) neat compounds, at 10.1 MHz and 303K δ_N(MP-2) = -330.1 δ_N(MP-3) = -327.6 δ_N(MP-4) = -329.4 ppm). For pyridine, 4-methylpyridine and 2,5-dimethylpyridine, it has been observed that additions of 0.0125 to 0.1 M $Cr(acac)_3$ do not change the $\delta^{15}N$ values (D10)

Table 6.15. ^{15}N chemical shifts of substituted pyridines

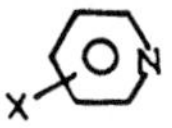

2	3	4	5	6	Pyridines $\delta^{15}N$ (a) (ppm/CH_3NO_2)	Pyridinium Salts $\delta^{15}N$ (g) (ppm/CH_3NO_2)	References of Shifts	Ref.
H	H	H	H	H	- 63.2(b,c)	-176.3(b)	HNO_3	L 8,D17
					- 60.6		CH_3NO_2 int.	D 16
						-168.6	NH_4Cl	D 16
CH_3	H	H	H	H	- 62.6		CH_3NO_2 int.	D 16
						-167.5	NH_4Cl	D 16
H	CH_3	H	H	H	- 61.7		CH_3NO_2 int.	D 16
						-178.4	NH_4Cl	D 16
H	H	CH_3	H	H	- 70.2		CH_3NO_2 int.	D 16
						-181.5	NH_4Cl	D 16
CH_3	CH_3	H	H	H	- 62.3		CH_3NO_2 int.	D 16
						-173.1	NH_4Cl	D 16
CH_3	H	CH_3	H	H	- 71.0		CH_3NO_2 int.	D 16
						-179.0	NH_4Cl	D 16
CH_3	H	H	CH_3	H	- 62.7		CH_3NO_2 int.	D 16
						-168.6	NH_4Cl	D 16
CH_3	H	H	H	CH_3	- 62.4		CH_3NO_2 int.	D 16
						-171.7	NH_4Cl	D 16
H	CH_3	CH_3	H	H	- 68.8		CH_3NO_2 int.	D 16
						-175.6	NH_4Cl	D 16
H	CH_3	H	CH_3	H	- 61.7		CH_3NO_2 int.	D 16
						-174.1	NH_4Cl	D 16
CH_3	H	CH_3	H	CH_3	- 68.0		CH_3NO_2	D 16
C_2H_5	H	H	H	H	- 64.0		CH_3NO_2 int.	D 16
iC_3H_7	H	H	H	H	- 67.3		CH_3NO_2 int.	D 16
tC_4H_9	H	H	H	H	- 64.7		CH_3NO_2 int.	D 16
H	H	C_2H_5	H	H	- 65.5		CH_3NO_2	D 16
H	H	iC_3H_7	H	H	- 64.8		CH_3NO_2	D 16
H	H	tC_4H_9	H	H	- 64.7		CH_3NO_2	D 16
tC_4H_9	H	H	H	tC_4H_9	- 70.4		CH_3NO_2 int.	D 16
tC_4H_9	H	tC_4H_9	H	tC_4H_9	- 76.4 (d)		CH_3NO_2 int.	D 16
H	H	$CH_3\underline{O}$	H	H	- 86.6 (c)		HNO_3	D 17
H	H	$CH_3C\underline{O}$	H	H	- 51.8 (c)		HNO_3	D 17
NH_2	H	H	H	H	-116.0 (e)		HNO_3	M 21
H	H	NH_2	H	H	-107.2 (f)		HNO_3	M 21
MP	H	H	H	H	- 63.6 (h)		CH_3NO_2	W 15
H	MP	H	H	H	- 60.9 (h)		CH_3NO_2	W 15
H	H	MP	H	H	- 64.5 (h)		CH_3NO_2	W 15

Table 6.16. ^{15}N chemical shifts of aminopyridines and aminopyrimidines in DMSO solutions measured at 10.1 MHz and 303-313K. The original reference was external NH_4NO_3, H_2O. 0.1M $Cr(acac)_3$ was added to observe tertiary N-atoms. (S 25)

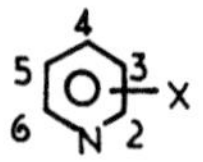

					$\delta^{15}N$ (ppm/CH_3NO_2)		
2	3	4	5	6	N_1	NH_2(a)	NH_2(b)
H	H	H	H	H	- 63.0		
NH_2	H	H	H	H	-113.8	-307.3	
H	NH_2	H	H	H	- 63.9	-325.3	
H	H	NH_2	H	H	-103.7	-312.0	
NH_2	NH_2	H	H	H	-114.5	-313.5	-330.4
NH_2	H	H	H	NH_2	-149.0	-309.1	-309.1
H	NH_2	NH_2	H	H	- 99.3	-336.9	-322.0

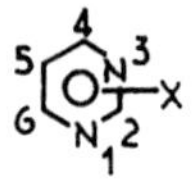

				$\delta^{15}N$ (ppm/CH_3NO_2)				
2	4	5	6	N_1	N_3	NH_2(a)	NH_2(b)	NH_2(c)
H	H	H	H	- 84.8	- 84.8			
NH_2	H	H	H	-129.9	-129.9	-297.9		
NH_2	NH_2	H	H	-164.5	-173.4	-301.6	-299.6	
H	NH_2	NH_2	H	-132.6(d)	-133.5(d)	-305.9	-338.0	
H	NH_2	H	NH_2	-149.4	-149.4	-309.1	-309.1	
(e)NH_2	NH_2	NH_2	H	-218.8	-218.8	-310.1(d)	-309.0(d)	-357.3
NH_2	NH_2	H	NH_2	-189.5	-189.5	-304.0	-306.0	-306.0
(e)H	NH_2	NH_2	NH_2	-181.7	-181.7	-301.5	-343.2	-301.5

(a) (b) (c) Refer to the NH_2 groups arranged following the increasing order of cycle substitution indices.

(d) Assignment may be reversed

(e) In DMSO + 0.2 eq. HCl

Table 6.17. ^{15}N chemical shifts of pyridine N-oxides, N-hydroxypyridinium salts

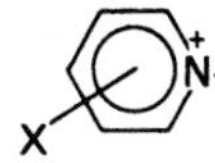
N–OH, CF_3COO^- and related compounds

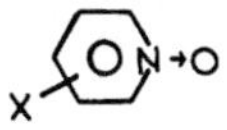

2	3	4	5	6	Pyridines N-Oxides $\delta^{15}N$ (ppm/ CH_3NO_2)	Conditions	N-Hydroxy Pyridinium Salts $\delta^{15}N$ (ppm/ CH_3NO_2)	Conditions	Reference of Shifts	Ref.
H	H	H	H	H	- 86.8	DMSO 2M	-135.7	TFA 2M	HNO_3 (a)	Y 9
					- 87.5	DMSO			CH_3NO_2 int.(b)	D 16
CH_3	H	H	H	H	- 90.9	DMSO 2M	-141.8	TFA 2M	HNO_3 (a)	Y 9
					- 89.6	DMSO			CH_3NO_2 int.(b)	D 16
H	CH_3	H	H	H	- 86.9	DMSO 2M	-141.8	TFA 2M	HNO_3 (a)	Y 9
					- 87.5	DMSO			CH_3NO_2 int.(b)	D 16
H	H	CH_3	H	H	- 96.6	DMSO 1M	-146.0	TFA 2M	HNO_3 (a)	Y 9
					- 96.3	DMSO			CH_3NO_2 int.(b)	D 16
CH_3	H	H	H	CH_3	- 92.9	DMSO 2M	-143.1	TFA 2M	HNO_3 (a)	Y 9
H	H	CH_3O	H	H	-106.4	DMSO 2M	-161.5	TFA 2M	HNO_3 (a)	Y 9
					-126.2	TFE 1M			HNO_3 (a)	Y 9
H	H	C_6H_5	H	H	-110.4	TFE 1M	-145.9	TFA 1M	HNO_3 (a)	Y 9
H	H	Cl	H	H	-106.2	TFE 1M	-141.7	TFA 1M	HNO_3 (a)	Y 9
H	H	NO_2	H	H	- 73.4	DMSO 1M	-112.9	TFA 1M	HNO_3 (a)	Y 9
					- 85.8	TFE 1M			HNO_3 (a)	Y 9
H	CH_3	NO_2	H	H	- 77.1	DMSO 1M	-122.0	TFA 1M	HNO_3 (a)	Y 9
quinoline N-oxide					-101.5	DMSO 1M	-150.3	TFA 1M	HNO_3 (a)	Y 9
8-hydroxyquinoline N-oxide					-111.0	DMSO 1M	-144.3	TFA 1M	HNO_3 (a)	Y 9

(a) At 18.25 MHz and room temperature (conversion factor = -6.2 ppm). The solvent effect on 2-methylpyridine N-oxide is high: DMSO: -90.9, CF_3CH_2OH: -103.8, H_2O: -109.2, CF_3COOH: -141.8 ppm

Table 6.18. ^{15}N chemical shifts of diazines and corresponding mono-N-oxides in DMSO solutions, measured at 10.1 MHz and 303-313K. The original reference was external NH_4NO_3, H_2O (0.1 M $Cr(acac)_3$ added) (S25)

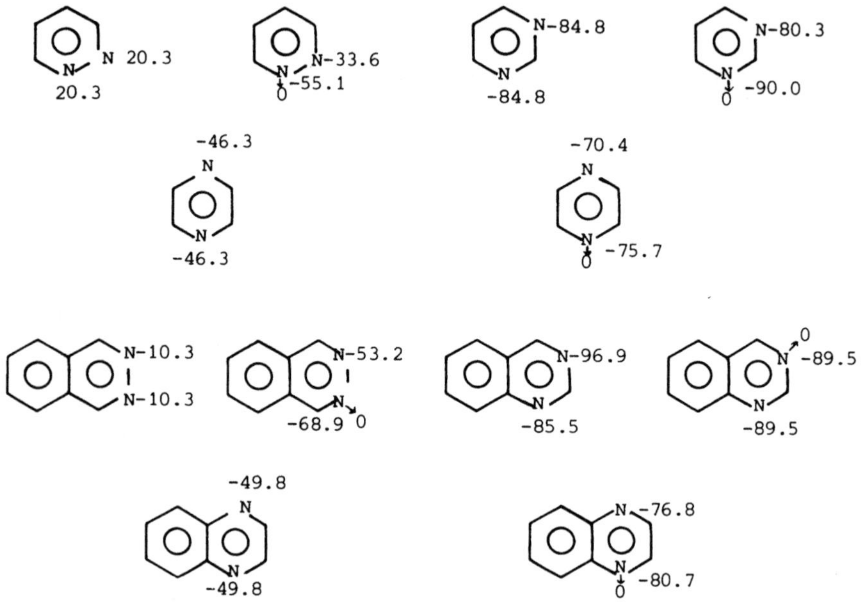

Table 6.19. ^{15}N chemical shifts of nucleosides and nucleotides. Numbering of atoms for purine and pyrimidine bases:

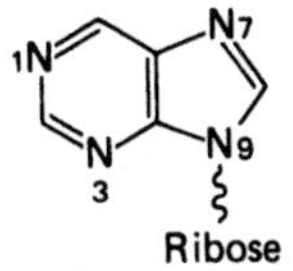

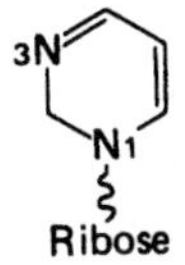

The ^{15}N chemical shifts of uracil and purine are respectively:

Uracil N_1 -250.0, N_3 -222.4 ppm in DMSO 0.8 M (H 10)

Purine	N_1	N_3	N_7	N_9		
	-130.8	-115.6	-191.8	-188.0	1.25M H_2O	(H10)
	-247.6	-266.7	-174.2	-174.2	DMSO	(H21)

Compounds	$\delta^{15}N$ (ppm/CH_3NO_2) N_1	N_3	N_7	N_9	NH_2	Conditions	Reference of Shifts	Ref.
Uridine	-237.8	-223.7				DMSO 0.8 M (b)	NH_4Cl	H 10
	-236.2	-223.9				H_2O-D_2 0.1M(b)	NH_4Cl	H 10
Thymidine	-237.5	-225.9				DMSO 0.8 M (b)	NH_4Cl	H 10
Cytidine	-228.4	-172.1			-287.7	DMSO 0.8 M (b)	NH_4Cl	H 10
	-229.5				-289.2	H_2O-D_2 0.1M(b)	NH_4Cl	H 10
Cytidine protonated	-227.3	-237.1			-275.2	DMSO (a)	NO_3D	M 21
Adenosine	-157.9 (c)	-145.0 (c)	-140.3	-211.9	-300.4	DMSO 0.4 M (b)	NH_4Cl	H 10
	-145.8	-158.9	-140.9	-211.8	-300.0	DMSO (a)	NO_3D	M 21
Guanosine	-234.1	-215.1	-134.4	-211.4	-308.4	DMSO 0.8 M (b)	NH_4Cl	H 10
	-234.2	-215.7	-134.8	-211.5	-308.2	DMSO (a)	NO_3D	M 21
Inosine	-207.1 (d)	-167.4	-132.6	-206.9 (d)		DMSO 0.8 M (b)	NH_4Cl	H 10
	-206.9	-167.4	-132.9	-206.9		DMSO (a)	NO_3D	M 21
Xanthosine	-227.7	-267.7	-133.4	-215.5		DMSO 0.8 M (b)	NH_4Cl	H 10
AMP	-158.3	-166.6	-151.0	-213.2	-303.2	H_2O (a)	NO_3D	M 21
ATP	-157.9	-166.5	-151.2	-213.4	-303.2	H_2O (a)	NO_3D	M 21
GMP	-235.3	-217.5	-147.4	-213.4	-309.7	H_2O (a)	NO_3D	M 21
UMP	-236.2	-222.7				H_2O (a)	NO_3D	M 21
TMP	-235.4	-225.7				H_2O (a)	NO_3D	M 21
CMP	-229.7	-181.4			-289.4	H_2O (a)	NO_3D	M 21

(a) At 18.25 MHz and 303 to 333 K (conversion factor = 6.2 ppm)

(b) At 18.24 MHz and 318K (or 301K, inverted gated decoupling) (conversion factor = -361.2 ppm)

(c) (d) The assignment may be reversed

Table 6.20. ^{15}N chemical shifts of pterin derivatives (S 15) for 0.7 M solutions

Pterin Derivatives	$\delta^{15}N$ (ppm/CH_3NO_2) (a) N(1)	N(3)	N(5)	N(8)	NH_2	N(10)	Amide	Solvent
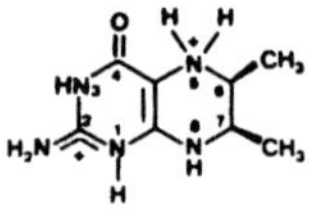	-269.4	-246.3	-333.2	-287.4	-296.5			TFA
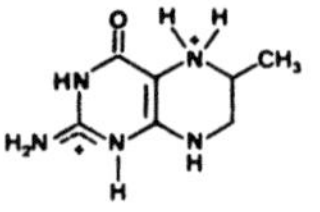	-269.0	-246.5	-332.9	-299.6	-297.5			TFA
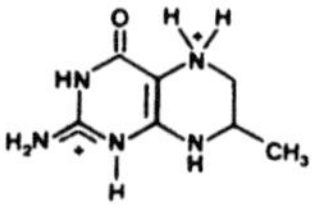	-268.1	-247.7	-346.3	-286.1	-297.3			TFA
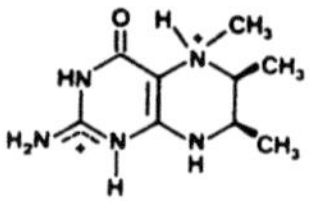	-270.1	-246.2	-330.6	-288.8	-295.9			TFA
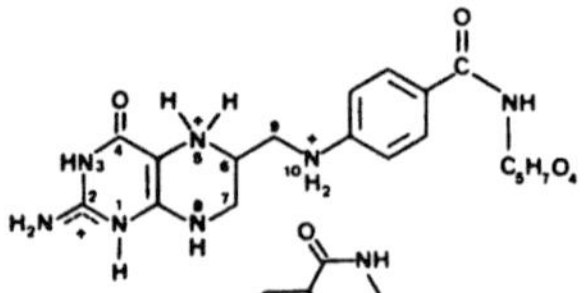	-267.7	-242.8	-335.5	-303.6	-295.5	-319.7	-261.4	6N HCl
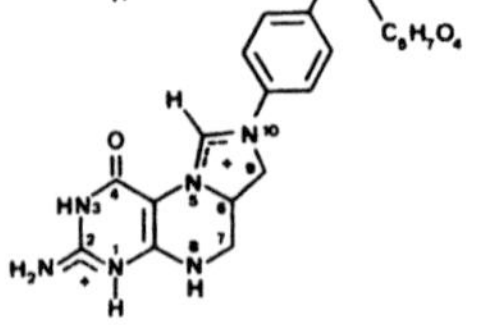	-268.3	-245.5	-251.3 (b)	-300.7	-298.1	-245.5 (b)	-262.1	TFA
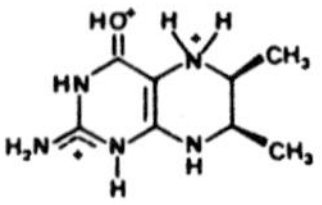	-263.2 (b)	-257.9 (b)	-337.2	-265.2 (b)	-288.1			FSO_3H
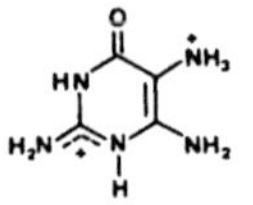	-255.8	-242.8	-350.0	-302.9	-298.3			H_2O

Table 6.20 (continued)

Pterin Derivatives	$\delta^{15}N$ (ppm/CH_3NO_2) (a)							Solvent
	N(1)	N(3)	N(5)	N(8)	NH_2	N(10)	Amide	
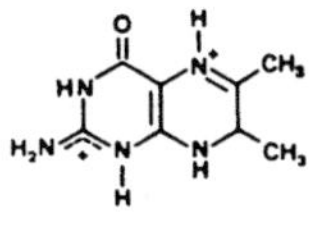	-266.5	-245.6	-194.1	-294.0	-296.3			H_2O
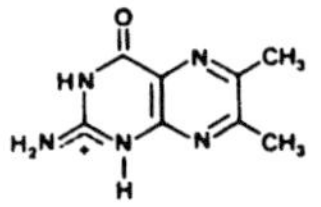	-262.3	-236.7	- 84.5 (b)	- 79.3 (b)	-296.7			TFA
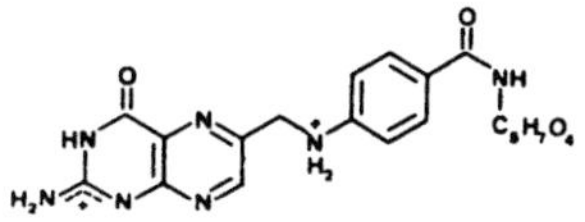	-264.1	-238.0	- 78.5	- 58.3	-294.8	-325.5	-262.2	TFA

(a) The shifts were measured with respect to $^{15}NO_3NH_4$ and converted to the CH_3NO_2 scale by the authors. At 10.1 MHz and room temperature

(b) Not identified

Footnote to Table 6.21

(a) The shifts were measured with respect to HNO_3 1M in D_2O at 28.25 MHz and 298K conversion factor = -6.2 ppm)

(b) From Ref. B44

(c) From Ref. R12

Table 6.21. ^{15}N chemical shifts of imidazoles and indoles. Solvent effect.: indole $CDCl_3$ δ = -255.4 DMSO sat. δ = -249.0 ppm. (See also K 43 for substituted pyrroles)

Compounds		δ ^{15}N (a) (ppm/CH_3NO_2) NH(1) (b)	N(2) (b)	Conditions
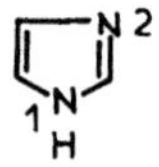	B	-177.2	-177.2	H_2O
	BH^+	-208.2	-208.2	H_3O^+
1 - methyl	B	-217.7	-134.7	H_2O
	BH^+	-210.3	-209.8	H_3O^+
4 - methyl	B	-179.0	-170.6	H_2O
	BH^+	-208.8	-204.8	H_3O^+
4-alanyl (histidine)	B	-202.7	-149.3	H_2O
	BH^+	-207.7	-205.2	H_3O^+
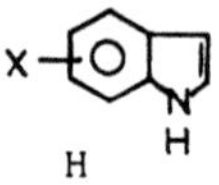		δNH (c)		
H		-247.3		DMSO 2M
2-COOH		-246.6		DMSO 2M
$2-CH_3$		-243.3		DMSO 2M
$3-CH_3$		-261.6		DMSO 2M
3-CHO		-234.1		DMSO 2M
$3-COCH_3$		-237.9		DMSO 1.6 M
$3-OCOCH_3$		-259.3		DMSO-d_6 sat.
$3-CH_2COOH$		-250.5		DMSO 1 M
$3-CH_2-CH_2COOH$		-253.0		DMSO-d_6 sat.
5-F		-257.0		$CDCl_3$ sat.
5-Cl		-256.0		$CDCl_3$ sat.
5-Br		-245.6		DMSO 1.3 M
$5-NO_2$		-240.4		DMSO 1.5 M
5-CN		-242.5		DMSO-d_6 sat.
5-COOH		-243.7		DMSO 2M
5-OH		-251.3		DMSO-d_6 sat.
$5-OCH_3$		-248.6		DMSO 1 M
$5-CH_3$		-248.1		DMSO 0.5 M
$5-NH_2$		-250.4		DMSO 1 M
$6-OCH_3$		-255.7		$CDCl_3$ sat.
$7-OCH_3$		-259.6		$CDCl_3$ sat.
$7-CH_3$		-247.9		DMSO 1M

Table 6.22: ^{15}N chemical shift of amides, N,N-dimethyl and N,N-diethylamides at 9.118 MHz and 301K (neat compounds) (M20d, M22). The shifts were measured with respect to NO_3Na, D_2O, H^+ (conversion factor = -1.5 ppm)

$R'_2N - CO - R$	R'		
R	H	CH_3 (a)	C_2H_5 (a)
H	- 267.6 - 266.1 (b)	- 275.2	- 248.7
CH_3	- 269.5	- 281.0	- 249.6
C_2H_5	- 271.7	- 285.8	- 256.3
nC_3H_7	- 271.0 (e)	- 282.9	- 254.8
iC_3H_7	- 275.1 (c)	- 285.8	- 256.3
tC_4H_9	- 275.2 (c)	- 287.1	
C_6H_5	- 279.3 (d)	- 278.3	- 252.7

(a) 0.08 M $Cr(acac)_3$ added

(b) Shifts referenced originally to NH_4NO_3 (B46) (conversion factor = -357.4 ppm)

(c) EtOH (5M)

(d) DMF (3M)

$(C_4H_9)_2N$-CHO (neat) δ = -256.5 ppm referenced originally to NMe_4Cl (conversion factor = -339.0 ppm) at 6.08 MHz and room temperature

Table 6.23. ^{15}N chemical shifts of N-substituted amides

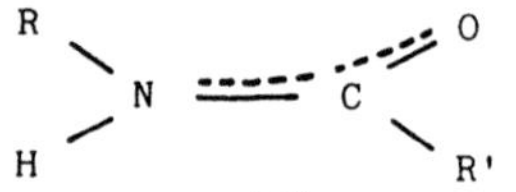

(Z)

R	R'	$\delta^{15}N$ (ppm/CH_3NO_2)	Conditions	Reference of Shifts	Ref.
CH_3	H	-270.1 -270.2 -270.2	neat (a) neat (b) neat (c)	NO_3Na NH_4NO_3 NMe_4Cl	D 18 B 46 W 21
CH_3	CH_3	-273.7 -274.4	neat (a) 40% $CHCl_3$ (d)	NO_3Na HNO_3	D 18 W 17
CH_3	C_2H_5	-276.5	neat (a)	NO_3Na	D 18
CH_3	nC_3H_7	-274.7	neat (a)	NO_3Na	D 18
CH_3	iC_3H_7	-278.4	neat (a)	NO_3Na	D 18
CH_3	C_6H_5	-280.1	25% $CDCl_3$ (a)	NO_3Na	D 18
CH_3	CH_2Cl	-273.9	neat (a)	NO_3Na	D 18
CH_3	CH_2OCH_3	-279.0	neat (a)	NO_3Na	D 18
CH_3	$CH_2N(CH_3)_2$	-280.3	neat (a)	NO_3Na	D 18
CH_3	$CH_2C_6H_5$	-273.7	neat (a)	NO_3Na	D 18
C_2H_5	CH_3	-254.8 -255.4	neat (a) 40% $CHCl_3$ (d)	NO_3Na HNO_3	D 18 W 17
$C_6H_5CH_2$	CH_3	-258.9	neat (a)	NO_3Na	D 18
C_6H_5	iC_3H_7	-250.7	30% CH_3CN (a)	NO_3Na	D 18
CH_3OCOCH_2	C_2H_5 (Z) (E)	-272.8 -271.5	neat (a) neat (a)	NO_3Na NO_3Na	D 18 D 18
$CH_3OCO-CH(CH_3)$	C_2H_5	-258.0	25% $CDCl_3$ (a)	NO_3Na	D 18
$HOCO-CH_2$	CH_3	-271.8	DMSO (e)	NH_4NO_3	G 13
$HO-CO-CH(C_4H_9i)$	CH_3	-259.5	DMSO (e)	NH_4NO_3	G 13
iC_3H_7	CH_3	-244.1	40% $CHCl_3$ (d)	HNO_3	W 17
tC_4H_9	CH_3	-241.8	40% $CHCl_3$ (d)	HNO_3	W 17
nC_3H_7	CH_3	-260.4	40% $CHCl_3$ (d)	HNO_3	W 17
sec. C_4H_9	CH_3	-262.4	40% $CHCl_3$ (d)	HNO_3	W 17
nC_4H_9	CH_3	-260.5	40% $CHCl_3$ (d)	HNO_3	W 17
nC_5H_{11}	CH_3	-260.2	40% $CHCl_3$ (d)	HNO_3	W 17
iC_4H_9	CH_3	-246.5	40% $CHCl_3$ (d)	HNO_3	W 17
$(CH_3)_3C-CH_2$	CH_3	-245.5	40% $CHCl_3$ (d)	HNO_3	W 17
$C_6H_5CH_2CH_2$	CH_3	-262.3	40% $CHCl_3$ (d)	HNO_3	W 17
$pCH_3O-C_6H_4$	CH_3	-249.1	15% DMSO (d)	HNO_3	W 17
$pCH_3C_6H_4$	CH_3	-247.8	15% DMSO (d)	HNO_3	W 17

Table 6.23 (continued)

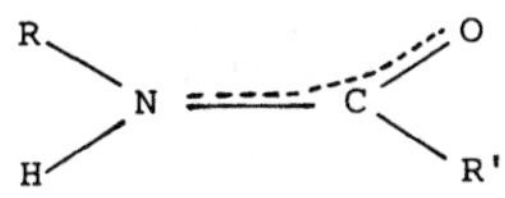

(Z)

R	R'	$\delta^{15}N$ (ppm/CH_3NO_2)	Conditions	Reference of Shifts	Ref.
C_6H_5	CH_3	-247.0	15% DMSO (d)	HNO_3	W 17
$pCl-C_6H_4$	CH_3	-248.2	15% DMSO (d)	HNO_3	W 17
nC_7H_{14}	nC_8H_{17}	-262.7	2M $CHCl_3$ (d)	HNO_3	W 18

(a) At 9.118 MHz and 301K (conversion factor = -1.5 ppm)

(b) At 9.2 MHz and room temperature (conversion factor = -357.4 ppm)

(c) At 6.08 MHz and ∿ 323 K (conversion factor = -336.7 ppm)

(d) At 18.25 MHz and room temperature (conversion factor = -6.2 ppm)

(e) At 9.2 MHz and room temperature (conversion factor = -361.2 ppm) 5M NH_4NO_3 in 2M HNO_3

Table 6.24. ^{15}N chemical shifts of Z, E diastereomeric amides (neat compounds)

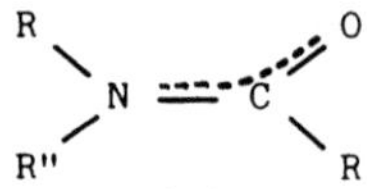

(E)

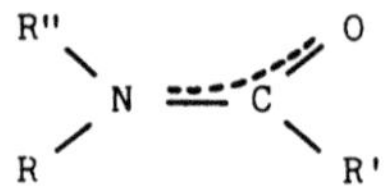

(Z)

R	R''	R'	(E/Z) % (b)	δ ^{15}N (ppm/CH_3NO_2) (a)
H	CH_3	H	(E) 10	-271.9
			(Z) 90	-270.1
				-270.2 (c)
H	C_2H_5	H	(E) 10	-254.0
			(Z) 90	-253.0
H	tC_4H_9	H	(E) 30	-231.1
			(Z) 70	-233.4
H	C_6H_5	H	(E) 40	-239.0 (d)
			(Z) 60	-241.6 (d)
H	$o-C_6H_4OCH_3$	H	(E) 33	-244.5
			(Z) 67	-247.5
CH_3	nC_4H_9	H	(E) 62	-265.7
			(Z) 38	
CH_3	$C_6H_5CH_2$	H	(E) 53	-264.0
			(Z) 47	
CH_3	nC_4H_9	CH_3	(E) 45	-269.0
			(Z) 55	-270.4
CH_3CH_2	CH_2CH_2OH	H	(E) 56	-254.5
			(Z) 44	-251.5
CH_3CH_2	CH_2CH_2O-CHO	H	(E) 55	-255.1
			(Z) 45	-254.8

(a) At 9.118 MHz and 301 K; the shifts were originally referenced to $NaNO_3$, D_2O, H^+ (conversion factor: -1.5 ppm (D18))

(b) Determined from ^1H spectroscopy

(c) Originally measured at 9.12 MHz, room temperature and referenced to NH_4NO_3 (B46) (conversion factor: -357.4 ppm)

(d) $CHCl_3$ 30%

Footnote to Table 6.25

(a) 0.08M $Cr(acac)_3$ was added. See also T.6-7 for acetyl derivatives of amino sugars

(b) From Ref. P14 (at 9.12 MHz and room temperature); the shifts were referenced to benzamide (conversion factor: -279.3 ppm)

Table 6.25. ^{15}N chemical shifts of N,N-dialkyl-amides, -carbamates and -benzamides (M22, M20d)

The spectra were recorded at 9.118 MHz and 301 K; the shifts were originally referenced to $NaNO_3$, D_2O, H^+ (conversion factor = -1.5 ppm)

$R_2N-CO-X$: R	X	$\delta^{15}N$ (ppm/CH_3NO_2)	Conditions
H	C_6H_5	-279.3	DMF 3M
H	$C_6H_4NO_2$-p	-276.0	DMSO 1.4 M (b)
H	C_6H_4CN-p	-276.6	DMSO 1.4 M (b)
H	$C_6H_4NO_2$-m	-277.8	DMSO 1.4 M (b)
H	C_6H_4Br-m	-278.4	DMSO 1.4 M (b)
H	C_6H_4F-m	-278.4	DMSO 1.4 M (b)
H	C_6H_4Br-p	-278.7	DMSO 1.4 M (b)
H	C_6H_4Cl-p	-278.8	DMSO 1.4 M (b)
H	$C_6H_4CH_3$ -m	-279.4	DMSO 1.4 M (b)
H	C_6H_4F-p	-279.4	DMSO 1.4 M (b)
H	$C_6H_4CH_3$ -p	-280.1	DMSO 1.4 M (b)
H	$C_6H_4OCH_3$-p	-281.0	DMSO 1.4 M (b)
CH_3	$CHCl_2$	-282.2	dioxane0.2 M (a)
CH_3	CCl_3	-286.0	dioxane0.2 M (a)
CH_3	Cl	-284.1	neat (a)
CH_3	OCH_3	-314.2	$CDCl_3$ 1M
CH_3	SCH_3	-286.5	$CDCl_3$ 3M
CH_3	$CH_2CON(CH_3)_2$	-278.2	neat (a)
CH_3	$CH_2C_6H_5$	-282.2	CH_3CN (a)
CH_3	$CH_2CH_2C_6H_5$	-282.0	neat (a)
CH_3	$C_6H_4OCH_3$-p	-282.8	dioxane0.2 M
CH_3	$C_6H_4CH_3$-p	-282.5	dioxane0.2 M
CH_3	C_6H_4Cl-p	-280.9	dioxane0.2 M
CH_3	C_6H_4CN-p	-279.9	dioxane0.2 M
CH_3	$C_6H_4NO_2$-p	-280.5	dioxane0.2 M
CH_3	$C_6H_4OCH_3$-o	-278.5	dioxane0.2 M
CH_3	$C_6H_4CH_3$-o	-279.1	dioxane0.2 M
CH_3	$C_6H_4NO_2$-o	-278.9	dioxane0.2 M
C_2H_5	$CH=CH-CH_3$	-254.8	neat (a)
C_2H_5	$CH_2COOC_2H_5$	-251.5	neat (a)
C_2H_5	Cl	-256.6	neat

Table 6.26. ^{15}N chemical shifts of N,N-dimethylthioamides and -carbamates (M22, M20d) The spectra were recorded at 9.118 MHz and 301 K using 0.08 M $Cr(acac)_3$ and the shifts were originally referenced to $NaNO_3$, D_2O, H^+ (conversion factor = -1.5 ppm)

$R_2N-CS-X$ R	X	$\delta^{15}N$ (ppm/CH_3NO_2)	Conditions
CH_3	H	-225.6	neat
CH_3	CH_3	-234.9	TCE 8M
CH_3	Cl	-233.8	TCE 8M
CH_3	SCH_3	-244.3	$CDCl_3$ 3M
CH_3	$S-S-CS-N(CH_3)_2$	-242.6	$CDCl_3$ 3M
CH_3	OCH_3	-269.8	$CDCl_3$ 4M
C_2H_5	SCH_3	-215.8	$CDCl_3$ 4M

Footnote to Table 6.27

(a) At 9.118 MHz and 301 K; shifts originally referenced to 2M $NaNO_3$, D_2O, 2M HNO_3 (conversion factor = -2.6 ppm)

(b) At 10.09 MHz and room temperature; 10-20 10^{-3}g $Cr(acac)_3$ added to N,N-disubstituted amides

(c) At 6.08 MHz and ~ 323 K (conversion factor = -336.7 ppm)

(d) At 18.25 MHz and room temperature (conversion factor = -6.2 ppm)

Table 6.27. ^{15}N chemical shifts of ureas

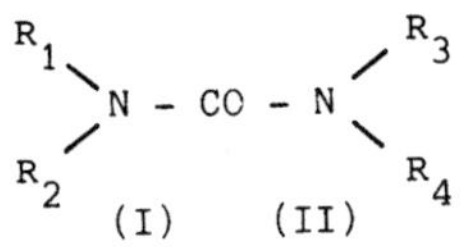

R_1	R_2	R_3	R_4	δN(I) (ppm/ CH_3NO_2)	δN(II) (ppm/ CH_3NO_2)	Conditions	Reference of Shifts	Ref.
H	H	H	H	-305.0 -305.2 -303.6 -305.0	-305.0 -305.2 -303.6 -305.0	H_2O/EtOH 4M DMF H_2O	(a) NO_3Na (b) NO_3H (c) NMe_4Cl (d) NO_3H	M 9 S 16 W 21 Y 16
H	H	H	CH_3	-307.4 -307.3 -307.7	-308.6 -311.1 -307.9	H_2O/EtOH 4M DMF H_2O	(a) NO_3Na (b) NO_3H (d) NO_3H	M 9 S 16 Y 16
H	H	CH_3	CH_3	-306.2 -307.5	-309.0 -314.6	H_2O/EtOH 4M DMF	(a) NO_3Na (b) NO_3H	M 9 S 16
H	CH_3	H	CH_3	-312.8 -313.6 -310.5	-312.8 -313.6 -310.5	EtOH 4M DMF H_2O	(a) NO_3Na (b) NO_3H (d) NO_3H	M 9 S 16 Y 16
H	CH_3	CH_3	CH_3	-311.6 -312.6	-318.3 -317.3	H_2O/EtOH 4M DMF	(a) NO_3Na (b) NO_3H	M 9 S 16
CH_3	CH_3	CH_3	CH_3	-317.9 -316.7 -317.8	-317.9 -316.7 -317.8	neat 4M DMF	(a) NO_3Na (b) NO_3H	M 9 S 16 W 21
H	H	H	C_2H_5	-307.3	-293.0	4M DMF	(b) NO_3H	S 16
H	C_2H_5	H	C_2H_5	-293.9 -295.3	-293.9 -295.3	H_2O/EtOH 4M DMF	(a) NO_3Na (b) NO_3H	M 9 S 16
H	H	C_2H_5	C_2H_5	-306.8	-283.9	H_2O/EtOH	(a) NO_3Na	M 9
CH_3	CH_3	C_2H_5	C_2H_5	-317.5	-292.5	H_2O/EtOH	(a) NO_3Na	M 9
C_2H_5	C_2H_5	C_2H_5	C_2H_5	-291.0	-291.0	neat	(a) NO_3Na	M 9
H	H	H	iC_3H_7	-279.6 -279.0	-308.3 -310.6	4M DMF H_2O	(b) NO_3H (d) NO_3H	S 16 Y 16
H	iC_3H_7	H	iC_3H_7	-279.6	-279.6	H_2O/EtOH	(a) NO_3Na	M 9
CH_3	CH_3	H	iC_3H_7	-312.6	-278.3	H_2O/EtOH	(a) NO_3Na	M 9
H	H	H	nC_3H_7	-307.3	-296.5	4M DMF	b) NO_3H	S 16
H	CH_3	H	nC_3H_7	-296.1	-310.7	H_2O	(d) NO_3H	Y 16
H	H	H	nC_4H_9	-307.3	-296.4	4M DMF	(b) NO_3H	S 16
H	H	H	iC_4H_9	-307.1	-297.9	4M DMF	(b) NO_3H	S 16
H	H	H	tC_4H_9	-306.1	-276.3	4M DMF	(b) NO_3H	S 16
H	CH_3	H	tC_4H_9	-278.2	-313.5	DMSO	(d) NO_3H	Y 16
H	H	H	C_6H_5	-302.7	-274.5	4M DMF	(b) NO_3H	S 16
H	CH_3	H	C_6H_5	-276.2	-308.9	DMSO	(d) NO_3H	Y 16
H	H	H	$CH_2CH=CH_2$	-306.9	-299.5	4M DMF	(b) NO_3H	S 16
H	CH_3	H	$CH_2C_6H_5$	-298.5	-313.5	DMSO	(d) NO_3H	Y 16
H	nC_4H_9	H	nC_4H_9	-297.8	-297.8	4M DMF	(b) NO_3H	S 16
H	C_6H_5	H	C_6H_5	-272.5	-272.5	4M DMF	(b) NO_3H	S 16

Table 6.28. ^{15}N chemical shifts of thioureas, recorded at 9.118 MHz and 301 K (M9) The shifts were originally referenced to 2M $NaNO_3$, D_2O, H^+ (conversion factor = -2.6 ppm)

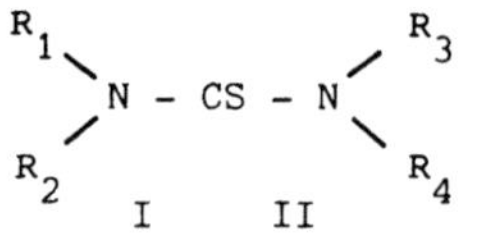

R_1	R_2	R_3	R_4	δN(I) (ppm/CH_3NO_2)	δN(II) (ppm/CH_3NO_2)	Conditions
H	H	H	H	-273.3	-273.3	H_2O/EtOH
H	H	H	CH_3	-277.3	-277.8	EtOH
H	H	CH_3	CH_3	-276.0	-277.8	H_2O/EtOH
H	H	H	$COCH_3$	-258.8	-220.2	$CDCl_3/Me_2CO$
H	CH_3	H	CH_3	-278.9	-278.9	EtOH
H	CH_3	CH_3	CH_3	-281.4	-287.3	$CDCl_3$
CH_3	CH_3	CH_3	CH_3	-287.6	-287.6	$CDCl_3$
CH_3CH_2	CH_3CH_2	CH_3CH_2	CH_3CH_2	-258.8	-258.8	$CDCl_3$
H	tC_4H_9	H	tC_4H_9	-243.9	-243.9	$CDCl_3$
CH_3	CH_3	H	$C_6H_4CH_3$-p	-279.3	-256.8	$CDCl_3$

Table 6.29. ^{15}N chemical shifts of lactams. Solvent effect: n = 7 cis, $CDCl_3$ 2M -259.6, DMSO 2M -260.6, C_2H_5OH 2M -257.7, TFA -239 ppm. A series of pyrrolin-2-ones and related compounds have also been studied (F 16)

$(CH_2)_n$-CO-NH	$\delta^{15}N$ (ppm/CH_3NO_2)	Conditions	Reference of Shifts	Ref.
n = 3	-265.5	$CHCl_3$ 2M (a)	HNO_3	W 18
	-247.0	TFA (b)	NO_3NH_4	K 13
4	-265.9	$CHCl_3$ 2M (a)	HNO_3	W 18
	-244.2	TFA (b)	NO_3NH_4	K 13
5	-262.6	$CHCl_3$ 2M (a)	HNO_3	W 18
	-240.2	TFA (b)	NO_3NH_4	K 13
6	-262.5	$CHCl_3$ 2M (a)	HNO_3	W 18
	-240.3	TFA (b)	NO_3NH_4	K 13
7 cis	-259.6	$CHCl_3$ 2M (a)	HNO_3	W 18
	-237.7	TFA (b)	NO_3NH_4	K 13
7 trans	-262.6	$CHCl_3$ 2M (a)	HNO_3	W 18
11	-261.7	$CHCl_3$ 2M (a)	HNO_3	W 18
	-236.9	TFA (b)	NO_3NH_4	K 13

(a) At 18.25 MHz and room temperature (conversion factor = -6.2 ppm)

(b) At 9.2 MHz and 305 K (conversion factor = -4.6 ppm)

Table 6.30. ^{15}N chemical shifts of penicillins and cephalosporins (P 9) (L 54)

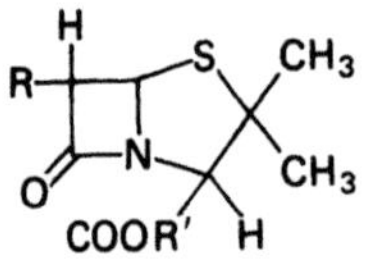
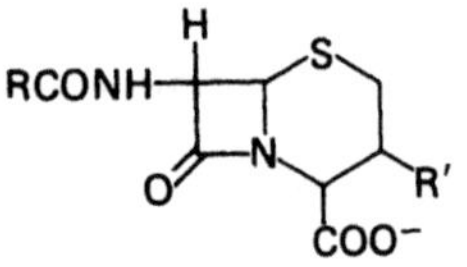

Th 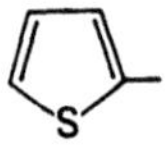 Tet 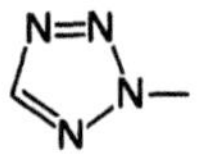 Thd 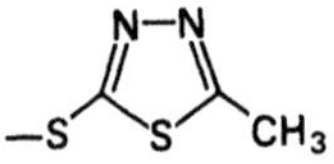

AAA HOCOCH(NH_2)$CH_2CH_2CH_2$-

The spectra were recorded at 10.09 MHz and room temperature; the shifts were originally referenced to 2.9M NH_4Cl in 1M HCl (conversion factor = -355.3 ppm)

Penicillins and Cephalosporins R		$\delta^{15}N$ (ppm/CH_3NO_2) N(cyclic)	N(R)	Conditions
(Penicillins)				
$C_6H_5CH_2CONH$-	CH_3 K	-221.0 -211.6	-270.2 -264.6	C_6H_6 H_2O
$C_6H_5OCH_2CONH$-	CH_3 Na	-220.5 -212.2	-274.8 -269.8	dioxane H_2O
2,6(CH_3O)$_2C_6H_3$-CONH -	CH_3 Na	-218.5 -210.8	-271.9 -259.5	C_6D_6 H_2O
$C_6H_5CH(NH_2)CONH$-	CH_3 Na	-219.9 -210.8	-277.1 -267.7	C_6D_6 H_2O
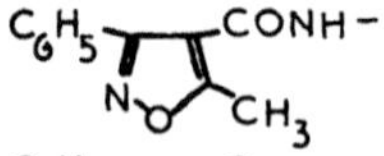	CH_3 Na	-219.9 -211.2	-267.6 -263.2	$CH_2Cl_2-C_6D_6$ H_2O
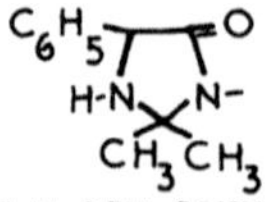	CH_3 K	-230.4 -223.2	-276.9 -269.3	H_2O H_2O
$C_6H_5OCH_2CONH$-	CH_3	-238.0	-276.4	dioxane
α-sulfoxide	Na	-229.4	-272.8	H_2O

Table 6.30 (continued)

Penicillins and Cephalosporins R	R'	$\delta^{15}N$ (ppm/CH_3NO_2) N(cyclic)	N(R)	Conditions
(Cephalosporins)				
H	$OCOCH_3$	-226.0	-265.7	(a)
CH_3	$OCOCH_3$	-225.6	-267.8	(a)
$C_6H_5CH_2$	$OCOCH_3$	-225.6	-268.6	(a)
$C_6H_5OCH_2$	$OCOCH_3$	-226.0	-274.2	(a)
$C_6H_5C(OH)H$	$OCOCH_3$	-225.7	-273.9	(a)
$ThCH_2$	$OCOCH_3$	-226.0	-269.8	(a)
$TetCH_2$	$OCOCH_3$	-226.2	-271.8	(a)
AAA	$OCOCH_3$	-225.9	-268.2	(a)
$ThCH_2$	Pyr	-224.5	-269.4	(a)
$TetCH_2$	Thd	-224.6	-271.6	(a)
$C_6H_5C(NH_2)H$	H	-225.4	-271.6	(a)

(a) In H_2Osolution, pH: 4 to 7

Table 6.31. ^{15}N chemical shifts of dipeptides

Peptide	δNH_2(a) (ppm/CH_3NO_2)	δ NHCO (a) (ppm/CH_3NO_2)	Conditions	Ref.
Gly - Gly	-355.1 -353.3	-266.6 -265.0	H_2O 0.2 M (a) H_2O pH 5.8(b)	P 10 G 13
Ala - Gly	-341.2	-266.6	H_2O pH 5.8(a)	P 10
Val - Gly	-346.8	-262.0	H_2O pH 5.8(a)	P 10
Leu - Gly	-343.0	-263.3	H_2O pH 5.8(a)	P 10
Phe - Gly	-343.6	-263.5	H_2O pH 4.9(a)	M 23
Gly - Ala	-354.6	-252.1	H_2O 0.2 M (a)	P 10
Ala - Ala	-341.6	-252.3	H_2O 0.2 M (a)	P 10
Val - Ala	-346.8	-248.8	H_2O 0.2 M (a)	P 10
Leu - Ala	-342.9	-250.2	H_2O 0.2 M (a)	P 10
Phe - Ala	-343.1	-256.5	H_2O pH 9 (a)	M 23
Gly - Val	-354.8	-258.4	H_2O 0.2 M (a)	P 10
Ala - Val	-341.3	-258.4	H_2O 0.2 M (a)	P 10
Val - Val	-346.8	-254.0	H_2O 0.2 M (a)	P 10
Leu - Val	-343.0	-256.2	H_2O 0.2 M (a)	P 10
Gly - Leu	-355.1	-254.3	H_2O 0.2 M (a)	P 10
Ala - Leu	-341.3	-254.9	H_2O 0.2 M (a)	P 10
Val - Leu	-347.1	-251.1	H_2O 0.2 M (a)	P 10
Leu - Leu	-343.4	-252.4	H_2O 0.2 M (a)	P 10
Phe - Leu	-353.0	-252.8	H_2O pH 12 (a)	M 23
Gly - Phe	-355.3	-256.9	H_2O pH 9 (a)	M 23
Gly - Tyr	-355.2	-256.8	H_2O pH 11 (a)	M 23

(a) At 18.23 MHz and room temperature; the shifts were originally referenced to 1M DNO_3 (conversion factor = -6.2 ppm)

(b) 9.12 MHz and room temperature; the shifts were originally referenced to 5M NH_4NO_3 in 2M HNO_3 (conversion factor = -359.0 ppm)

(*) In fact, the NH_4NO_3 and DNO_3 scales were claimed to be related through the conversion factor = -355.0 ppm) (G13)

Two acetyldipeptides were also studied:

H_3CCO—Gly—Gly and H_3CCO—Leu—Gly (in DMSO)

-269.9 -257.0 -256.8 -274.4 ppm

Table 6.32. ^{15}N chemical shifts of cyclodipeptides (2.5-dioxopiperazines) in TFA at 9.12 MHz and 303 K. The shifts were referenced to a D_2O solution of NH_4NO_3 (conversion factor = -4.6 ppm) (K26)

Homo-Cyclodipeptides		Hetero-Cyclodipeptides		
c(Gly - Gly)	-268.0	c(Gly - Ala)	-270.5	-253.0
c(Ala - Ala)	-255.1	c(Gly - Leu)	-270.1	-256.5
c(Leu - Leu)	-258.0	c(Gly - Val)	-268.1	-261.3
c(Val - Val)	-260.9	c(Gly - Phe)	-266.4	-257.7
c(Phe - Phe)	-256.8	c(Gly - Phg)	-269.6	-253.9
c(Tyr - Tyr)	-257.3	c(Ala - Leu)	-254.5	-258.2
c(Sar - Sar)	-266.2	c(Ala - Val)	-252.3	-263.4
c(Pro - Pro)	-242.1	c(Ala - Phe)	-251.2	-258.2
c(L-Ala - D-Ala)	-255.5	c(Ala - Sar)	-254.5	-270.1
c(L-Leu - D-Leu)	-258.4	c(L-Ala - D-Leu)	-254.6	-259.0(a)
c(L-Val - D-Val)	-261.5	c(L-Ala - D-Val)	-252.5	-263.9(a)
c(L-Phe - D-Phe)	-257.1			

(a) Measured at 20.28 MHz and calculated for 9.12 MHz.

Table 6.33. ^{15}N chemical shifts of tripeptides, measured at 18.25 MHz and room temperature in H_2O solutions (0.1 to 0.2M); the pH of the solution were those of the isoelectric points and the shifts were referenced to 1M DNO_3 (M23) (conversion factor = -6.2 ppm)

(1)		(2)		(3)	δNH_2 (1)	δ NH-CO (2)	δ NH-CO (3)
Gly	-	Gly	-	Gly	-355.1	-272.9 -270.4(a)	-266.9 -264.2(a)
Ala	-	Gly	-	Gly	-340.9	-272.8	-266.8
Leu	-	Gly	-	Gly	-342.6	-269.9	-266.7
Val	-	Gly	-	Gly	-346.3	-268.5	-266.5
Gly	-	Gly	-	Ala	-355.1	-272.8	-252.7
Ala	-	Ala	-	Ala	-341.6	-258.0	-253.2
Gly	-	Ala	-	Gly	-355.8	-258.4	-267.9
Gly	-	Leu	-	Gly	-355.3	-260.0	-266.1

(a) At 9.12 MHz and room temperature. The shifts were originally referenced to 5M NH_4NO_3 in 2M HNO_3 at pH = 8 (G13) (conversion factor = -359.0 ppm); (*) see Table 6.31)

One acetyl tripeptide was also studied:

CH_3CO—Gly—Gly—Gly

(in H_2O) -265.8 -270.2 -265.0 (4M NH_4Cl in 2M HCl) (L19)

(in DMSO) -269.3 -275.0 -275.0 (5M NH_4NO_3 in 2M HNO_3) (G13)

Table 6.34. ^{15}N chemical shifts of diastereomeric oligopeptides measured at 36.47 MHz and room temperature in DMSO/acetone 1/1 solutions, and referenced originally to external 2M NH_4NO_3 in D_2O (conversion factor = -4.6 ppm). Some aminoacids were used at the enriched level of the ^{15}N isotope (K 17). The tacticity of polypeptides was also investigated by ^{15}N spectroscopy (K 44).

Diastereomeric	Oligopeptides
BOC-L-Ala-L-Ala-OH	BOC-L-Ala-D-Ala-OH
-288.3 -262.6	-288.3 -263.0
BOC-L-Val-L-Val-OH	BOC-L-Val-D-Val-OH
-294.2 -265.8	-294.9 -266.5
BOC-L-Val-L-Val-L-Val-OH	BOC-L-Val-L-Val-D-Val-OH
-293.3 -264.1 -264.3	-293.3 -265.2 -265.2
BOC-L-Val-L-Val-L-Val-OCH_3	BOC-L-Val-L-Val-D-Val-OCH_3
-293.3 -265.0 -258.1	-293.7 -265.0 -258.5
BOC-L-Val-L-Val-L-Phe-OCH_3	BOC-L-Val-L-Val-D-Phe-OCH_3
-293.6 -264.8 -261.6	-293.8 -265.6 -262.0
BOC-L-Val-L-Val-L-Val-Gly-OEt	BOC-L-Val-D-Val-L-Val-Gly-OEt
-293.6 -264.3 -263.5 -274.2	-294.6 -264.3 -263.8 -274.2
BOC-L-Ala-L-Ala-D-Ala-OH	BOC-L-Ala-D-Ala-L-Ala-OH
-287.7 -263.1 -262.1	-287.5 -262.5 -262.1
BOC-L-Val-L-Val-D-Val-OCH_3	BOC-L-Val-D-Val-L-Val-OCH_3
-293.3 -265.2 -265.0	-294.4 -266.2 -265.1

Table 6.35. ^{15}N chemical shifts of polypeptides in TFA solutions, measured at 9.12 or 18.24 MHz and room temperature. The shifts were referenced to external NH_4NO_3 in D_2O (conversion factor = -4.6 ppm) (K19, K20, K26)

Homopolypeptides	Mixed Polypeptides
(Gly)n	(Ala - Ala - Gly)n
-272.0	-256.0 -256.8 -272.7
(Ala)n	(Ala - Gly - Gly)n
-256.9	-256.1 -272.6 -272.2
(Leu)n	(Leu - Gly - Gly)n
-255.3	-257.8 -271.2 -271.7
(Val)n	(Val - Gly - Gly)n
-254.6	-260.1 -268.5 -271.7
(Phe)n	(Phe - Gly - Gly)n
-255.4	-259.1 -269.1 -272.2
(Pro)n	(Pro - Gly - Gly)n
-239.5	-242.5 -270.8 -271.7
(Sar)n	(γAbu - βAla - Gly)n
-272.9	-258.7 -247.1 -266.5
(Phg)n	(βAla - γAbu - Gly)n
-255.4	-260.5 -247.8 -265.2
(βAla)n	(βAla - Gly - Gly)n
-252.5	-259.6 -266.2 -272.7
(γAbu)n	(γAbu - Gly - Gly)n
-248.3	-257.2 -265.7 -271.6
(δAva)n	(δAva - Gly - Gly)n
-244.6	-254.8 -265.2 -271.8
(εAca)n	(εAca - Gly - Gly)n
-240.8	-252.9 -264.9 -272.2
	(γAbu - βAla - Gly)n
	-257.1 -247.1 -272.5
	(Ala - γAbu - Gly)n
	-255.9 -258.1 -262.6
	(Val - Ala - Gly)n
	-260.3 -252.5 -272.1
	(Ala - Val - Gly)n
	-255.6 -258.3 -268.5

Abu: aminobutyryl Aca: aminocaproyl
Ava: aminovaleryl

Table 6.36. ^{15}N chemical shifts of synthetic polyamides in TFA solutions measured at 9.2 MHz and room temperature, and referenced to external NH_4NO_3 (conversion factor: -4.6 ppm) (K13, K18)

$-HN(CH_2)_n-NH-CO-(CH_2)_p-CO$		$\delta^{15}N$ (ppm/CH_3NO_2)	$H_2N-(B)-NH_2+HOCO-(A)COOH$		$\delta^{15}N$ (ppm/CH_3NO_2)
n	p		B	A	
2	2	-254.1	$(CH_2)_2$	C_6H_4	-261.7
2	3	-256.1	$(CH_2)_3$	C_6H_4	-255.9
2	4	-253.7	$(CH_2)_4$	C_6H_4	-252.0
2	6	-252.6	$(CH_2)_6$	C_6H_4	-249.6
2	8	-252.1	$(CH_2)_8$	C_6H_4	-247.1
2	10	-251.9	C_6H_4	$(CH_2)_2$	-241.1
3	2	-249.6	C_6H_4	$(CH_2)_3$	-241.2
4	2	-248.1	C_6H_4	$(CH_2)_4$	-241.3
4	6	-242.2	C_6H_4	$(CH_2)_6$	-240.7
6	2	-246.9	C_6H_4	$(CH_2)_8$	-241.5
6	3	-244.0	$(CH_2)_2$	(E)CH=CH	-256.0
6	4	-240.9	$(CH_2)_4$	(E)CH=CH	-244.1
6	6	-239.0	$(CH_2)_{10}$	(E)CH=CH	-243.0
6	8	-239.2			
8	2	-246.2			
10	2	-246.1			

Table 6.37. ^{15}N chemical shifts of nylon homopolymers (K13) and polysulfamides (K24), in TFA solutions, measured at 9.12 MHz and room temperature. The shifts were referenced to external NH_4NO_3 (conversion factor = -4.6 ppm). Mixed polymers containing sulfamide and peptide fragments were also studied:

	$\delta(SO_2NH)$	$\delta(NH\omega 1)$	$\delta(NH\omega 2)$
$-NH-(CH_2)_2-SO_2-NH$ - Gly - Gly	-265.8	-296.3	-271.8
- Gly - βAla	-259.8	-296.5	-257.7
-βAla - Gly	-265.3	-292.5	-266.0

$CO-NH(CH_2)_n-CO$	$\delta^{15}N$ (ppm/CH_3NO_2)	$-NH-(A)SO_2NH-(B)-CO$		$\delta^{15}N$ (ppm/CH_3NO_2)	
n		(A)	(B)	$NHSO_2$	NH(ω)
1	-271.9	$(CH_2)_3$	$(CH_2)_2$	-249.3	-292.2
2	-252.6	$(CH_2)_3$	$(CH_2)_5$	-246.2	-292.9
3	-248.3	C_6H_4	$(CH_2)_5$	-243.7	-312.8
4	-244.6	$(CH_2)_2$	CH_2	-264.9	-295.5
5	-240.8	$(CH_2)_2$	$(CH_2)_2$	-257.2	-291.5
6	-239.3	$(CH_2)_2$	$(CH_2)_3$	-256.1	-291.1
7	-238.3	$(CH_2)_2$	$(CH_2)_5$	-254.4	-290.7
11	-237.8				

Table 6.38. ^{15}N chemical shifts of imines

R (R\C = N/R'' with R')	R'	R"	$\delta^{15}N$ (ppm/ CH_3NO_2)	Conditions		Reference of Shifts	Ref.
$CH_3-N-C_6H_4-S$ (N and S both bonded to C)		H	-192.3	$CHCl_3$ 30%	(a)	NO_3Na	N4
CH_3	C_2H_5	CH_3	- 73.8	neat	(b)	NO_3Na	N5
H	iC_3H_7	CH_3	- 65.2	neat	(b)	NO_3Na	N5
H	$C_6H_5CH=CH$	CH_3	- 49.6	$CHCl_3$ 50%	(b)	NO_3Na	N5
H	iC_3H_7	nC_3H_7	- 52.1	neat	(b)	NO_3Na	N5
CH_3	C_2H_5	$C_6H_5CH_2$	- 67.0	neat	(b)	NO_3Na	N5
H	CH_3	iC_3H_7	- 32.3	neat	(b)	NO_3Na	N5
H	nC_3H_7	iC_3H_7	- 32.0	neat	(b)	NO_3Na	N5
H	iC_3H_7	iC_3H_7	- 34.6	neat	(b)	NO_3Na	N5
H	$CH_3CH=CH$	iC_3H_7	- 29.5	neat	(b)	NO_3Na	N5
H	C_6H_5	CH_3	- 56.9	neat	(b)	NO_3Na	N5
			- 62.1	$CHCl_3$ 36 mol %	(c)	HNO_3	W19
H	C_6H_5	C_2H_5	- 46.8	$CHCl_3$ 36 mol %	(c)	HNO_3	W19
H	C_6H_5	iC_3H_7	- 34.7	$CHCl_3$ 36 mol %	(c)	HNO_3	W19
H	C_6H_5	tC_4H_9	- 26.8	$CHCl_3$ 36 mol %	(c)	HNO_3	W19
H	C_6H_5	nC_3H_7	- 49.1	$CHCl_3$ 36 mol %	(c)	HNO_3	W19
H	C_6H_5	iC_4H_9	- 49.2	$CHCl_3$ 36 mol %	(c)	HNO_3	W19
H	C_6H_5	neo C_5H_{11}	- 49.3	$CHCl_3$ 36 mol %	(c)	HNO_3	W19
H	C_6H_5	tC_5H_{11}	- 28.0	$CHCl_3$ 36 mol %	(c)	HNO_3	W19
H	C_6H_5	C_6H_{11}	- 36.4	$CHCl_3$ 20 mol %	(c)	HNO_3	W19
H	$pCH_3O-C_6H_4$	C_6H_{11}	- 44.7	$CHCl_3$ 20 mol %	(c)	HNO_3	W19
H	$pCH_3-C_6H_4$	C_6H_{11}	- 40.1	$CHCl_3$ 20 mol %	(c)	HNO_3	W19
H	$pCl-C_6H_4$	C_6H_{11}	- 34.3	$CHCl_3$ 20 mol %	(c)	HNO_3	W19
H	$pNO_2-C_6H_4$	C_6H_{11}	- 20.9	$CHCl_3$ 20 mol %	(c)	HNO_3	W19
H	iC_3H_7	C_6H_5	- 55.1	neat	(b)	NO_3Na	N5
CH_3	C_2H_5	C_6H_5	- 52.9	neat	(b)	NO_3Na	N5
H	$C_6H_5CH=CH$	C_6H_5	- 47.4	$CHCl_3$ 50 %	(b)	NO_3Na	N5
H	C_6H_5	C_6H_5	- 52.2	$CHCl_3$ 50 %	(b)	NO_3Na	N5
			- 54.1	$CHCl_3$ 20 mol %	(c)	HNO_3	W19
H	$p(CH_3)_2N-C_6H_4$	C_6H_5	- 72.9	$CHCl_3$ 20 mol %	(c)	HNO_3	W19
h	$pCH_3O-C_6H_4$	C_6H_5	- 62.6	$CHCl_3$ 20 mol %	(c)	HNO_3	W19
H	$pCH_3-C_6H_4$	C_6H_5	- 58.1	$CHCl_3$ 20 mol %	(c)	HNO_3	W19
H	pF - C_6H_4	C_6H_5	- 55.5	$CHCl_3$ 20 mol %	(c)	HNO_3	W19
H	$pCl-C_6H_4$	C_6H_5	- 52.8	$CHCl_3$ 20 mol %	(c)	HNO_3	W19

Table 6.38 (continued)

R	R'	R"	$\delta^{15}N$ (ppm/ CH_3NO_2)	Conditions	Reference of Shifts	Ref.
H	$pNO_2-C_6H_4$	C_6H_5	-41.3	$CHCl_3$ 20 mol %(c)	HNO_3	W19
H	C_6H_5	$pCH_3O-C_6H_4$	- 57.6	$CHCl_3$ 20 mol %(c)	HNO_3	W19
H	C_6H_5	$pCH_3-C_6H_4$	- 55.2	$CHCl_3$ 20 mol %(c)	HNO_3	W19
H	C_6H_5	$pCl-C_6H_4$	- 58.2	$CHCl_3$ 20 mol %(c)	HNO_3	W19
H	C_6H_5	$pNO_2-C_6H_4$	- 57.9	$CHCl_3$ 20 mol %(c)	HNO_3	W19
H	$(CH_3)_2N$	C_6H_5	-147.3	neat (b)	NO_3Na	N5
$(CH_3)_2C=CH$	$(CH_3)_2N$	C_6H_5	-137.0	neat (b)	NO_3Na	N5
$(CH_3)_2N$	$(CH_3)_2N$	C_6H_5	-173.2	neat (b)	NO_3Na	N5
C_6H_5	$S-CH_2-C(COCH_3) = CH-$		- 87.6	2M $CDCl_3$ (d)	NO_3Na	R16
$pCH_3O-C_6H_4$	$S-CH_2-C(CHO) = CH$		- 88.2	2M $CDCl_3$ (d)	NO_3Na	R16
C_6H_5	$S-CH_2-C(COOCH_3) = CH$		- 90.8	2M $CDCl_3$ (d)	NO_3Na	R16
$pCH_3O-C_6H_4$	$S-CO-CD = CH$		- 94.4	2M $CDCl_3$ (d)	NO_3Na	R16
$pCH_3O-C_6H_4$	$S-CS-CH = CH$		- 90.9	2M $CDCl_3$ (d)	NO_3Na	R16

(a) At 6.08 MHz and 304 K with ^{15}N enriched material; (conversion factor = -1.5 ppm)

(b) At 9.118 MHz and 301 K; (conversion factor = -1.5 ppm)

(c) At 18.25 MHz and room temperature (conversion factor = -6.2 ppm)

(d) At 9.118 MHz and 301 K (conversion factor = -3.1 ppm)

0.05 M $Cr(acac)_3$ was added to the solution

Table 6.39. ^{15}N chemical shifts of oximes

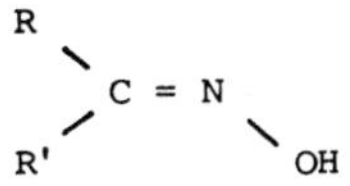

R	R'	$\delta^{15}N$ (ppm/CH_3NO_2) Z	E	Conditions		Reference of Shifts	Ref.
H	H	- 2.2	- 2.2	H_2O 20% mol	(a)	NO_3D	B47
H	CH_3	-30.3	-34.6	$CHCl_3$ 36% mol	(a)	NO_3D	B47
			-31.7	$CHCl_3$ 50% v/v	(b)	NO_3Na	N5
H	C_2H_5	-32.8	-34.4	$CHCl_3$ 36% mol	(a)	NO_3D	B47
H	nC_3H_7	-31.6	-33.5	$CHCl_3$ 36% mol	(a)	NO_3D	B47
H	iC_3H_7	-36.2	-36.3	$CHCl_3$ 36% mol	(a)	NO_3D	B47
			-33.0	$CHCl_3$ 50% v/v	(b)	NO_3Na	N5
H	$secC_4H_9$	-32.2	-32.7	$CHCl_3$ 36% mol	(a)	NO_3D	B47
H	iC_4H_9	-35.1	-35.0	$CHCl_3$ 36% mol	(a)	NO_3D	B47
H	$Et(Me)CHCH_2$	-32.2	-32.2	$CHCl_3$ 36% mol	(a)	NO_3D	B47
H	$(Et)_2CH$	-32.0	-34.2	$CHCl_3$ 36% mol	(a)	NO_3D	B47
H	C_6H_{11}	-34.5	-34.7	$CHCl_3$ 36% mol	(a)	NO_3D	B47
H	C_6H_5		-26.3	$CHCl_3$ 36% mol	(a)	NO_3D	B47
			-22.5	$CHCl_3$ 50% v/v	(b)	NO_3Na	N5
H	$pCH_3C_6H_4$		-29.3	$CHCl_3$ 36% mol	(a)	NO_3D	B47
H	$pCH_3OC_6H_4$		-30.9	$CHCl_3$ 36% mol	(a)	NO_3D	B47
H	$pClC_6H_4$		-23.5	$CHCl_3$ 36% mol	(a)	NO_3D	B47
CH_3	CH_3	-45.9	-45.9	$CHCl_3$ 36% mol	(a)	NO_3D	B47
		-42.0	-42.0	$CHCl_3$ 50% v/v	(b)	NO_3Na	N5
CH_3	C_2H_5	-47.6	-44.5	$CHCl_3$ 36% mol	(a)	NO_3D	B47
			-45.5	$CHCl_3$ 50% v/v	(b)	NO_3Na	N5
CH_3	nC_3H_7	-45.1	-42.2	$CHCl_3$ 36% mol	(a)	NO_3D	B47
CH_3	iC_3H_7	-49.6	-44.5	$CHCl_3$ 36% mol	(a)	NO_3D	B47
CH_3	tC_4H_9		-43.2	$CHCl_3$ 36% mol	(a)	NO_3D	B47
CH_3	iC_4H_9	-45.3	-42.3	$CHCl_3$ 36% mol	(a)	NO_3D	B47
CH_3	$secC_4H_9$		-54.7	neat	(c)	$N(Me)_4^+$	L27
CH_3	$neoC_5H_{11}$		-37.6	$CHCl_3$ 36% mol	(a)	NO_3D	B47
CH_3	$cycloC_3H_5$	-47.8	-47.8	$CHCl_3$ 36% mol	(a)	NO_3D	B47
C_2H_5	C_2H_5	-46.3	-46.3	$CHCl_3$ 36% mol	(a)	NO_3D	B47
C_2H_5	iC_3H_7	-46.5	-44.3	$CHCl_3$ 36% mol	(a)	NO_3D	B47
C_2H_5	tC_4H_9		-43.7	$CHCl_3$ 36% mol	(a)	NO_3D	B47
iC_3H_7	iC_3H_7	-43.5	-43.5	$CHCl_3$ 36% mol	(a)	NO_3D	B47
C_6H_5	C_6H_5	-34.2	-34.2	$CHCl_3$ 36% mol	(a)	NO_3D	B47
$CH_2 - CH_2 - CH_2$		-52.4	-52.4	$CHCl_3$ 36% mol	(a)	NO_3D	B47

Table 6.39 (continued)

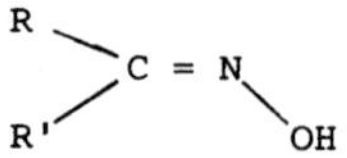		$\delta^{15}N$ (ppm/CH_3NO_2)		Conditions	Reference of Shifts	Ref.
R	R'	Z	E			
$\overset{|}{C}H_2-(CH_2)_2-\overset{|}{C}H_2$		-53.0	-53.0	$CHCl_3$ 36% mol (a)	NO_3D	B47
$\overset{|}{C}H_2-(CH_2)_3-\overset{|}{C}H_2$		-52.6	-52.6	$CHCl_3$ 36% mol (a)	NO_3D	B47
		-57.2	-57.2	neat (d)	NH_4Cl	L20
$\overset{|}{C}H_2-(CH_2)_4-\overset{|}{C}H_2$		-46.8	-46.8	$CHCl_3$ 36% mol (a)	NO_3D	B47
		-34.6	-34.6	neat (d)	NH_4Cl	L20
$\overset{|}{C}H_2-(CH_2)_5-\overset{|}{C}H_2$		-45.7	-45.7	$CHCl_3$ 36% mol (a)	NO_3D	B47
		-31.8	-31.8	neat (d)	NH_4Cl	L20
$CH_3-\overset{|}{C}H-(CH_2)_3-\overset{|}{C}H_2$		-55.2	-50.6	$CHCl_3$ 36% mol (a)	NO_3D	B47
		-57.0	-57.0	neat (d)	NH_4Cl	L20
$\overset{|}{C}H_2-CH(CH_3)-(CH_2)_2-\overset{|}{C}H_2$		-52.3	-52.3	$CHCl_3$ 36 % mol (a)	NO_3D	B47
bicyclo(2.2.1) heptanone		-59.6	-57.6	$CHCl_3$ 36% mol (a)	NO_3D	B47
nortricyclanone		-70.1	-65.2	$CHCl_3$ 36% mol (a)	NO_3D	B47

(a) At 18.25 MHz and room temperature (conversion factor = -6.2 ppm)

(b) At 9.118 MHz and 301 K (conversion factor = -1.5 ppm)

(c) At 27.36 MHz and room temperature (conversion factor = -336.7 ppm)

(d) At 27.36 MHz and room temperature (conversion factor = -355.3 ppm) solvent not specified

Table 6.40. ^{15}N chemical shifts of isocyanates and isothiocyanates

R - N = C = A: R	A	$\delta^{15}N$ (ppm/CH_3NO_2)	Conditions		Reference of Shifts	Ref.
CH_3	O	-365.3	neat	(a)	HNO_3	Y11
C_2H_5	O	-348.6	neat	(a)	HNO_3	Y11
nC_4H_9	O	-352.1	neat	(a)	HNO_3	Y11
iC_3H_7	O	-335.5	neat	(a)	HNO_3	Y11
tC_4H_9	O	-326.0	neat	(a)	HNO_3	Y11
C_6H_{11}	O	-338.3	neat	(a)	HNO_3	Y11
C_6H_5	O	-333.7	neat	(a)	HNO_3	Y11
$pCl-C_6H_4$	O	-334.2	DMSO 3M	(a)	HNO_3	Y11
1-naphthyl	O	-338.0	neat	(a)	HNO_3	Y11
CH_3	S	-289.9	DMSO 3M	(a)	HNO_3	Y11
C_2H_5	S	-277.0	neat	(a)	HNO_3	Y11
nC_4H_9	S	-279.0	neat	(b)	CH_3NO_2	M20e
C_6H_{11}	S	-268.2	neat	(a)	HNO_3	Y11
C_6H_5	S	-273.1	neat	(a)	HNO_3	Y11
$CH_2=C(C_6H_{13})$	S	-266.8	neat	(b)	CH_3NO_2	M20e
Li	S	-163.1	0.1 DMF	(b)	NO_3Na	V1
K	S	-161.1	0.1 DMF	(b)	NO_3Na	V1
NH_4	S	-168.3	5M HCl 2N	(c)	NH_4NO_3	B46
$(C_4H_9)_4N$	S	-166.9	2M dioxane	(b)	CH_3NO_2	M20e
iC_3H_7	$N-C_3H_7i$	-276.6	$CHCl_3$ 20% v/v	(a)	HNO_3	Y10
		-276.2	DMSO 20% v/v	(a)	HNO_3	Y10
		-277.1	neat	(a)	HNO_3	Y10
C_6H_{11}	$N-C_6H_{11}$	-280.7	$CHCl_3$ 10% v/v	(a)	HNO_3	Y10
iC_3H_7	$N-C_4H_9t$	-275.8	C_6H_{12} 10% v/v	(a)	HNO_3	Y10
tC_4H_9	$N-C_3H_7i$	-267.6	C_6H_{12} 20% v/v	(a)	HNO_3	Y10
iC_3H_7	$N-C_6H_5$	-270.8	DMSO 20% v/v	(a)	HNO_3	Y10
C_6H_5	$N-C_3H_7i$	-283.7	DMSO 20% v/v	(a)	HNO_3	Y10
C_2H_5	$N-(CH_2)_3N(Me)_2$	-297.2 (d)	neat	(a)	HNO_3	Y10
$(Me)_2N(CH_2)_3$	$N-C_2H_5$	-294.4 (d)	neat	(a)	HNO_3	Y10

(a) At 18.25 MHz and room temperature; (conversion factor = -6.2 ppm)

(b) At 9.118 MHz and 301 K; (conversion factor = -1.5 ppm)

(c) At 9.2 MHz and room temperature; (conversion factor = -357.4 ppm)

(d) Assignment may be reversed

Table 6.41. ^{15}N chemical shifts of compounds containing the cyano group

Compounds Containing the Cyano Group X-C≡N		$\delta^{15}N$ (ppm/CH_3NO_2)	Conditions		Reference of Shifts	Ref.
CH_3		-137.1	neat	(a)	NH_4NO_3	B46
		-138.4	neat	(b)	NO_3Na	M20
CH_3CH_2		-138.8	neat	(c)	NO_3Na	I1
$ON-N(C_3H_7i)-CH_2$	Z	-130.0	neat	(b)	NO_3Na	G20
	E	-126.0	neat	(b)	NO_3Na	G20
nC_4H_9-S		-102.5	neat	(b)	NO_3Na	M20e
$CH_2=CH-S$		- 95.3	neat	(b)	NO_3Na	M20e
$CH_2=C(C_4H_9n)-S$		- 95.2	neat	(b)	NO_3Na	M20e
C_6H_5		-121.5	DMSO 1.4 M	(d)	NH_4Cl	P14
$pNO_2-C_6H_4$		-112.7	DMSO 1.4 M	(d)	NH_4Cl	P14
$pCH_3-C_6H_4$		-121.8	DMSO 1.4 M	(d)	NH_4Cl	P14
$pCH_3O-C_6H_4$		-124.2	DMSO 1.4 M	(d)	NH_4Cl	P14

(a) At 9.12 MHz and 300 K (conversion factor = -357.4 ppm)

(b) At 9.118 MHz and 301 K (conversion factor = -1.5 ppm)

(c) At 9.12 MHz and room temperature; the shifts were referenced to 5M NH_4Cl in 2M HCl (conversion factor = -352.5 ppm)

Table 6.42. ^{15}N chemical shifts of iminium, amidinium and guanidinium salts measured at 9.118 MHz, 301 K and originally referenced to $NaNO_3$, D_2O, H^+ (conversion factor = -1.5 ppm); 0.08M $Cr(acac)_3$ was added to the solutions. (R13)

Iminium Salts

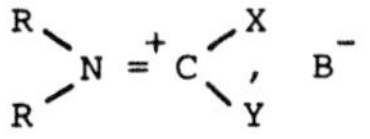

R	X	Y	B	$\delta^{15}N$ (ppm/CH_3NO_2)	Conditions	
CH_3	H	H	CF_3COO	-156.5	$CDCl_3$	2M
CH_3	H	OCH_3	OSO_2F	-236.0	CD_2Cl_2	2M
CH_3	H	OCH_3	CH_3SO_4	-241.0	CH_2Cl_2	2.5M
CH_3	CH_3	OCH_3	OSO_2F	-239.0	CH_2Cl_2	2M
CH_3	H	Cl	Cl	-171.4	CH_2Cl_2	1.5M
CH_3	H	Cl	$OPOCl_2$	-174.4	CH_2Cl_2	1.5M
CH_3	H	SCH_3	OSO_2F	-205.7	CH_3CN	3M
CH_3	H	SCH_3	$MeSO_4$	-204.5	CH_3CN	4M
CH_3	H	SCH_3	Br	-205.9	CH_3CN	4M
CH_3	SCH_3	Cl	Cl	-258.0	CH_2Cl_2	2M
CH_3	SCH_3	Cl	CF_3COO	-260.5	CH_2Cl_2	2M

Amidinium Salts

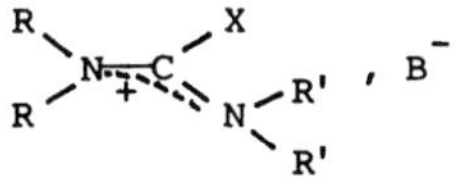

R	R'	X	B	$\delta N(R)_2$ (ppm/CH_3NO_2)	$\delta N(R')_2$ (ppm/CH_3NO_2)	Conditions	
CH_3	CH_3	H	Cl	-275.9	-275.9	CH_3CN	2M
CH_3	CH_3	CH_3	CH_3SO_4	-273.3	-273.3	CH_2Cl_2	2M
CH_3	CH_3	OMe	OSO_2F	-294.2	-294.2	CH_3CN	2M
CH_3	CH_3	SMe	OSO_2F	-269.1	-269.1	CH_3CN	2M
CH_3	CH_3	Cl	Cl	-269.0	-269.0	CH_3CN	2M
				-268.1	-268.1	CH_2Cl_2	2M
CH_3	CH_3CH_2	H	Cl	-275.6	-250.6	CH_3CN	2M
CH_3	CH_3CH_2	OMe	OSO_2F	-291.8	-272.1	CH_3CN	2M
CH_3CH_2	CH_3CH_2	H	Cl	-249.7	-249.7	CH_3CN	2M
CH_3CH_2	CH_3CH_2	OMe	OSO_2F	-270.2	-270.2	CH_3CN	2M

Table 6.42 (continued)

Guanidinium Salts 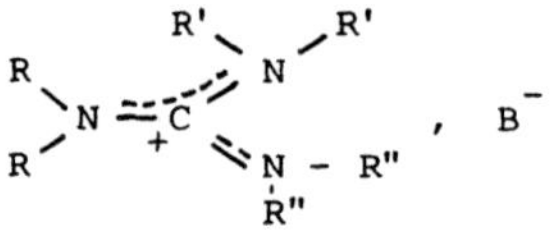 (a)

R	R'	R"	B	$\delta N(R)_2$ (ppm/CH_3NO_2)	$\delta N(R')_2$ (ppm/CH_3NO_2)	$\delta N(R")_2$ (ppm/CH_3NO_2)
CH_3	CH_3	CH_3	Cl	-305.4	-305.4	-305.4
CH_3	CH_3	CH_3CH_2	Cl	-300.5	-300.5	-283.7
CH_3	CH_3	H, C_6H_5	Cl	-303.2	-303.2	-270.3
CH_3	CH_3	H, C_6H_5	CF_3COO	-302.8	-302.8	-268.9
CH_3	CH_3CH_2	CH_3CH_2	Cl	-304.3	-280.1	-280.1
CH_3CH_2	CH_3CH_2	CH_3CH_2	Cl	-277.9	-277.9	-277.9

(a) In CH_2Cl_2 (2 M)

Table 6.43. ^{15}N chemical shifts of vinamidinium salts and cyclobutene cyanines. The assignment of N_I and N_{II} is ambiguous

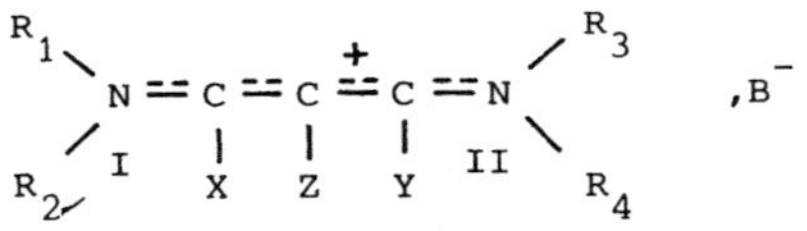

R_1	R_2	R_3	R_4	X	Y	Z	B	$\delta^{15}N$ (ppm/CH_3NO_2) N_I	N_{II}	Conditions	Reference for Shifts	Ref.
CH_3	CH_3	CH_3	CH_3	H	H	H	ClO_4	-257.7	-257.7	DMSO 0.6 M (a)	NH_4Cl	J7
CH_3	CH_3	CH_3	CH_3	H	Cl	H	Cl	-248.1	-258.3	$CDCl_3$ 2 M (b)	NO_3Na	R13
CH_3	CH_3	CH_3	CH_3	H	Cl	CH_3	$OPOCl_2$	-244.2	-262.5	$CDCl_3$ 2 M (b)	NO_3Na	R13
CH_3	CH_3	CH_3	CH_3	CH_3	Cl	H	Cl	-239.7	-264.1	$CDCl_3$ 2 M (b)	NO_3Na	R13
CH_3	CH_3	CH_3	CH_3	C_2H_5	Cl	CH_3	$OPOCl_2$	-191.4	-322.8	$CDCl_3$ 2 M (b)	NO_3Na	R13
CH_3	CH_3	CH_3	CH_3	$C_6H_5CH_2$	Cl	C_6H_5	$OPOCl_2$	-210.0	-273.5	$CDCl_3$ 2 M (b)	NO_3Na	R13
CH_3	CH_3	CH_3	CH_3	$-CH(CH_3)-$		CH_3	$OPOCl_2$	-272.8	-272.8	$CDCl_3$ 2 M (b)	NO_3Na	R13
CH_3	CH_3	CH_3	CH_3	$-CH(C_2H_5)-$		C_2H_5	$OPOCl_2$	-271.6	-271.6	$CDCl_3$ 2 M (b)	NO_3Na	R13
CH_3	CH_3	CH_3	CH_3	$-CH(C_6H_5)-$		C_6H_5	$OPCl_2$	-269.8	-269.8	$CDCl_3$ 2 M (b)	NO_3Na	R13
C_6H_5	H	CH_3	CH_3	H	H	H	ClO_4	-237.5	-247.4	DMSO 0.6 M (a)	NH_4Cl	J7
C_6H_5	H	C_6H_5	H	H	H	H	ClO_4	-229.7	-229.7	DMSO 0.6 M (a)	NH_4Cl	J7

(a) At 27.35 MHz and room temperature; (conversion factor = -360.8 ppm). (The shifts were originally referenced to 5M NH_4NO_3 in 2M HNO_3 and converted to NH_4^+ using the conversion factor 356.2_5)

(b) At 9.118 and 301 K; (conversion factor = -1.5 ppm)

Table 6.44. ^{15}N chemical shifts of hydrazines, aziridines and related compounds

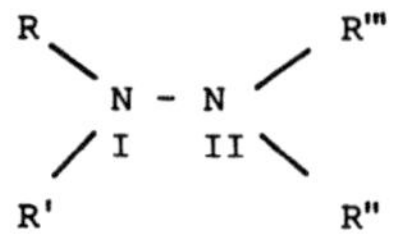

R	R'	R"	R"'	$\delta^{15}N$ (ppm/CH_3NO_2) $N_{(I)}$	$N_{(II)}$	Conditions		Reference of Shifts	Ref.
H	H	H	H	-334.9	-334.9	neat	(a)	HNO_3	L28
CH_3	H	H	H	-328.1	-305.6	neat	(a)	HNO_3	L28
CH_3	CH_3	H	H	-322.8	-281.5	neat	(a)	HNO_3	L28
CH_3	H	CH_3	H	-306.7	-306.7	neat	(a)	HNO_3	L28
CH_3	CH_3	CH_3	H	-307.8	-285.4	neat	(a)	HNO_3	L28
CH_3	CH_3	CH_3	CH_3	-277.7	-277.7	neat	(a)	HNO_3	L28
C_6H_5	H	H	H	-294.3	-319.6(d)	neat	(a)	HNO_3	L28
				-324.3	-299.4(d)	neat	(b)	NH_4NO_3	B46
C_6H_5	H	C_6H_5	H	-287.7	-287.7	dioxane sat.	(a)	HNO_3	L28
C_6H_5	C_6H_5	H	H		-293.3	neat	(a)	HNO_3	L28
H	$C(CH_3)_2$		CH_3	-304.3		CH_3OH	(c)	NO_3Na	M20b
H	$C(CH_3)_2$		C_3H_7i	-289.5	-277.6	CH_3OH	(c)	NO_3Na	M20b
	$=C(NH_2)-S-C(NH_2)=$			- 87.9	- 87.9	H_2O	(e)	HNO_3	W14
	$=C(NH_2)-S-C(CH_3)=$			- 79.0	- 79.0	1M DMSO	(e)	HNO_3	W14
	$=C(SH)-S-C(SH)=$			-114.2	-114.2	1.2M C_2H_5OH	(e)	HNO_3	W14
	$=C(NH_2)-S-CS-$		H	-116.8	-166.7	2M DMSO	(e)	HNO_3	W14

(a) At 6.07 MHz and room temperature; (conversion factor -6.2 ppm)

(b) At 9.2 MHz and 300 K; (conversion factor -361.2 ppm)

(c) At 9.118 MHz and 301 K; (conversion factor = -1.5 ppm)

(d) It should be noted that the assignments of the NH and NH_2 signals are different in Refs. L 28 and B 46

(e) At 18.25 MHz and room temperature (conversion factor = -6.2 ppm)

Table 6.45. ^{15}N chemical shifts of nitrosamines, measured at 9.118 MHz and 310 - 360 K (G14)

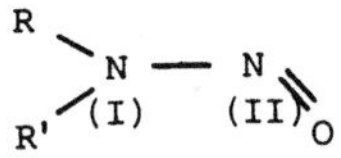				$\delta^{15}N$ (a) (ppm/CH_3NO_2)	
R	R'	Configuration		N_I	N_{II}
CH_3	CH_3			-148.2	+154.8
C_2H_5	C_2H_5			-123.8	+157.1
n-C_3H_7	n-C_3H_7			-127.7	+161.0
i-C_3H_7	i-C_3H_7			-108.7	+164.5
n-C_4H_9	n-C_4H_9			-127.7	+159.7
C_6H_5	C_6H_5			-111.3	+174.4
CH_2 - $(CH_2)_2$ - CH_2				-122.8	+154.2
CH_2 - $(CH_2)_3$ - CH_2				-133.1	+153.1
CH_2 - $(CH_2)_4$ - CH_2				-123.2	+157.7
CH_2-CH_2-O-CH_2-CH_2				-140.5	+153.2
CH_3	C_2H_5	Z	29	-136.7	+154.8
C_2H_5	CH_3	E	71		
CH_3	n-C_4H_9	Z	21	-138.6	+159.0
n-C_4H_9	CH_3	E	79		
t-C_4H_9	CH_3	E	100	-120.2	+160.3
C_6H_5	CH_3	E	100	-129.3	+163.8
CH_3	$CH_2C_6H_5$	Z	29	-139.4	+155.0
$C_6H_5CH_2$	CH_3	E	71	-137.3	+157.9
CH_3	CH_2CN	Z	56	-155.1	+159.2 (c)
$CNCH_2$	CH_3	E	44	-150.1	+163.5 (c)
CH_3CH_2	C_6H_5	Z	5	(b)	(b)
C_6H_5	CH_2CH_3	E	95	-116.1	+165.4
CH_3CH_2	$CH_2C_6H_5$	Z	50	-127.0	+156.1
$C_6H_5CH_2$	CH_2CH_3	E	50	-124.4	+158.4
CH_3CH_2	CH_2CH_2OH	Z	50	-126.1	+156.4
$HOCH_2CH_2$	CH_2CH_3	E	50		
t-C_4H_9	$CH(CH_3)_2$	E	100	-105.2	+175.4
$(CH_3)_2CH$	$CH_2C_6H_5$	E	82	-118.0	+158.4
$CH_2C_6H_5$	$CH(CH_3)_2$	Z	18	-115.9	+163.3
$(CH_3)_2CH$	CH_2CN	Z	93	-132.6	+161.4 (c)

Table 6.45 (continued)

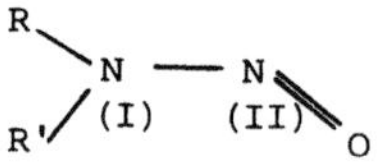

R	R'	Configuration		N_I	N_{II}
				$\delta^{15}N$ (a) (ppm/CH_3NO_2)	
$CNCH_2$	$CH(CH_3)_2$	E	7	(b)	(b)
$(CH_3)_2CH$	$CH(CN)CH_3$	Z	80	-124.3	+163.7 (c)
$CH_3(CN)CH$	$CH(CH_3)_2$	E	20	-117.5	+170.6 (b)
$CH_3 - \overset{\mid}{C}H - (CH_2)_3 - \overset{\mid}{C}H_2$		E Z	67 33	-124.4	+155.5

(a) Neat compounds containing 0.1 M $Cr(acac)_3$; shifts referenced originally to $NaNO_3$, D_2O, H^+ (conversion factor = -1.5 ppm)

(b) Signals not detected

(c) The ^{15}N signals of the C≡N groups are: -131.0 ppm for the Z isomers and -126.2 to -124.3 ppm for the E isomers.

Footnote to Table 6.46

(a) At 9.118 MHz and 301 K; (conversion factor = -1.5 ppm)

(b) At 9.08 MHz and 304 K; (conversion factor = -1.5 ppm)

(c) At 18.25 MHz and room temperature; (conversion factor = -6.2 ppm)

Miscellaneous polynitrogen compounds were also studied:

i) $Ar-N{=}\overset{+}{N}{=}\overset{-}{N}$ (I) (II) δ_{N_I} = -250.8 $\delta_{N_{II}}$ = -128.8 ppm/CH_3NO_2 at 10.14 MHz ($Cr(acac)_3$ added) (T7)

ii) Nitrogen (liquid) δ_N = -70.0 neat at 18.25 MHz and 77 K; originally referenced to NH_4NO_3 (conversion factor = -4.6 ppm) (B64)

iii) M-N≡N (I II)

	M = CH_2	M = camphor	M = fluorene	
δ_{N_I}	= 23.5	-103.5	-86.5	in 0.1M $CDCl_3$ solution
$\delta_{N_{II}}$	= 23.5	30.5	65.5	($Cr(acac)_3$ added)

at 9.12 MHz and 300 K; originally referenced to NH_4^+ (conversion factor = -352.5 ppm) (M37)

Table 6.46. ^{15}N chemical shifts of hydrazones and diazines (mainly E isomers)

R	R'	R''	R'''	$\delta^{15}N$ (ppm) N(I)	N(II)	Conditions	Reference of Shifts	Ref.
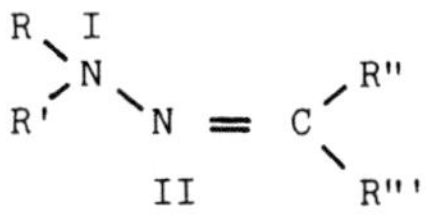								
H	H	$S-C_6H_4-N(CH_3)$		-283.4	-150.0	$CDCl_3$ 25% (b)	NO_3Na	N4
CH_3	CH_3	H	CH_3	-279.8	- 24.4	neat (a)	NO_3Na	G14,N5
CH_3	CH_3	H	C_2H_5	-280.8	- 25.9	neat (a)	NO_3Na	G14,N5
CH_3	CH_3	H	nC_3H_7	-280.4	- 24.0	neat (a)	NO_3Na	G14,N5
CH_3	CH_3	H	iC_3H_7	-281.8	- 28.5	neat (a)	NO_3Na	G14,N5
CH_3	CH_3	H	iC_4H_9	-279.5	- 22.7	neat (a)	NO_3Na	G14,N5
CH_3	CH_3	H	C_6H_5	-274.7	- 25.3	neat (a)	NO_3Na	G14,N5
CH_3	CH_3	H	$CH_2C_6H_5$	-278.5	- 24.6	neat (a)	NO_3Na	G14,N5
CH_3	CH_3	CH_3	CH_3	-287.5	- 27.5	neat (a)	NO_3Na	G14,N5
CH_3	CH_3	CH_3	C_2H_5	-288.8	- 23.0	neat (a)	NO_3Na	G14,N5
$= CH-CH_3$		H	CH_3	- 15.0	- 15.0	neat (a)	NO_3Na	G14,N5
$= CH-C_3H_7n$		H	nC_3H_7	- 14.3	- 14.3	neat (a)	NO_3Na	G14,N5
$= CH-C_3H_7i$		H	iC_3H_7	- 19.2	- 19.2	neat (a)	NO_3Na	G14,N5
$= CH-C_6H_5$		H	C_6H_5	- 17.2	- 17.2	$CDCl_3$ 50% (a)	NO_3Na	G14,N5
$= CH-CH_2C_6H_5$		H	$CH_2C_6H_5$	- 17.9	- 17.9	$CDCl_3$ 50% (a)	NO_3Na	G14,N5
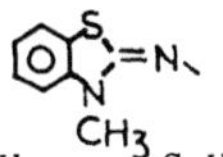		H	CH_3	-109.9	- 23.2	$CDCl_3$ 25% (b)	NO_3Na	N4
		CH_3	H	-114.7	- 22.0	$CDCl_3$ 25% (b)	NO_3Na	N4
H	C_6H_5	H	C_6H_5	-237.0	- 54.0	DMSO 20% (c)	NO_3D	W19
H	C_6H_5	H	$C_6H_4OCH_3p$	-239.0	- 60.0	DMSO 20% (c)	NO_3D	W19
H	C_6H_5	H	$C_6H_4CH_3p$	-237.9	- 56.4	DMSO 20% (c)	NO_3D	W19
H	C_6H_5	H	C_6H_4Clp	-236.1	- 52.8	DMSO 20% (c)	NO_3D	W19
H	C_6H_5	H	$C_6H_4NO_2p$	-230.3	- 43.4	DMSO 20% (c)	NO_3D	W19

Table 6.47. ^{15}N chemical shifts of nitramines, nitrourethane derivatives and azoxybenzenes

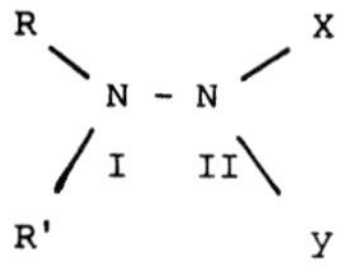

R	R'	X	Y	$\delta^{15}N$ (ppm/CH_3NO_2) N(I)	N(II)	Conditions		Reference of Shifts	Ref.
H	CH_3	O	O	-221.9	-23.9	CH_2Cl_2	(a)	NO_3Na	I1
CH_3	CH_3	O	O	-217.3	-25.0	CH_2Cl_2	(a)	NO_3Na	I1
C_2H_5	C_2H_5	O	O	-192.8	-27.8	neat	(b)	NO_3Na	G14
H	CH_3OCO	O	O	-188.3	-44.6	H_2O	(a)	NO_3Na	I1
CH_3	CH_3OCO	O	O	-183.3	-40.3	CH_2Cl_2	(a)	NO_3Na	I1
CH_3	C_2H_5OCO	O	O	-184.0	-40.6	CH_2Cl_2	(a)	NO_3Na	I1
CH_3	$SiMe_3$	O	O	-201.4	-19.4	CH_2Cl_2	(a) 243K	NO_3Na	I1
EtOCO	$SiMe_3$	O	O	-170.4	-38.2	CH_2Cl_2	(a)	NO_3Na	I1
CH_3	NO_2	O	O	- 91.4	-40.5	CH_2Cl_2	(a)	NO_3Na	I1
	CH_3	O^-	OCH_3	-109.8	-66.1	CH_2Cl_2	(a)	NO_3Na	I1
	CH_3	OCH_3	O^-	-102.4	-52.1	CH_2Cl_2	(a)	NO_3Na	I1
	CH_3	$OSiMe_3$	O^-	- 93.1	-54.8	CH_2Cl_2	(a) 243K	NO_3Na	I1
	CH_3OCO	OCH_3	O^-	-106.6	-45.9	CH_2Cl_2	(a)	NO_3Na	I1
	C_2H_5OCO	OC_3H_7i	O^-	-100.7	-49.7	CH_2Cl_2	(a)	NO_3Na	I1
	C_6H_5	C_6H_5	O^-	- 46.7	-54.1	$CDCl_3$	(c)	CH_3NO_2	S9
	C_6H_5	O^-	C_6H_5	-19.8	-36.0	$CDCl_3$	(c)	CH_3NO_2	S9
	CH_3	O	O (d)	-119.3	-27.0	D_2O	(a)	NO_3Na	I1
	CH_3OCO	O	O (d)	-136.2	- 8.9	D_2O	(a)	NO_3Na	I1

(a) At 9.12 MHz and 300 K; the shifts were originally referenced to $C_2H_5C{\equiv}N$ which is 135.1 ppm upfield from $NaNO_3$ (conversion factor of the published data to the CH_3NO_2 scale: -3.0 ppm)

(b) At 9.118 MHz and 301 K; (conversion factor = -1.5 ppm)

(c) At 10.1 MHz and 253 K

(d) These salts have the following structure: $R\text{-}\bar{N}\text{-}NO_2$, NH_4^+

Table 6.48. ^{15}N chemical shifts of nitro derivatives and furazanes

Nitro Derivatives and Furazanes		$\delta^{15}N$ (ppm/CH_3NO_2)	Conditions		Reference of Shifts	Ref.
$pXC_6H_4NO_2$	X = H	- 9.9	neat	(a)	CH_3NO_2	L30
	NH_2	- 9.2	neat	(a)	CH_3NO_2	L30
	NO_2	- 14.1	neat	(a)	CH_3NO_2	L30
	CH_3	- 10.0	neat	(a)	CH_3NO_2	L30
	F	- 11.7	neat	(a)	CH_3NO_2	L30
	CH_3O	- 10.3	neat	(a)	CH_3NO_2	L30
	CH_3CO	- 10.0	neat	(a)	CH_3NO_2	L30
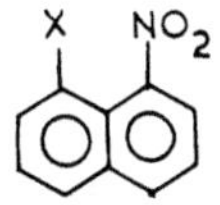	X = H	- 5.9	$CHCl_3$	(b)	HNO_3	S29
	NO_2	- 7.9	$CHCl_3$	(b)	HNO_3	S29
	CN	- 8.2	$CHCl_3$	(b)	HNO_3	S29
	I	- 5.7	$CHCl_3$	(b)	HNO_3	S29
	Cl	- 2.2	$CHCl_3$	(b)	HNO_3	S29
	NH_2	+ 2.8	$CHCl_3$	(b)	HNO_3	S29
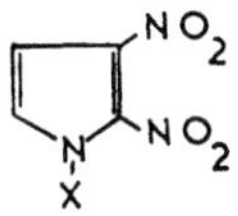	X = H	{ - 13.5 - 23.1	20 to 30% H_2O	(c)	CH_3NO_2	L46
	X = Na	{ - 16.3 - 28.3	20 to 30% H_2O	(c)	CH_3NO_2	L47
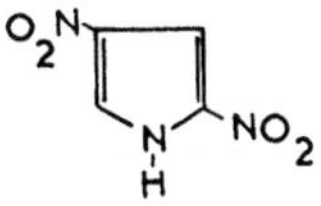		{ - 18.3 - 25.7	20 to 30% H_2O	(c)	CH_3NO_2	L46
O = N — NO_2 (I) (II)	N(I) N(II)	+292.1 + 63.1	CH_2Cl_2 173K	(g)	NH_4Cl	S1
BF_4^- NO_2^+		-130.2	SO_2 liq.	(d)	NO_3Na	07
BF_4^- NO^+		- 2.0	SO_2 liq.	(d)	NO_3Na	07
BF_4^- $ON\text{-}O\text{-}S(CH_3)_2^+$		+618.1	SO_2 liq.	(d)	NO_3Na	07
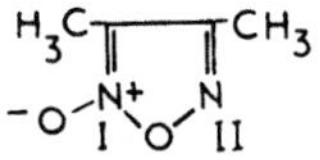	N(I) N(II)	- 13.2 - 25.3	acetone 20% v/v	(e)	HNO_3	Y12
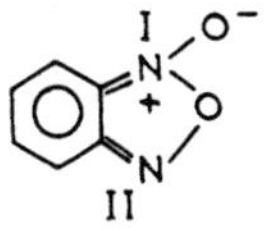	N(I) N(II)	- 4.5 - 18.0	acetone 2M	(f)	HNO_3	Y12

Table 6.48 (continued)

Nitro Derivatives and Furazanes	$\delta^{15}N$ (ppm/CH_3NO_2)	Conditions	Reference of Shifts	Ref.
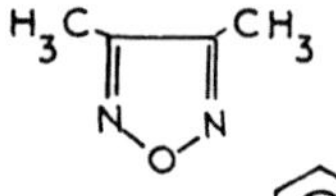	+ 24.8	acetone 20% v/v (e)	HNO_3	Y12
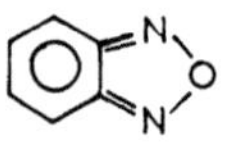	+ 36.3	acetone 2M (e)	HNO_3	Y12

(a) Not specified

(b) At 18.25 MHz and 308 K, $Cr(acac)_3$ added (conversion factor = -6.2 ppm)

(c) At 6.08 MHz; obtained from double resonance spectra

(d) At 8.059 MHz and 213 K (conversion factor = -2.4 ppm). The shifts were originally referenced to pyridine dissolved in CH_2Cl_2, which is at -66.3 ppm from CH_3NO_2

(e) At 18.25 MHz and room temperature (conversion factor = -6.2 ppm)

(f) At 18.25 MHz and 263 K (conversion factor = -6.2 ppm)

(g) At 9.12 MHz and 173 K; (conversion factor = -352.9 ppm)

Footnote to Table 6.49

(a) At 27.35 MHz and room temperature; (conversion factor = -360.8 ppm) measured with respect to 5M NH_4NO_3 in 2M HNO_3 and converted by the authors to NH_4^+: -356.2 ppm

(b) At 9.125 MHz and room temperature; (conversion factor = -352.9 ppm)

(c) At 9.125 MHz and 263 K; (conversion factor = -352.9)

(d) At 9.12 MHz and room temperature; (conversion factor = -336.7 ppm) The shifts were originally referenced to urea and converted to the Me_4NCl scale using the factor: 32.5 ppm. $Cr(acac)_3$ added

(e) At 10.1 MHz and 293 K; (conversion factor = -353.6 ppm). The observed discrepancy could be explained by folding back of the signals due to a maladjusted sweep width in A 12

(f) At 9.12 MHz and room temperature; shifts were referenced to KNO_2 and converted to an acidified saturated NH_4Cl solution (590.7 ppm); (conversion factor to the CH_3NO_2 scale = -353.6 ppm)

MBTA 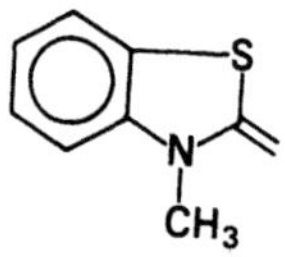

Table 6.49. ^{15}N chemical shifts of polyazenium salts

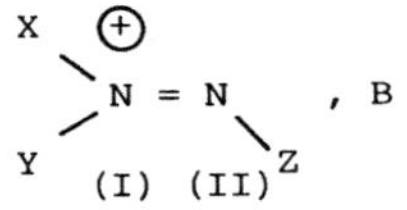

X	Y	Z	B	$\delta^{15}N$ (ppm/CH_3NO_2) N(I)	N(II)	N(Z)	Conditions		Reference of Shifts	Ref.
CH_3	CH_3	$CH=CH-N(Me)_2$	ClO_4^-	-205.2	+ 28.4	-230.1	DMSO d_6 0.6 M	(a)	NH_4Cl	J7
C_6H_5	H	$CH=CH-N(Me)_2$	ClO_4^-	-173.5	+ 7.0	-216.3	DMSO d_6 0.6 M	(a)	NH_4Cl	J7
CH_3	CH_3	$CH=CH-NHC_6H_5$	ClO_4^-	-194.6	+ 32.9	-215.1	DMSO d_6 0.6 M	(a)	NH_4Cl	J7
C_6H_5	H	$CH=CH-NHC_6H_5$	ClO_4^-	-184.8	+ 10.6	-200.4	DMSO d_6 0.6 M	(a)	NH_4Cl	J7
O^-	O^-	O^-	2 Na^+	- 27.9	- 43.4	-	D_2O	(b)	NH_4Cl	S1
O^-	SO_3^-	O^-	2 K^+	- 46.9	+ 58.1	-	D_2O	(b)	NH_4Cl	S1
$^-O-N=N(O^-)-CH_2$	O^-	O^-	2 Na^+	- 82.4	+ 62.6	-	D_2O	(b)	NH_4Cl	S1
$(Et)_2N$	O^-	O^-	$(Et)_2NH_2^+$	-137.9	+ 33.6	-	$CDCl_3$	(c)	NH_4Cl	S1
p-$O_2NC_6H_4$	C_2H_5	N = MBTA	BF_4^-	-151.1	+ 77.0	- 54.0	CF_3COOH	(f)	NH_4Cl	F1
	$CHCOOC_2H_5$			-165.8	- 49.5		$CDCl_3$ 1/2 M	(d)	$N(Me)_4Cl$	A12
				-110.9	+ 5.3			(e)	NH_4Cl	L2
	$C(COOC_2H_5)_2$			-177.5	- 58.2		$CDCl_3$ 1/2 M	(d)	$N(Me)_4Cl$	A12
	$C(C_6H_5)_2$			-130.7	+ 5.4		$CDCl_3$ 1/2 M	(d)	$N(Me)_4Cl$	A12
	$CH-CO-C_6H_5$			-165.2	- 59.6		$CDCl_3$ 1/2 M	(d)	$N(Me)_4Cl$	A12

Table 6.50. ^{15}N chemical shifts of triazenes and azides

$R(R')N-N=N-X$ (I, II, III)

R	R'	X	$\delta^{15}N$ (ppm/CH_3NO_2) $N_{(I)}$	$N_{(II)}$	$N_{(III)}$	Conditions		Reference of Shifts	Ref.
CH_3	CH_3	C_6H_5	-226.3	+ 70.5	- 21.5	$CDCl_3$	(a)	NH_4Cl	A11
CH_3	CH_3	$C_6H_4NO_2p$	-215.6	+ 74.4	- 33.6	$CDCl_3$	(a)	NH_4Cl	A11
CH_3	CH_3	C_6H_4Clp	-224.7	+ 70.0	- 26 3	$CDCl_3$	(a)	NH_4Cl	A11
CH_3	CH_3	$C_6H_4OCH_3p$	-229.2	+ 67.5		$CDCl_3$	(a)	NH_4Cl	A11
CH_3	CH_3	$C_6H_4CH_3p$	-231.4	+ 67.8	- 18.5	neat	(f)	NO_3Na	G14
MBTA (Z)		C_6H_5	- 83.4	+ 85.4	+ 33.6	pyridine	(b)	NH_4Cl	F1
MBTA (E)		C_6H_5	- 83.3	+113.2	+ 51.3	pyridine	(b)	NH_4Cl	F1
MBTA (Z)		$C_6H_4OCH_3p$	- 84.0	+ 79.0	+ 32.0	pyridine	(b)	NH_4Cl	F1
MBTA (E)		$C_6H_4OCH_3p$	- 84.5	+103.3	+ 51.6	pyridine	(b)	NH_4Cl	F1
MBTA (E)		$C_6H_4NO_2p$	- 48.5	+ 76.6	-165.9	CF_3COOH	(b)	NH_4Cl	F1
$PN-N_{(I)}\cdots N_{(II)}(C_6H_5)\cdots N_{(III)}\cdots C_6H_5$ Z				- 63.7	- 59.6	$CDCl_3$	(c)	CH_3NO_2	S9
$PN-N_{(I)}\cdots N_{(II)}(C_6H_5)\cdots N_{(III)}\cdots C_6H_5$ E				- 64.7	- 60.5	$CDCl_3$	(c)	CH_3NO_2	S9

$R-N=\overset{+}{N}=\overset{-}{N}$ (I, II, III)

R	$N_{(I)}$	$N_{(II)}$	$N_{(III)}$	Conditions		Reference of Shifts	Ref.
CH_3	-321.2	-129.7	-171.0	C_6D_6 30%	(d)	NO_3Na	M24
C_6H_5	-287.9	-136.2	-146.9	acetone-d_6 25%	(d)	NO_3Na	M24
$C_6H_4NO_2$	-281.5	-139.5	-143.6	DMSO-d_6 10%	(d)	NO_3Na	M24
$C_6H_2(NO_2)_3$	-288.8	-150.6	-142.2	DMSO-d_6 30%	(d)	NO_3Na	M24
C≡N	-314.8	-149.2	-147.0	CD_3CN 5%	(e)	NO_3Na	M24
$C-(C_3N_3)$	-260.6	-145.1	-134.3	CD_2Cl_2 10%	(d)	NO_3Na	M24
Li	-279.9	-131.3	-279.9	H_2O	(g)	NO_3Na	M10
Na	-280.9	-130.4	-280.9	H_2O	(g)	NO_3Na	M10
$(CH_3)_2P(O)$	-294.3	-146.4	-175.0	$CDCl_3$	(g)	NO_3Na	M10
$(CH_3)_2P(S)$	-289.4	-143.3	-169.8	acetone-d_6	(g)	NO_3Na	M10
$(CH_3)_2P(Se)$	-290.7	-142.3	-168.0	C_6D_6	(g)	NO_3Na	M10
$(Et)_2P(O)$	-299.0	-144.9	-176.8	C_6D_6	(g)	NO_3Na	M10
$(Et)_2P(S)$	-293.7	-142.9	-172.3	C_6D_6	(g)	NO_3Na	M10
$(CH_3O)_2P(O)$	-304.1	-148.2	-173.7	CH_3CN	(g)	NO_3Na	M10

Table 6.51. ^{15}N chemical shifts of aminophosphines P(III) with various amine fragments, measured at 9.118 MHz and 301 K, and referenced to 2M $NaNO_3$, D_2O, H^+ (conversion factor = -1.5 ppm) (G10)

Aminophosphine

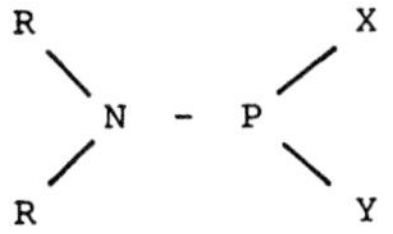

R	X	Y	$\delta^{15}N$ (ppm/CH_3NO_2)
CH_3	Cl	C_6H_5	-334.2
C_2H_5	Cl	C_6H_5	-306.8
nC_3H_7	Cl	C_6H_5	-310.2
$(CH_3)_2CH$	Cl	C_6H_5	-287.5
nC_4H_9	Cl	C_6H_5	-310.9
CH_3-CH_2-$CH(CH_3)$	Cl	C_6H_5	-287.0
$C_6H_5CH_2$	Cl	C_6H_5	-305.2
$C_6H_5CH_2$, CH_3	Cl	C_6H_5	-321.2
CH_3	Cl	Cl	-321.5
C_2H_5	Cl	Cl	-293.4
$(CH_3)_2CH$	Cl	Cl	-273.9
$CH_3CH_2CH(CH_3)$	Cl	Cl	-275.8
$C_6H_5CH_2$, CH_3	Cl	Cl	-308.2

Footnote to Table 6.50

(a) At 9.12 MHz and room temperature; (conversion factor = -353.6 ppm) (sat. NH_4Cl in HCl)

(b) At 9.12 MHz and room temperature; (conversion factor = -353.6 ppm) (see note (f) Table 6.49)

(c) At 10.1 MHz and 253 K

(d) At 10.15 MHz and 308 K ($NaNO_3$ 1M); (conversion factor = -3.0 ppm)

(e) At 10.15 MHz and 253 K ($NaNO_3$ 1M); (conversion factor = -3.0 ppm)

(f) 9.118 MHz and 340 K; (conversion factor = -1.5 ppm)

(g) At 10.14 MHz and room temperature; (conversion factor = -3.0 ppm)

MBTA 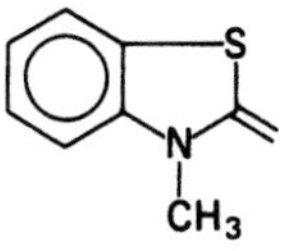

PN 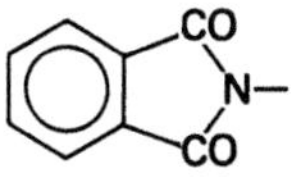

Table 6.52. ^{15}N chemical shifts of aminophosphines P(III), with various phosphorous fragments, measured at 9.118 MHz and 301 K, and referenced to 2M $NaNO_3$, D_2O, H^+ (conversion factor = -1.5 ppm) (G10)

Aminophosphine

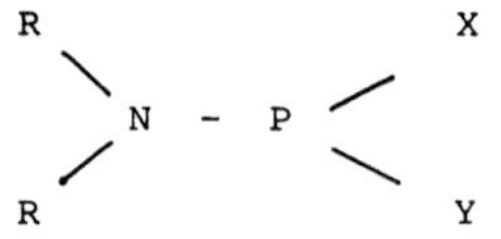

R	X	Y	$\delta^{15}N$ (ppm/CH_3NO_2)
CH_3	CH_3	CH_3	-374.5
CH_3	C_6H_5	C_6H_5	-372.2
CH_3	Cl	CH_3	-331.5
CH_3	Cl	C_6H_5	-334.2
CH_3	Cl	Cl	-321.5
CH_3	$N(Me)_2$	C_6H_5	-348.2
CH_3	$N(Me)_2$	Cl	-331.6
CH_3	$N(Me)_2$	$N(Me)_2$	-350.9
CH_3	$N(Me)_2$	$N(Et)_2$	-353.4
CH_3	$N(Me)_2$	$N(iPr)_2$	-359.0
CH_3	$N(Me)_2$	OEt	-343.0
CH_3	$N(Me)_2$	NHC_6H_5	-297.5 (a)
CH_3	$N(Et)_2$	Cl	-327.5
CH_3	$N(Et)_2$	$N(Et)_2$	-355.1
CH_3	$N(Et)_2$	$N(iPr)_2$	-359.3
CH_3	$N(iPr)_2$	Cl	-327.5
CH_3	NHC_6H_5	NHC_6H_5	-297.2 (a)
C_2H_5	Cl	C_6H_5	-306.8
C_2H_5	Cl	Cl	-293.4
C_2H_5	$N(Me)_2$	Cl	-307.0
C_2H_5	$N(Me)_2$	$N(Me)_2$	-319.9
C_2H_5	$N(Me)_2$	$N(Et)_2$	-322.9
C_2H_5	$N(Me)_2$	$N(iPr)_2$	-332.3
C_2H_5	$N(Et)_2$	Cl	-304.5
C_2H_5	$N(Et)_2$	$N(Et)_2$	-323.8
C_2H_5	$N(Et)_2$	$N(iPr)_2$	-332.4
C_2H_5	$N(iPr)_2$	Cl	-305.2
$(CH_3)_2CH$	Cl	C_6H_5	-287.5
$(CH_3)_2CH$	Cl	Cl	-273.9

Table 6.52 (continued)

Aminophosphine 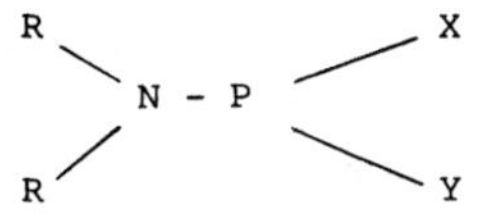

R	X	Y	$\delta^{15}N$ (ppm/CH_3NO_2)
$(CH_3)_2CH$	$N(Me)_2$	Cl	-288.8
$(CH_3)_2CH$	$N(Me)_2$	$N(Me)_2$	-299.2
$(CH_3)_2CH$	$N(Me)_2$	$N(Et)_2$	-300.5
$(CH_3)_2CH$	$N(Et)_2$	Cl	-286.6
$(CH_3)_2CH$	$N(Et)_2$	$N(Et)_2$	-302.1
$(CH_3)_2CH$	$N(iPr)_2$	Cl	-285.2

(a) From INDOR spectra in C_6H_6; reference $N(Me)_4$ I, H_2O (conversion factor = -337.3 ppm) (M26)

Table 6.53. ^{15}N chemical shifts of aminophosphorous P(IV) derivatives

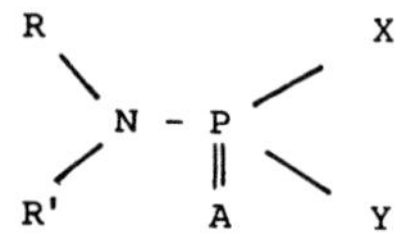

R	R'	X	Y	A	$\delta^{15}N$ (ppm/ CH_3NO_2)	Conditions		Reference for Shift	Ref.
CH_3	CH_3	$N(Me)_2$	$N(Me)_2$	O	-357.1	neat	(a)	NO_3Na	M20c
CH_3	CH_3	$N(Me)_2$	$N(Me)_2$	S	-333.1	neat	(a)	NO_3Na	M20c
C_2H_5	C_2H_5	$N(Et)_2$	$N(Et)_2$	O	-315.4	neat	(a)	NO_3Na	M20c
CH_3	CH_3	$N(Me)_2$	Cl	O	-343.4	$CHCl_3$ 30%	(a)	NO_3Na	P11
CH_3	CH_3	C_6H_5	Cl	O	-337.0	neat	(a)	NO_3Na	M20c
CH_3	CH_3	Cl	Cl	O	-324.5	neat	(a)	NO_3Na	M20c
C_6H_5	H	CH_3	CH_3		-309.1	C_6H_6	(b)	$N(Me)_4I$	M26
C_6H_5	H	CH_3	CH_3	O	-293.6	DMSO d_6	(b)	$N(Me)_4I$	M26
C_6H_5	H	CH_3	CH_3	S	-301.0	dioxane	(b)	$N(Me)_4I$	M26
C_6H_5	H	CH_3	CH_3	Se	-304.6	CH_2Cl_2	(b)	$N(Me)_4I$	M26
C_6H_5	H	CH_3	CH_3	Te	-308.0	$CH_2Cl_2+C_6H_6$	(b)	$N(Me)_4I$	M26
C_6H_5	H	tBu	tBu		-321.6	TMB	(b)	$N(Me)_4I$	M26
C_6H_5	H	tBu	tBu	O	-316.7	$TMB+CH_2Cl_2$		$N(Me)_4I$	M26
C_6H_5	H	tBu	tBu	S	-321.3	$TMB+CHCl_3$		$N(Me)_4I$	M26
C_6H_5	H	tBu	tBu	Se	-325.8	$TMB+CHCl_3$		$N(Me)_4I$	M26
C_6H_5	$SnMe_3$	tBu	tBu		-340.3	$C_6H_6+CH_2Cl_2$		$N(Me)_4I$	M26
C_6H_5	$SnMe_3$	CH_3	CH_3		-332.8	C_6H_6	(b)	$N(Me)_4I$	M26
C_6H_5	$SnMe_3$	CH_3	CH_3	S	-307.8	C_6H_6	(b)	$N(Me)_4I$	M26
H	H	F	F		-315.9	C_6H_6	(c)	$N(Me)_4I$	A17
H	PF_2	F	F		-251.0	C_6H_6	(d)	$N(Me)_4I$	A16
PF_2	PF_2	F	F		-198.3	C_6H_6	(d)	$N(Me)_4I$	A16

(a) At 9.118 MHz and 301 K; 0.08M $Cr(acac)_3$ added; (conversion factor = -1.5 ppm)

(b) From INDOR spectra at room temperature; $N(Me)_4I$: ν = 10133351 Hz ($\nu_{o,TMS}$ = 100.0 MHz) (conversion factor = -337.3 ppm) TMB 1,3,5 - trimethylbenzene

Table 6.54. ^{15}N chemical shifts of cyclic amino phosphorous derivatives

Compound	$\delta^{15}N$ (ppm/CH_3NO_2)		Conditions		Reference of Shifts	Ref.
Cyclophosphazenes						
$(NPCl_2)n$						
n = 3	-258.8		$CDCl_3$	(a)	CH_3NO_2	T 5
n = 4	-248.0		$CDCl_3$	(a)	CH_3NO_2	T 5
n = 5	-253.3		$CDCl_3$	(a)	CH_3NO_2	T 5
Diazaphospholanes						
CH_3-N-CH_2-CH_2-N(CH_3)-P-R	N endo	N exo				
R						
$N(CH_3)_2$	-341.3	-331.1	neat	(b)	NO_3Na	G 11
$N(Et)_2$	-342.8	-299.7	neat	(b)	NO_3Na	G 11
$N(C_3H_7i)_2$	-349.6	-281.4	neat	(b)	NO_3Na	G 11
	-353.8	-286.0	neat	(e)	CH_3NO_2	H 8
	-339.8	-301.5	neat	(b)	NO_3Na	G 11
NHC_6H_5		-282.4	C_6H_6	(c)	$N(Me)_4I$	M 26
Cl	-319.7		neat	(b)	NO_3Na	G 11
Phosphoramidates						
 N - $P(O)(OCH_3)_2$						
	-367.4		C_6D_6 50%	(d)	CH_3NO_2	G 6
	-352.2		C_6D_6 50%	(d)	CH_3NO_2	G 6
	-340.9		C_6D_6 50%	(d)	CH_3NO_2	G 6
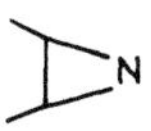 cis	-338.4		C_6D_6 50%	(d)	CH_3NO_2	G 6
trans	-338.1		C_6D_6 50%	(d)	CH_3NO_2	G 6
	-327.2		C_6D_6 50%	(d)	CH_3NO_2	G 6
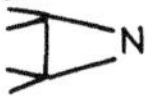	-320.6		C_6D_6 50%	(d)	CH_3NO_2	G 6
	-346.2		C_6D_6 50%	(d)	CH_3NO_2	G 6
	-334.4		C_6D_6 50%	(d)	CH_3NO_2	G 6
	-333.7		C_6D_6 50%	(d)	CH_3NO_2	G 6
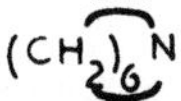	-333.6		C_6D_6 50%	(d)	CH_3NO_2	G 6
	-337.0		C_6D_6 50%	(d)	CH_3NO_2	G 6
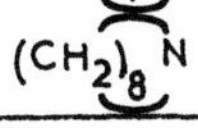	-335.4		C_6D_6 50%	(d)	CH_3NO_2	G 6

Table 6.55. ^{15}N chemical shifts of phosphonium salts

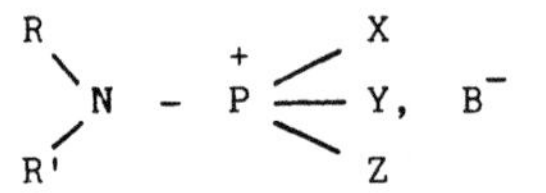

R	R'	X	Y	Z	B	$\delta^{15}N$ (ppm/ CH_3NO_2)	Conditions	Reference of Shifts	Ref.
C_6H_5	H	CH_3	CH_3	CH_3	I	-322.3	CH_2Cl_2 (a)	$N(Me)_4I$	M 26
C_6H_5	H	CH_3	CH_3	SMe	I	-315.9	$CHCl_3$ (a)	$N(Me)_4I$	M 26
C_6H_5	H	CH_3	CH_3	SeMe	I	-315.7	CH_2Cl_2 (a)	$N(Me)_4I$	M 26
C_6H_5	H	CH_3	tBu	tBu	I	-339.6	DMSO d_6(a)	$N(Me)_4I$	M 26
C_6H_5	H	tBu	tBu	SeMe	I	-323.7	DMSO d_6(a)	$N(Me)_4I$	M 26
C_6H_5	H	CH_3	$-N(Me)-(CH_2)_2-N(Me)-$		I	-370.7	CH_2Cl_2 (a)	$N(Me)_4I$	M 26
CH_3	CH_3	NMe_2	NMe_2	Cl	Cl	-346.6	$CHCl_3$ (b)	NO_3Na	P 11
CH_3	CH_3	NMe_2	NMe_2	Cl	$OPOCl_2$	-346.9	$CHCl_3$ (b)	NO_3Na	P 11

(a) From INDOR spectra $N(Me_4)I$: ν = 10.133351 Hz ($\nu_{o,TMS}$ = 100.0 MHz) (conversion factor = -337.3 ppm)

(b) At 9.118 MHz and 301 K; (conversion factor = -1.5 ppm)

Footnote to Table 6.54

(a) At 9.1 MHz and room temperature

(b) At 9.118 MHz and 301 K; 0.1 M $Cr(acac)_3$ added; (conversion factor = -1.5 ppm)

(c) From INDOR $^1H\{^{15}N\}$ spectra; (conversion factor = -337.3 ppm)

(d) At 10.14 MHz and room temperature

(e) At 18.25 MHz and room temperature (CD_3NO_2)

Cyclic P(IV) derivatives have comparable $\delta^{15}N$ (exo) values (F3)

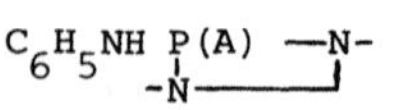 A=S $\delta^{15}N$ = -289.5 A=Se $\delta^{15}N$ = -236.4 ppm

Other diazaphospholanes have also been studied (H17):

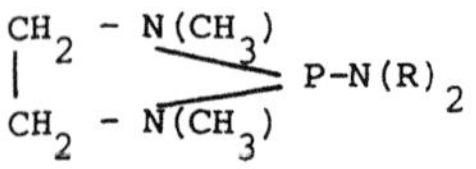

R	CH_3	C_2H_5	$i-C_3H_7$	$CH_3, i-C_3H_7$
solvent	C_6D_6	C_6D_6	CH_2Cl_2	C_6D_6
$\delta^{15}N_{endo}$	-343.4	-345.7	-353.8	-345.5
$\delta^{15}N_{exo}$	-332.8	-302.7	-286.0	-308.7

The chemical shift of the labelled cyclic nitrogen atom in a diazaphosphetidin is shown to be -309.7 ppm (in C_6D_6) (A 31)

Table 6.56. ^{15}N chemical shifts of sulfonamides, measured at 18.25 MHz and 292-300 K with 9 mol-% solution in DMSO. The shifts were originally referenced to 1M DNO_3 (conversion factor = -6.2 ppm) (S12)

$RR'N-SO_2-R''$			$\delta^{15}N$ (ppm/CH_3NO_2)
R	R'	R''	
H	H	$pC_6H_4NH_2$	-284.3
H	H	C_6H_5	-285.9
H	C_2H_5	C_6H_5	-281.6
H	$CH_2C_6H_5$	C_6H_5	-283.4
H	C_6H_5	C_6H_5	-259.7
H	$p-C_6H_4-OCH_3$	C_6H_5	-263.0
H	$p-C_6H_4-F$	C_6H_5	-261.4
H	$p-C_6H_4-CH_3$	C_6H_5	-261.2
H	$p-C_6H_4-Br$	C_6H_5	-259.7
H	$p-C_6H_4-CN$	C_6H_5	-254.5
H	$p-C_6H_4-NO_2$	C_6H_5	-253.3
H	iC_5H_{11}	C_6H_5	-272.6
H	C_6H_{11}	C_6H_5	-272.2
H	$-CH_2-CH_2-$	$-CH_2-CH_2-$	-284.4
C_2H_5	C_2H_5	C_6H_5	-280.2
$-CH_2-CH_2-$	$-CH_2-CH_2-$	C_6H_5	-280.0
nC_3H_7	nC_3H_7	C_6H_5	-283.8
$-CH_2-(CH_2)_3-$	CH_2-	C_6H_5	-278.9
$CH_3-CH-(CH_2)_3-$	$CH-CH_3$	C_6H_5	-276.2
$C_6H_5CH_2$	$C_6H_5CH_2$	C_6H_5	-279.3
H	$-CO-$	C_6H_4-	-221.6
H	CH_3CO	C_6H_5	-210.8
$C_6H_5CH_2$	CH_3CO	C_6H_5	-216.1
$m-NO_2-C_6H_4$	CH_2-CH_2-	CH_2-	-261.7

Table 6.57. ^{15}N chemical shifts of various compounds containing N-S bonds

Compounds with N - S Bond	$\delta^{15}N$ (ppm/ CH_3NO_2)	Conditions	Reference of Shifts	Ref.
Sulfamates:				
NH_2SO_3H	-287.9	H_2O pH 0.15 (a)	NO_3Na	N 6
$NH_2SO_3NH_4$	-288.6	H_2O pH 5 (a)	NO_3Na	N 6
$NH(SO_3NH_4)_2$	-227.3	H_2O, NH_4OH pH 8.3 (a)	NO_3Na	N 6
$N(SO_3NH_4)_3$	-154.8	H_2O, NH_4OH pH 9 (a)	NO_3Na	N 6
Sulfinylamines				
$C_6H_{11}N = S = O$	- 28.8	neat (b)	HNO_3	Y 11
$C_6H_5 - N = S = O$	- 63.8	neat (b)	HNO_3	Y 11
$pCH_3OC_6H_4 - N = S = O$	- 64.4	neat (b)	HNO_3	Y 11
$pNO_2\ C_6H_4 - N = S = O$	- 70.9	DMSO 3M (b)	HNO_3	Y 11
Various N-S Derivatives				
$S-N-S-N-S-N-S^+$, Cl^- I II I	I + 2.0 II - 10.0	HNO_3 70 % 3M (c)	NH_4Cl	L 4
$(CH_3)_2S^+ - NO_2$, BF_4^-	+258.5	SO_2 liq. 213 K (d)	NO_3Na	O 7
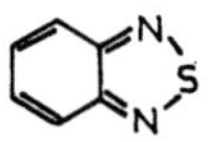	- 50.5	DMSO 2M (b)	HNO_3	Y 12
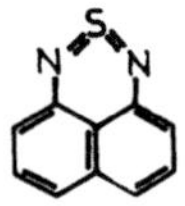	- 89.3	DMSO 2M (b)	HNO_3	Y 12
$C_6H_5-N=S=N-C_6H_5$	-119.9	DMSO 2M (b)	HNO_3	Y 12

(a) At 9.118 MHz and 301 K; (conversion factor = -1.5 ppm)

(b) At 18.25 MHz and room temperature; (conversion factor = -6.2 ppm)

(c) At 6.08 MHz and room temperature 5.7M NH_4Cl; 0.1M HCl, 10^{-4}M Fe^{3+} (conversion factor = -353 ppm)

(d) At 8.06 MHz and low temperature; (conversion factor = -3.0 ppm) (the shifts were originally referenced to pyridine in CH_2Cl_2, which is at -63.9 ppm from 2M $NaNO_3$ in H_2O)

Table 6.58. ^{15}N chemical shifts of silylamines, measured at 9.118 MHz and 301 K with neat compounds. The shifts were originally referenced to 2M $NaNO_3$, D_2O, H^+ (conversion factor = -1.5 ppm) (F8)

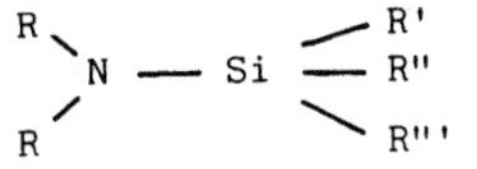				δ ^{15}N (ppm/CH_3NO_2)
R	R'	R"	R"'	
CH_3	H	CH_3	CH_3	-400.5
CH_3	CH_3	CH_3	CH_3	-378.5
CH_3	C_6H_5	CH_3	CH_3	-381.2
CH_3	C_6H_5	C_6H_5	CH_3	-381.5
CH_3	C_6H_5	C_6H_5	Cl	-368.4
CH_3	C_6H_5	Cl	Cl	-361.9
CH_3	Cl	Cl	Cl	-354.9
C_2H_5	H	CH_3	CH_3	-374.5
C_2H_5	CH_3	CH_3	CH_3	-351.5
C_2H_5	CH_3	CH_3	$C(CH_3)_3$	-351.4
C_2H_5	C_6H_5	CH_3	CH_3	-349.5
C_2H_5	C_6H_5	C_6H_5	CH_3	-342.5
C_2H_5	C_6H_5	CH_3	$CH=CH_2$	-337.5
C_2H_5	C_6H_5	C_6H_5	Cl	-338.9
C_2H_5	C_6H_5	Cl	Cl	-331.5
C_2H_5	C_6H_5	C_6H_5	C_6H_5	-302.5
C_2H_5	Cl	Cl	Cl	-325.4
iC_3H_7	CH_3	CH_3	CH_3	-315.5
iC_3H_7	C_6H_5	C_6H_5	Cl	-316.4
iC_3H_7	C_6H_5	Cl	Cl	-311.3
iC_3H_7	Cl	Cl	Cl	-304.5
H,Si(Me)$_3$	CH_3	CH_3	CH_3	-352.9(a)

(a) At 6.08 MHz; shifts referred to 12M Me_4NCl, D_2O (conversion factor = -336.7 ppm) (W21)

Some spectra of silatranes have also been measured (ch.5, M20f)

Table 6.59. ^{15}N chemical shifts of molecular complexes of Mo, W, Cr, Ti, Zr, Co(e)

Molecular Complexes Mo, W, Cr, Ti, Zr	$\delta^{15}N$ (ppm/CH_3NO_2)		Conditions	Reference of Shift	Ref.
trans $MoF(NNH_2)$ $dppe_2$, BF_4 (I, II)	- 83.3	-243.9	CH_2Cl_2	(a) CD_3NO_2, T_1R	C21
Mo (NNH_2)(8-quin), $(PMe_2Ph)_3Cl$ (I, II)	- 64.3	-220.8	CH_2Cl_2	(a) CD_3NO_2, T_1R	C21
$MoCl(NNH_2)(C_5H_5N)$, $(PMe_2Ph)_3Cl$ (I, II)	- 72.8	-227.0	CH_2Cl_2	(a) CD_3NO_2, T_1R	C21
Mo(NMe=CH-CH=NMe), $(CO)_4$ (I, I)	- 74.7		$CDCl_3$	(b) DMF	S13
	- 76.7		C_6D_6	(b) DMF	S13
$Mo(\eta^5-C_5H_5)(CO)_2(NO)$ (I)	+ 37.4		$CHCl_3$	(c) HNO_3	B48
$Mo(\eta^5-C_5H_5(CO)(PPh_3)(NO)$ (I)	+ 35.2		$CHCl_3$	(c) HNO_3	B48
$Mo(\eta^5-C_5H_5)(NO)_2Cl$ (I)	+185.2		$CHCl_3$	(c) HNO_3	B48
trans$[Mo(N{=}N)_2$ $(dppe)_2]$ (I II)	- 46.5	- 46.2	THF	(a) CD_3NO_2, T_1R	C 1
cis $[Mo(N{=}N)_2(PMe_2Ph)_4]$ (I II)	- 42.6	- 34.9	THF	(a) CD_3NO_2, T_1R	C 1
trans$[W(N{=}N)_2$ $(dppe)_2]$ (I II)	- 63.5	- 52.0	THF	(c) CD_3NO_2, T_1R	C 1
cis $W(N{=}N)_2(PMe_2Ph)_4$ (I II)	- 61.2	- 35.9	THF	(c) CD_3NO_2, T_1R	C 1
trans $WF(NNH_2)$ $(dppe)_2$, BF_4 (I II)	-101.4	-255.1	CH_2Cl_2	(a) CD_3NO_2, T_1R	C21
$W(NNH_2)$(8-quin)$(PMe_2Ph)_3$ Cl (I II)	- 82.1	-241.6	CH_2Cl_2	(a) CD_3NO_2, T_1R	C21
W $Cl(NNH_2)$ $(C_5H_5N)(PMe_2Ph)_3Cl$ (I II)	- 90.6	-240.6	CH_2Cl_2	(a) CD_3NO_2, T_1R	C21
$W(\eta^5-C_5H_5)(CO_2)(NO)$	+ 16.3		$CHCl_3$	(c) HNO_3	B48
$W(\eta^5-C_5H_5)(NO)_2Cl$	+172.7		$CHCl_3$	(c) HNO_3	B48
$Cr(\eta^5-C_5H_5)(CO)_2(NO)$	+ 48.8		$CHCl_3$	(c) HNO_3	B48
$Cr(\eta^5-C_5H_5)(NO)_2Cl$	+184.4		$CHCl_3$	(c) HNO_3	B48
$[Cr(\eta^5-C_5H_5)(NO)_2]_2$	+121.5		$CHCl_3$	(c) HNO_3	B48
$Ti(C_5Me_5)_2$ (N≡N) (I II)	+165.0 (N_2 = +299)	+214.4	$C_6D_5CD_3$ 212K	$N(Et)_4Cl$ (d)	B29
$[Zr(C_5Me_5)_2$ $(N{\equiv}N)]_2$ N_2 (I II)	+ 10.8 (N_2 = + 178.8)	+ 79.8	$C_6D_5CD_3$ 245K	HNO_3 (c)	M 8

Table 6.60. ^{15}N chemical shifts of molecular complexes of Pt, Pd and Rh

Molecular Complexes	$\delta^{15}N$ (ppm/CH_3NO_2) I	II	III	Conditions		Reference of Shifts	Ref.
L-Pt(Cl_2)(NH_2-C_6H_{13}n) (I)							
L : P(nC_4H_9)$_3$	-352.2			$CDCl_3$	(a)	NH_4Cl	M 7
P(CH_3)(C_6H_5)$_2$	-353.3			$CDCl_3$	(a)	NH_4Cl	M 7
P($pCH_3C_6H_4$)$_3$	-353.1			$CDCl_3$	(a)	NH_4Cl	M 7
As(C_4H_9n)	-357.7			$CDCl_3$	(a)	NH_4Cl	M 7
As(CH_3)(C_6H_5)$_2$	-359.7			$CDCl_3$	(a)	NH_4Cl	M 7
As($pCH_3C_6H_4$)$_3$	-359.2			$CDCl_3$	(a)	NH_4Cl	M 7
$NH_2C_6H_{13}$n	-397.5			$CDCl_3$	(a)	NH_4Cl	M 7
C_2H_4	-356.6			$CDCl_3$	(a)	NH_4Cl	M 7
Cl(cis C_2H_4)	-385.8			$CDCl_3$	(a)	NH_4Cl	M 7
Pt(NH_3)$_2$(H_2O)$_2$ $^{2+}$ cis (I)	+ 67(b)			H_2O		(b)	A 22
Pt(en)(H_2O)$_2$ $^{2+}$ (I)	+ 45.9(b)			H_2O		(b)	A 22
Pt(NH_3)$_2$(MeIm)$_2$ $^{2+}$ cis (I) (II,III)	+ 46.4(b)	- 4.7(b)	+91.9(b)	H_2O		(b)	A 22
Pt(en)(H_2O) (MeIm) $^{2+}$ (I) (II,III)	+ 44.8(b) + 29.7(b)	- 4.8(b)	+93.1(b)	H_2O		(b)	A 22
Pt(en)(MeIm)$_2$ (I) (II,III)	+ 28.9(b)	- 4.6(b)	+91.1(b)	H_2O		(b)	A 22
Pt(Me Im)$_2$(H_2O)$_2$ $^{2+}$ cis (II,III)		- 5.4(b)	+121 (b)	H_2O		(b)	A 22
L - Pd(Cl_2)(NH_2-C_6H_{13}n)							
L : P(nC_4H_9)$_3$	-357.6			$CDCl_3$	(a)	NH_4Cl	M 7
P(CH_3)(C_6H_5)$_2$	-358.8			$CDCl_3$	(a)	NH_4Cl	M 7
P($pCH_3C_6H_4$)$_3$	-358.8			$CDCl_3$	(a)	NH_4Cl	M 7
As(nC_4H_9)$_3$	-360.7			$CDCl_3$	(a)	NH_4Cl	M 7
As(CH_3)(C_6H_5)$_2$	-362.6			$CDCl_3$	(a)	NH_4Cl	M 7
As($pCH_3C_6H_4$)$_3$	-362.1			$CDCl_3$	(a)	NH_4Cl	M 7
$NH_2C_6H_{13}$n	-386.1			$CDCl_3$	(a)	NH_4Cl	M 7

Table 6.60 (continued)

Molecular Complexes	$\delta^{15}N$ (ppm/CH_3NO_2) I	II	III	Conditions	Reference of Shifts	Ref.
$Rh(Cl_3)(en)_3$	-384.7			H_2O 1.5M(c)	NH_4Cl	B 34
$Rh(Cl_3)(pn)_3$	-368.4		-387.4	H_2O 1.5M(c)	NH_4Cl	B 34
$Rh(Cl_3)(tn)_3$	-369.9			H_2O 1.5M(c)	NH_4Cl	B 34
$Rh(Cl_3)(bpy)_3$	-166.8			H_2O 1.5M(c)	NH_4Cl	B 34
$Rh(Cl_3)(phen^3)_3$	-172.1			H_2O 1.5M(c)	NH_4Cl	B 34
trans $Rh(Cl_2)(en)_2Cl$	-381.5			H_2O 1.5M(c)	NH_4Cl	B 34
trans $Rh(Cl_2)(py)_4Cl$	-161.9			H_2O 1.5M(c)	NH_4Cl	B 34

(a) At 9.12 MHz and 303 K (conversion factor = -352.9 ppm) (aqueous NH_4Cl)

(b) At 10.16 MHz and room temperature; the shift reference is the uncomplexed amine which should resonate at a lower frequency than the complexed amine:

en: ethylenediamine
pn: 1,2-diaminopropane
tn: 1,3-diaminopropane
phen: 1,10-phenanthroline
MeIm: N-methylimidazole Me-N(II)
Py : pyridine
bpy : 2,2 bipyridine

N(I) bound-Pt complex with (5-Cl,2-OH)C_6H_3 $\underset{II}{-N}$ = $\underset{I}{N}$ - C_6H_2(OH-2, CH_3-3, tBu-5) gave the following ^{15}N signals:

free ligand: $\delta^{15}N(I)$ = +71.7 ppm

complex: $\delta^{15}N(I)$ = -107.9 $\delta N(II)$ = +7.9 (P4)

(originally referenced to NH_4Cl, in $CDCl_3$ solutions, 10^{-2}M Cr(acac)$_3$, at 9.12 MHz and $\sim$ 300 K)

(c) At 10.09 MHz and room temperature; (conversion factor = -352.9 ppm)

Footnote to Table 6.59

(a) At 18.24 MHz and 303 K

(b) At 27.3 MHz and room temperature (conversion factor = -275.2 ppm)

(c) At 18.25 MHz and room temperature; (conversion factor = -6.2 ppm)

(d) At 6.08 MHz (the high frequency signal is +615 ppm of external $(Et)_4N$ Cl, 11M in H_2O) (conversion factor = -316 ppm)

(e) Co(III) complexes such as $Co(gly)_3$ have been investigated (F 17) (Y 18)

pH = C_6H_5, dppe = $(C_6H_5)_2PCH_2CH_2P(C_6H_5)_2$, 8-quin = quinolin-8-olate

Table 6.61. ^{15}N chemical shifts of ^{15}N enriched samples of diamagnetic porphyrin derivatives

Porphyrin Derivatives	$\delta^{15}N$ (ppm/CH_3NO_2)	Conditions	Reference of Shifts	Ref.
$(C_6H_5)_4Po(Zn)$(meso)	-179.6	0.02 M $CHCl_3$	NO_3Na	(a) G 15
$(C_6H_5)_4Po(Zn)L$				
L				
Piperidine(meso)	-178.4	0.02 M $CHCl_3$	NO_3Na	(a) G 15
3-CN pyridine	-179.4	0.02 M $CHCl_3$	NO_3Na	(a) G 15
4-CN pyridine	-179.2	0.02 M $CHCl_3$	NO_3Na	(a) G 15
3-CHO pyridine	-178.9	0.02 M $CHCl_3$	NO_3Na	(a) G 15
4-CHO pyridine	-178.8	0.02 M $CHCl_3$	NO_3Na	(a) G 15
3-$COCH_3$ pyridine	-178.6	0.02 M $CHCl_3$	NO_3Na	(a) G 15
4-$COCH_3$ pyridine	-178.6	0.02 M $CHCl_3$	NO_3Na	(a) G 15
Pyridine	-178.1	0.02 M $CHCl_3$	NO_3Na	(a) G 15
3-CH_3 pyridine	-177.9	0.02 M $CHCl_3$	NO_3Na	(a) G 15
4-CH_3 pyridine	-177.9	0.02 M $CHCl_3$	NO_3Na	(a) G 15
3-NH_2 pyridine	-177.8	0.02 M $CHCl_3$	NO_3Na	(a) G 15
4-NH_2 pyridine	-177.3	0.02 M $CHCl_3$	NO_3Na	(a) G 15
$(C_6H_5)_4Po(Fe)L_2$				
L				
pyridine (meso)	-175.9	0.015 M C_5D_5N-D_2O	HNO_3	(b) M 12
morpholine(meso)	-150.2	0.015 M $CDCl_3$	HNO_3	(b) M 12
pyrrolidine(meso)	-172.0	0.015 M $CDCl_3$	HNO_3	(b) M 12
$(C_2H_5)_8Po(R)_2$(d)				
$(R)_2$				
H_2	-253.1 -143.7	0.019 M $CDCl_3$,220K	NO_3NH_4	(c) K 14 a,b
H,CH_3	-260.0 -247.3 -129.8	0.018 M $CDCl_3$	NO_3NH_4	(c) K 14 a,b
$(C_2H_5)_8Po(R)_4^{2+}$(d)				
$(R)_4$				
H_4	-257.7	TFA	NO_3NH_4	(c) K 14 a,b
$(CH_3)_2$, H_2	-264.8 -258.5	TFA	NO_3NH_4	(c) K 14 a,b
CH_3, H_3	-263.0 -256.9 -255.3	TFA	NO_3NH_4	(c) K 14 a,b
$(C_2H_5)_8Po(M)$				
M				
Ni	-261.0	$CDCl_3$	NO_3NH_4	(c) K 14 a,b
Mg	-184.4	$CDCl_3$	NO_3NH_4	(c) K 14 a,b
Zn	-183.2	$CDCl_3$+pyridine	NO_3NH_4	(c) K 14 a,b
Cd	-177.1	pyridine	NO_3NH_4	(c) K 14 a,b

Table 6.61 (continued)

Porphyrin Derivatives	$\delta^{15}N$ (ppm/CH_3NO_2)	Conditions	Reference of Shifts	Ref.
$(C_2H_5)_8Po(M)L_2$				
M L_2				
Fe II $(Pyr)_2$	-191.9	0.03 M pyridine	NO_3NH_4	(c) O 8
CO,Pyr	-235.3	0.03 M pyridine	NO_3NH_4	(c) O 8
$(\gamma Pic)_2$	-189.5	0.03 M γ-picoline	NO_3NH_4	(c) O 8
Co(III)Br,Pyr	-264.0	0.03 M pyridine	NO_3NH_4	(c) O 8
$(\gamma Pic)_2$	-264.3	0.03 M γ-picoline	NO_3NH_4	(c) O 8

(a) At 9.12 MHz and 300 K; Ref. 1M $NaNO_3$ in D_2O (conversion factor = -3.6 ppm)

(b) At 10.14 MHz (conversion factor = -6.2 ppm)

(c) At 10.09 MHz and 301 K; reference NH_4NO_3 in DMSO-d_6 (conversion factor = -4.0 ppm)

(d) R are the substituents of the nitrogen atoms. Po is the porphyrin parent compound

Other ^{15}N data concern exchange processes (G8, I5, K14 see Sect.7) and natural products (B32)

	dihydropyrrol	methylpyrrol	methyl-ethyl-pyrrol	vinylpyrrol
chlorophyll:	-128.5	-186.1	-169.0	-188.9
pheophytin :	- 79.7	-241.6	-134.0	-250.0

(at 9.12 MHz and 313 K; Ref. 5M NH_4Cl in 2N HCl; conversion factor = -352.5 ppm)

Table 6.62. Substituent parameters of main groups

σ_P Hammett constant (monosubstituted benzenes) (F14)

σ^* Charton aliphatic constant (C26)

σ_I, σ_R Taft inductive and mesomeric constants (mainly from ^{19}F-NMR) (E5)

F,R Swain-Lupton field and resonance constants (F14)

E_s Taft steric parameter (U1)

Substituent	σ_P	σ^*	σ_I	σ_R	F	R	E_s
H	0	0.49	0	0	0	0	/
CH_3	-0.17	0	-0.08	-0.15	-0.04	-0.13	-1.24
C_2H_5	-0.15	-0.10	-0.03	-0.14	-0.05	-0.10	-1.31
nC_3H_7	-0.15	-0.11	-0.02	-0.11	-0.06	-0.08	-1.60
iC_3H_7	-0.15	-0.19	-0.03	-0.12	-0.05	-0.10	-1.71
tC_4H_9	-0.20	-0.30	0.02	-0.17	-0.05	-0.10	-2.37
C_6H_5	-0.01	0.60	0.08	-0.09	0.08	-0.08	-1.01(a) -3.82(b)
OH	-0.37	1.34	0.25	-0.43	0.29	/	-0.55
OCH_3	-0.27	1.81	0.25	-0.43	0.26	-0.51	-0.55
OC_2H_5	-0.24	1.68	0.26	-0.44	0.22	-0.44	-0.55
OC_6H_5	-0.32	2.43	0.37	-0.31	0.34	-0.35	-0.55
$OCOCH_3$	0.31	/	0.27	-0.21	/	/	/
NH_2	-0.66	0.62	0.01	-0.48	/	/	-0.61
$N(CH_3)_2$	/	0.32	0.10	-0.54	/	/	-1.05
F	0.06	3.21	0.52	-0.32	0.43	-0.34	-0.46
Cl	0.23	2.96	0.47	-0.18	0.41	-0.15	-0.97
Br	0.23	2.84	0.44	-0.19	0.44	-0.17	-1.16
I	0.18	2.46	0.42	-0.14	0.40	-0.19	-1.40
CF_3	0.54	2.61	0.41	0.10	0.38	/	-2.40
CCl_3	/	2.65	0.30	0.03	/	/	-3.30
CHO	0.42	/	0.25	0.27	/	/	/
$COCH_3$	0.50	1.65	0.18	0.19	/	/	/
COOH	0.45	2.08	0.32	0.29	/	/	/
$COOCH_3$	0.45	2.00	0.11	0.19	/	/	/
SH	0.15	1.68	0.19	-0.15	/	/	-1.07
SCH_3	0.00	1.56	0.13	-0.16	/	/	-1.07
SO_2CH_3	0.72	3.63	0.55	0.16	/	/	/

Table 6.62 (continued)

Substituent	σ_p	σ^*	σ_I	σ_R	F	R	E_s
CN	0.66	3.30	0.48	0.22	0.51	0.19	-0.51
NO_2	0.78	4.00	0.57	0.20	0.67	0.16	-1.01(a) -2.52(b)

(a) Minimum steric hindrance

(b) Maximum steric hindrance

7 $^{n}J^{15}N \sim X$ Coupling Constants

A number of coupling constants between ^{15}N and other nuclei, e.g. $X \equiv {}^{1}H$, ^{19}F, ^{31}P, can be obtained using CW spectroscopy and ^{15}N enriched samples. However, with the availability of commercial FT spectrometers, the direct determination of $^{n}J^{15}N{\sim}X$ coupling constants from X-undecoupled ^{15}N spectra, at the natural abundance level of the ^{15}N isotope, is possible and our knowledge of the coupling interactions involving nitrogen is rapidly increasing. The first section of this chapter is concerned with the presentation of coupling constant data. We have only considered the results which involve the ^{15}N isotope. No transformation of $J^{14}N{\sim}X$ coupling constants into $J^{15}N{\sim}X$ coupling constants has been attempted. Moreover we shall not discuss the theoretical analysis of coupling here since this aspect has been treated elsewhere (W 29), and, rather than successively examining the coupling constants in given series of compounds, whe shall present the results in terms of the more or less quantitative correlations which have been discussed.

7.1 Tables of $^{n}J^{15}N{\sim}X$ Coupling Constants

-$^{n}J^{15}N{\sim}^{1}H$ Coupling Constants

$-{}^{n}J^{15}N{\sim}^{183}W$

Table 7.65 Miscellaneous tungsten derivatives

$-{}^{n}J^{15}N{\sim}^{195}Pt$

Table 7.66 Miscellaneous platinum derivatives

$-{}^{n}J^{15}N{\sim}^{199}Hg$

Table 7.67 Miscellaneous mercury derivatives

$-{}^{n}J^{15}N{\sim}^{207}Pb$

Table 7.68 Miscellaneous lead derivatives

$-D^{15}N{\sim}X$ in nematic phase

Table 7.69 Miscellaneous derivatives

7.2 Correlation Analysis of ${}^{n}J^{15}N{\sim}X$ Coupling Constants

Before embarking on the discussion of the coupling constants, it should be recalled that the gyromagnetic ratio of the ^{15}N isotope has a negative value. The sign of a coupling constant involving ^{15}N will therefore be the opposite of that of the corresponding coupling involving ^{14}N for which γ is positive. In order to facilitate the comparison between the effects of electronic structure on the coupling interactions, it is often convenient to determine a reduced coupling constant K. The sign of K then depends only on the electron distribution:

$$K_{N{\sim}X} = \frac{4\pi^2 J_{N{\sim}X}}{h\gamma_N\gamma_X} \tag{7.1}$$

In SI units, K is expressed in newton ampere^2 meter^{-3} (equivalent to 10 cm^{-3}).

The behaviour of the coupling constants involving nitrogen is frequently discussed by analogy with the case of the ^{13}C-H coupling constants. Thus, to the extent that the Fermi contact mechanism is considered as governing J_{N-X}, correlations with bond hybridization, similar to those which have been employed for $^{1}J^{13}$C-H, may be proposed. Moreover, as in the case of $^{1}J^{13}$C-H, attention is also drawn to the possible effect of variations in the effective nuclear charge. Hence an attempt has been made to establish certain relationships between J^{15}N-X and parameters such as substituent electronegativities or Hammett constants. However in reality, mechanisms other than the Fermi contact mechanism may contribute, namely interactions with the orbital motion of the electrons or with the spin of the extra-nuclear electrons. The Behaviour of J_{N-X} is therefore expected to be complex and semiempirical correlations may give some insight into the nature of the coupling interactions and help to specify the possible applications to structural determinations.

7.2.1 $^{n}J_{N-H}$ Coupling Constants

7.2.1.1 Influence of the s Character of the Bonding Orbitals

It was soon recognized that $^{1}J^{15}$N-H strongly depends on the hybridization of the nitrogen since $^{1}J_{N-H}$, which equals 73 Hz in the tetrahedral ammonium ion (Table 7.2), increases when the sp_2 character increases (see for example substituted anilines Table 7.9, amides Table 7.16 or pyrroles Table 7.13) to reach 135 Hz in protonated nitriles (Table 7.28). The following relationships with the percentage s character of the nitrogen orbital engaged in the N-H bond have been proposed:

(B 24) $$\%s = 0.43 \left|{}^{1}J_{N-H}\right| - 6 \qquad (7.2)$$

(B 26) $$\%s = 0.34 \left|{}^{1}J_{N-H}\right| \qquad (7.3)$$

(W 7) $$\%s = -0.59 \left|{}^{1}J_{N-H}\right| - 17.5 \qquad (7.4)$$

These equations are based on different test compounds and correspond to different solvent properties. They can be useful in discussing nitrogen hybridization provided that the rough nature of such a treatment is kept in mind. The degree of planarity at the nitrogen atom of amino-phosphines, -arsines and -sulfenamides for example has been examined from this point of view (C 11) (C 13).

Similarly, the $^3J_{N-CC-H}$ coupling constant has been related, on a theoretical basis, to the s character of the nitrogen hybrid orbital a^2, and to the value of $^3J_{H\sim H}$, for a corresponding fragment H-C-C-H' with an identical dihedral angle (K 11).

$$^3J^{15}{}_{N\sim H}\ (\theta) = -1.514\ a^2\left(^3J_{H\sim H'}(\theta)\right) \tag{7.5}$$

Hence the s character a^2 of the hybrid orbital should be the dominant element in determining the variation of the ratio $J_{N-H}/J_{H-H'}$ in different molecules.

7.2.1.2 Relationship between $^nJ^{15}{}_{N\sim H}$ and Electronegativities or Hammett Constants of the Substituents

As concluded from the theoretical and experimental examination of $^1J^{13}$C-H coupling constants, the percentage s character is not the only factor which conditions the Fermi contact contributions for the J_{NH} interactions. Variations in the effective nuclear charges of the coupled atoms may also play a part. Although a complete breakdown of the electronic properties of a substituent is not possible, it can be shown that good correlations between J^{15}N-H and the electronegativity of the substituents exist in certain series of compounds. Thus, in spite of several exceptions an increase in $|^1J^{15}N\text{-}H|$ accompanies the increase in group electronegativities of X for compounds XNH_2 or X_2NH (X = SiH_3 to SCF_3) (C 11) (C 13).

Since electronic effects are also well defined by the Hammett parameters, σ, it is not surprising that reasonable correlations between $^nJ_{N\sim H}$ and σ are observed, in certain families of compounds. This is the case for *meta* and *para* substituted anilines for which two linear curves $^1J_{NH} = f(\sigma)$ have been drawn corresponding to two different solvents DMSO and $CHCl_3$ (A 1) (A 4) (see Sect. 5.3). The following relationships are derived from these results:

$$\text{(DMSO)}\quad ^1J^{15}N\text{-}H = 6.4\ \sigma + 82.3 \quad (R = 0.93,\ \sigma = 0.85,\ N = 17) \tag{7.6}$$

$$(CDCl_3)\quad ^1J^{15}N\text{-}H = 6.3\ \sigma + 78.4 \quad (R = 0.93,\ \sigma = 1.0,\ N = 18) \tag{7.7}$$

7.2.1.3 Geometrical Dependency of $^nJ^{15}{}_{N\sim H}$: Bond Angles and Lone-Pair Orientation

The *dihedral angle dependence* of the $^3J_{H-C-C-H}$ coupling constants has been extensively applied to structural investigations. Indeed the existence of an analogous geometrical behaviour for $^3J_{N-C-C-H}$ is expected to be of prime importance in the investigation of the conformation of the backbone chain of peptides for example. Unfortunately, the measured $^3J_{N\sim H}$ coupling constants are usually small in magnitude and it is difficult to find model compounds with a well-defined geometry. However, the theoretical expression (7.5), for example, suggests that $^3J_{N-H}$, like $^3J_{H-H}$, should have a dihedral dependence (K 11). Based on the values, $^3J_{N-H}$ = (-) 1.5 Hz and $^3J_{N\sim H}$ = 2.85 Hz measured in acetylglycine and acetaldehyde respectively, the *trans* and *gauche* H-C-C-N coupling constants are tentatively estimated at -4.3 Hz and -0.05 Hz for the considered fragment, N-CO-C-H. Several correlations between $^3J_{N-H}$ and the dihedral angles θ of the peptide structures, have been proposed on theoretical (B 70) (S 37) and experimental bases (B 8) (D 8).

(B 70) $$^3J_{N-C(O)-C-H} = \begin{cases} -5.15\cos^2\theta + 2.30\cos\theta + 0.98 \; (cis) & (7\text{-}8a) \\ -5.34\cos^2\theta + 2.15\cos\theta + 0.87 \; (trans) & (7\text{-}8b) \end{cases}$$

(B 8) $$^3J_{N-C(O)-C-H} = -4.6\cos^2\theta + 3.0\cos\theta + 0.8 \qquad (7\text{-}9)$$

(D 8) $$^3J_{N-H} = -4.4\cos^2\theta + 1.2\cos\theta + 0.15 \qquad (7\text{-}10)$$

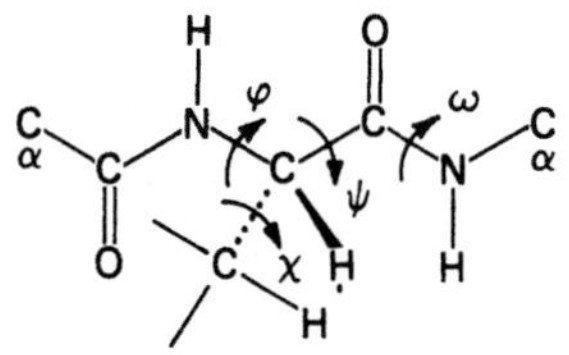

Equations (7.8a) and (7.8b) show that the *cis* or *trans* arrangement of the amide bond has little influence on the dihedral angle dependency. Eq. (7.10) is derived from values measured in a rigid model of known geometry, alumichrome, which is analogous to the ferric cyclohexapeptide ferrichrome A (D 8). This equation is convenient for estimating the conformational angle χ from the $^3J_{N-H_\beta}$ coupling constants. However, discrepancies are observed in the prediction of the ψ dihedral angle from ${}^{3}J_{N-CO-CH_\alpha}$ values and it is suggested that this latter coupling constant depends simultaneously on the conformational angle φ and eventually on ω. Moreover, since both positive and negative values are expected, a sign determination is, in principle, necessary for derivating angular information.

Orientation with respect to substituents, or more specially with respect to a lone pair at the nitrogen, is also shown to condition systematically the relative values of the coupling constants in a number of structures. Thus, in the amide fragment, the $|{}^{1}J_{N-H}|$ value appears to be higher for the proton in an *anti* position with respect to the oxygen than for the corresponding proton in a *syn* position (Table 7.16). This property can be used as an indicator of *cis, trans* isomerism about the peptide bond (B 16) (S 5) (W 28) (W 1) since $|{}^{1}J_{N-H}|$ varies between 92-94.5 Hz and between 89 to 91 Hz, respectively for a *trans* and for a *cis* arrangement in linear-dipeptides (see Table 7.21). A more refined treatment involves the dependency of ${}^{1}J{}^{15}N\text{-}H$ on the degree of planarity at the peptide bond which is defined by an angle ω (L 16). On going to a non planar amide configuration the nitrogen valence electrons shift from a partial sp^2 to an sp^3 hybridization and ${}^{1}J_{NH}$ should decrease. This suggests a cyclic dependence of ${}^{1}J_{NH}$ starting with $|{}^{1}J_{NH}| \simeq 95$ Hz for a planar *trans* structure, going through a minimum ($\simeq$ 67 Hz) for $\omega = 90^0$ and increasing back to about 90 Hz for a planar *cis* arrangement. These trends therefore enable the ${}^{1}J_{NH}$ values to be used for detecting distortion with respect to a *transoid* ($\omega = 0$) structure for example.

A considerable orientation effect is exhibited by compounds which possess a lone pair at the nitrogen. Thus, the *geminal* coupling constant ${}^{2}J_{N-C-H}$ may be of a high negative value in structures such as pyridine or oximes where the lone pair is in a *syn* relationship with respect to the coupled hydrogen. On the contrary, if there is no lone pair (Table 7.14), or if an *anti* relationship exists (Table 7.24), the value of ${}^{2}J_{N-C-H}$ is smaller and possibly positive.

Obviously, this property finds interesting applications in stereochemical determinations involving oximes (Tables 7.23 - 7.25), cyclic amines (Table 7.1), tetrahydro-1,3-oxazines (Table 7.1), nitrosamines (Table 7.31) or alkyl nitrites (Table 7.32) for example (G 21) (J 2) (J 4) (K7) (W 30) (O 3) (R 9) (S 20) (see also Sect. 5.3).

7.2.2 $^{n}J^{15}N^{13}C$ Coupling Constants

In general, $^{1}J_{N-C}$ is negative. However, a positive value has been determined in the strained oxaziridine ring (J 2) or in pyridine (B 12) for example. The absolute value is generally found between 0 and 20 Hz but a large value of 77.5 Hz, has been reported for 2,4,6-trimethylbenzonitrileoxide (C 14). Relatively small values are measured for $^{2}J_{CN}$ and $^{3}J_{CN}$ and both signs have been determined (Tables 7.38 to 7.50).

7.2.2.1 Influence of Hybridization

Although the relationship (7.11) proposed earlier (B 24)

$$(\%\ s_C) \cdot (\%\ s_N) = 80\ \left|^{1}J_{C-N}\right| \tag{7.11}$$

holds for some coupling constants it has been shown on both theoretical and experimental bases that no general relation with hybridization exists. Thus, the coupling $^{1}J_{N-C}$ between a trigonal nitrogen and an *sp* carbon in a fragment $\rangle$N-C≡C- behaves anomalously (Table 7.38) (B 22). More generally, deviations are found for compounds (C 14) (B 24) (B 21) (S 31) such as nitriles or isocyanides which involve a triple bond and this particular behaviour of multiple bonding is explained by an important participation of the orbital and dipolar mechanisms (K 39) (S 32). In addition, when present on a single or double bond, a lone pair with some *s*-character noticeably contributes to the Fermi contact term and no simple relationship with the percentages of *s* character may then be expected (S 31) (S 32) (S 33). Nevertheless, satisfactory correlations with hybridization are found in limited series as shown by the relatively good linear trend observed between $^{1}J_{N-C}$ and $^{1}J_{N-H}$ of a number of amine derivatives (W 10).

$$\left|^{1}J_{N-C}\right| \simeq 0.392\ \left|^{1}J_{N-H}\right| - 20.2 \tag{7.12}$$

By combining this equation with Eq.(7.2) the following relationship is deduced for the ^{13}C - $^{15}N\!<^{H}_{R}$ fragment.

$$\left|^{1}J_{N-C}\right| = 0.912\ (\%2sN) - 14.7 \tag{7.13}$$

In fact, $^{1}J_{CN}$ is considered a good criterion of the planarity of the amino group. Its magnitude increases from a pyramidal nitrogen to a planar amino group and a linear correlation with the bond order $p^2_{s_N s_C}$ is found (W 10). Although no general relation is expected (M 5), this behaviour is corroborated on a theoretical basis by the linear correlation which exists between the calculated Fermi contact contribution and INDO $p^2_{s_N s_C}$ values for compounds which do not exhibit the lone pair effect discussed above (S 32).

7.2.2.2 Relations involving Electronic Properties of the Substituents

As has been discussed for nitrogen-proton interactions, variations in J_{N-C} may also reflect the electronic properties of the substituents in appropriate series. In this respect, the magnitude of $^1J_{NC}$ in *meta* and *para* substituted anilines is determined primarily by substituent polar effects (A 14). Taking the values measured in this series into consideration we have observed that the changes in $^1J_{N-C}$ reasonably follow the variations in the chemical shift of the substituted carbon

$$^1J_{^{15}N-^{13}C} = 0.26\ \delta^{13}C_{(ppm)} - 26.5 \qquad (R = 0.93,\ \sigma = 0.35,\ N = 12) \qquad (7.14)$$

and are in good agreement with the variations of the *Hammett σ parameter*

$$^1J_{^{15}N-^{13}C} = 2.89\ \sigma + 11.95 \qquad (R = 0.93,\ \sigma = 0.4,\ N = 12) \qquad (7.15)$$

By contrast, changes in the effective nuclear charge at the substituted carbon, C_1, of anilinium ions do not constitute the predominant contributions to $^1J_{C-N}$. Steric interaction between the solvated ammonium moiety and the *ortho* ring substituents seems to be a determining factor since the magnitude of $^1J_{CN}$ varies from about 9 Hz in *meta* and *para* substituted anilinium ions to 13.5 Hz in a crowded *ortho* substituted structure (A 14).

- Variations of $^1J_{NC}$ in *para* subtituted anilines are also shown to follow the variations in the *NC π bond order*. Thus, the magnitude of $|^1J_{NC}|$, which is 10.9 Hz in aniline, increases when *para* substitution leads to an enhanced participation of a resonance structure involving a N=C double bond. It reaches a value of 15 Hz in paranitroaniline (W 10). Such variations with the π bond order have also been pointed out in the discussion on protonation effects (see Sect. 5.3).

- Steric effects can be used to interpret nitrogen carbon couplings in amides substituted by strained cyclic groups at the carbonyl group (L 14). Thus, $^{2}J_{CN}$, which could be negative, decreases with increasing substitution at the α carbon (Table 7.46). It is observed that only a rough correlation exists between $^{1}J_{CN}$ and $^{2}J_{CN}$ in this series. Related to this, it is noted that the absolute value of $^{1}J_{CN}$ decreases from 17.5 Hz to 15 Hz when the size of the substituent in nitrile derivatives increases (CH_3 to t-C_4H_9) (M 5) (Table 7.48).

Finally, nitrogen *lone pair orientation* can markedly influence $^{1}J_{NC}$ and $^{2}J_{NC}$ in oximes (B 17) (L 12). Thus, $|^{1}J_{NC}|$ = 4.0 Hz and 2.3 Hz in the E and Z isomers of acetaldoxime respectively. However, substantial differences only arise when one of the substituents is a proton. For $^{2}J_{N-C}$, the absolute values are greater when the lone pair is *cis* to the coupled carbon than in the cases where the lone pair is *trans*. Moreover, a sign reversal is observed and the results hence parallel those for $^{2}J_{NH}$ (see above) (B 17) (Table 7.44). Although $^{3}J_{C-N}$ is significant only in unsaturated oximes, the same relation to lone pair orientation seems to hold (L 12).

7.2.3 Coupling Constants between ^{15}N and other Hetero Nuclei

7.2.3.1 $^{n}J^{15}N\sim^{15}N$

Owing to the necessity of ^{15}N enrichment, relatively few $J^{15}N\sim^{15}N$ coupling constants have been obtained (Table 7.51 to 7.53) (see also (S 1)) for a documented review) and the structural features which influence this parameter are still not fully understood. However, although the range of measured $^{1}J_{NN}$ values is restricted to about 25 Hz, an increase in $|^{1}J_{NN}|$ accompanying the increase in sp^2 hybridization is supported by the values determined in phenylhydrazine (6.7 Hz), p-nitrobenzaldehyde phenylhydrazone (10.7 Hz) p-hydroxyazobenzene (15.0 Hz) and dibenzylnitrosamine (19.0 Hz) for example (B 7). From a theoretical investigation of hydrazine, it is concluded that for compounds with s hybridized lone pairs the dominant Fermi contact contribution varies with the dihedral angle ϕ between the nitrogen lone pairs according to the following relation (S 36)

$$J \simeq a_0 + a_1 \cos\phi + a_2 \cos^2\phi$$

with a_0 = -3.8 Hz, a_1 = -11.5 Hz, a_2 = +3.5 Hz for hydrazine. It is also predicted that most $^{1}J_{NN}$ coupling constants should be negative (K 40).

As regards the $^{2}J_{NN}$ parameters, a lone pair effect similar to those which have been discussed for J_{NH} and J_{NC} seems to intervene as shown by the changes resulting from the protonation of several nucleotides (Table 7.53) (B 36).

7.2.3.2 $J^{15}N{\sim}^{31}P$

The one-bond nitrogen-phosphorous coupling constant attains a large magnitude (Tables 7.56 to 7.59) and a sign change is observed from a tricoordinate phosphorus in $(CF_3)_2PNH_2$ ($^{1}J_{P-N}$ = + 52.6 Hz) to a penta-coordinate phosphorous in $F_3P(NH_2)_2$ ($^{1}J_{PN}$ = - 81.5 Hz) for example (S 7). Qualitatively, this sign reversal is associated with hybridization at the phosphorus atom going from about 0 % to 33.3 % 3 *s* character. An intermediate negative value of -53.2 Hz is determined for the tetracoordinate phosphorus of $F_3P{=}NPF_2$ which possesses about 25 % 3 *s* character. Related to this, a reasonable linear correlation has been drawn between $^{1}J_{P-N}$ and the calculated bond order $P^{2}_{3s(P),\ 2s(N)}$, corroborating the fact that $^{1}J_{PN}$ can be tentatively related to the percentage *s* character of the P-N bonds (G 4).

However, a discussion of the behaviour of $^{1}J_{PN}$ at a more refined level is difficult due to the complex variations which may occur in the P-N fragment. In addition to the determining effect of phosphorus hybridization the coupling is sensitive to the planarity at the nitrogen and when comparing substituent effects possible concomitant changes in the various conformational equilibria must be considered. In spite of these difficulties, satisfactory explanations of the behaviour of J_{PN} in limited series can be proposed and several types of correlation are of some help. Thus, in tricoordinate phosphorus derivatives, $^{1}J_{PN}$ varies between about + 50 Hz to + 96 Hz (Table 7.56) and depends on both electronic and steric effects. Qualitatively, large increases in the electronegativity of the substituents at the phosphorus are shown to lead to substantial increases in J_{PN} (G 5). A factor analysis of $^{1}J_{PN}$, based on a matrix of data for aminophosphines ordered as a function of the substitution at the nitrogen shows that, in contrast to the case of $\delta^{15}N$ (see Sect. 6.2), two factors are required to account for the J_{P-N} variations and a correlation with β and γ parameters is proposed (G 10).

$$^{1}J_{PN} = 76.6 + 1.3\ (\pm\ 0.2)n_{\beta} + 0.7\ (\pm\ 0.3)n_{\gamma} \qquad (R = 0.95,\ \sigma = 0.8,\ N = 8)$$

As already mentioned in Section 6.2, to the extent that no important steric crowding occurs, $^1J_{PN}$ is satisfactorily correlated with $\delta^{15}N$ (or $\delta^{31}P$) in certain series of aminophosphines $\frac{X}{Y}\!\!>P-NR_2$ (R = CH_3 or C_2H_5) (G 10)

$$R = CH_3 \quad {}^1J_{PN} = 400.3 + 0.98\ \delta^{15}N \qquad (R = 0.986,\ \sigma = 2.2,\ N = 13)$$

$$R = C_2H_5 \quad {}^1J_{PN} = 385.6 + 1.0\ \delta^{15}N \qquad (R = 0.98,\ \sigma = 2.8,\ N = 10)$$

The magnitude of the coupling interaction increases when $\delta^{15}N$ (and $\delta^{31}P$) undergo paramagnetic shifts and it is noticeable that the slopes are similar for the dimethyl- and diethylamino derivatives. Rough variations of J_{P-N} with the *inductive electron withdrawing* properties of the substituent, represented by the Taft parameter σ_I, can be deduced.

$$^1J_{PN} = 44(\pm\ 3) + 71(\pm\ 8)\ \sigma_I(X,Y) \qquad (R = 0.93,\ \sigma = 5,\ N = 13)$$

Similarly, the coupling constant, involving the endocyclic nitrogen of diazophospholanes $[CH_2N(CH_3)]_2P$-R varies with the inductive effect of the phosphorous substituents since $^1J_{PN}$ has the following values: 49.4 Hz for R = C_6H_5, 50 - 52 Hz for R = N< , 57.6 Hz for R = O, *sec*-but. and 63.2 Hz for R = Cl. However, several anomalies arise, and for diisopropylamino compounds of the type $R_2NP(X)(Y)$ for example, different behaviours are observed according to whether the phosphorus carries an halogen substituent or not.

It is also noticed that increases in $^1J_{PN}$ are observed either when the size of the substituents at the phosphorus decreases or when the size of the substituents at the nitrogen increases, as exemplified by the following results (G 10) (G 11).

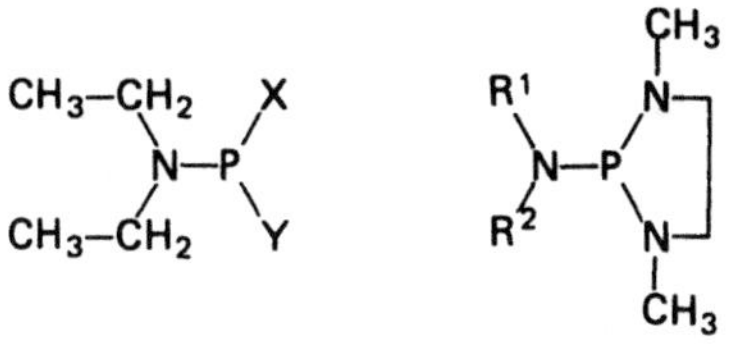

(I) (II)

X	y	$^{1}J_{PN}$ (Hz)	R^1	R^2	$^{1}J_{PN}$ (Hz)
NMe_2	$NiPr_2$	49.8	H	C_6H_5	84.2
NEt_2	NEt_2	61.5	CH_3	CH_3	91.2
NEt_2	NMe_2	66.2	C_2H_5	C_2H_5	92.6
NMe_2	NMe_2	72.1	$-(CH_2)_5-$		92.7
CH_3-N \ CH_2CH_2 /	$N-CH_3$	92.6	*i*C_3H_7	*i*C_3H_7	96

Very large variations in $^{1}J_{P-N}$ involving a tetra- or a pentacoordinate phosphorus atom are observed, but this coupling constant remains generally negative (Table 7.57). From investigation of heterocyclic phosphoramidates $\rangle N-P(O)(OCH_3)_2$ (G 6), it is shown that $^{1}J_{PN}$ experiences a large increase in going from a pyramidal-like (*sp*^{3}) nitrogen in a three membered ring (J = 9.3 Hz) to a trigonal *sp*^{2} nitrogen in the larger rings or in the acyclic compound $Et_2NPO(OCH_3)_2$ (J = 42.2 Hz). Similarly, attempts to rationalize the behaviour of J_{PN} in phosphazenes in terms of electronic distribution and bond orders have been carried out (T 4) (T 5) (T 6).

7.2.3.3 *Miscellaneous Couplings to Nitrogen*

Results obtained for coupling constants with nuclei other than H, C, N or P (Tables 7.54 - 7.55) (7.60 to 7.69) are scarce. We shall only mention here that a good correlation is noted for the variations in the $J^{15}N-^{111}Cd$ coupling constant of *meso*-tetraphenylporphyrin ($^{15}N_4$) as a function of the basicity (pKa) of the ring substituted pyridine ligand. The increased electron donation from the pyridine to the metal should strengthen this bond at the expense of the porphyrin-to-metal coordination, thus producing a decrease in $^{1}J^{15}N-^{111}Cd$ (D 22). Finally, it is pointed out that the $^{1}J^{15}N-^{195}Pt$ coupling constants measured in platinum- amine complexes, depend on the *s* character of the Pt and N hybrids used in the bonding (P 5) (A 20). The fact that $^{1}J_{N-Pt(II)}$ is about 1.5 times larger than the corresponding value $^{1}J_{N-Pt(IV)}$ is in accordance with a change of hybridization from *dsp*^{2} to *d*^{2}*sp*^{3} (J_{NPt} = 351 Hz in *cis* $(PtCl_2A_2)$ and 249 Hz in *cis* $(PtCl_4A_2)$ with A = n-dodecylamine) (P 5). Moreover, it is observed that $^{1}J_{NPt}$ is larger for *cis*-(PtX_2A_2) complexes than for the corresponding trans complexes (J_{NPt} = 334 Hz in *cis*- $(PtBr_2A_2)$ and 279 Hz in *trans*- $(PtBr_2A_2)$ with A = *n*-dodecylamine). In *cis*-diamine-platinum (II) complexes, it also appears that $^1J_{^{15}N-^{195}Pt}$ decreases from 388.6 Hz to 286.5 Hz when the H_2O ligands of *cis*-$(^{15}NH_3)_2Pt(H_2O)_2{}^{2+}$ for example are replaced by the stronger 1-methyl-imidazole ligands. This behaviour, which is consistent with a weakening of the *trans* Pt-N bond as a result of interaction with a stronger ligand, is potentially of biological use (A 22) (K 2). It can also be noted that a linear correlation is observed between $^1J_{N-Pt}$ in *trans* hexylamine complexes of platinum *trans*-$(PtCl_2AL)$ (with A = $NH_2(CH_2)_5CH_3$ and L a variable ligand) and $^1J_{Pt-P}$ in analogous phosphine complexes (*trans*-$(PtCl_2A'L)$ with A' = PBu_3). These changes are governed by the variations in the percentage of *s*-character at the platinum (a^2_{Pt}) (M 7).

Table 7.1. $^{n}J_{^{15}N-^{1}H}$ Coupling constants of amines

Amines		^{1}J	^{2}J	Conditions	Ref.
NH_3		61		/	B 42
		64 ± 2		neat	O 2
		61.2 ± 0.9		neat	B 2
		61.8 ± 0.5		neat	L 57
		61.2 ± 0.3		gaz (a)	A 5
NH_2D		61.8 ± 0.5		neat	L 10
NHD_2		61.8 ± 0.5		neat	L 10
CH_3NH_2		64.0		gaz	P 2
		64.5 ± 0.3		gaz (a)	A 5
			1.0	/	B 24
		-65.0 ± 0.2	-1.0 ± 0.1	gaz	P 1
H_2N-CH_2-R (1, 1')			≃0.5 (1) ≃0.5 (1')	$CDCl_3$ (b)	C 6
HN-$(CH_3)_2$		67.0 ± 0.3		gaz (a)	A 5
HN-R_2	(c)	86.7	3.7	DMSO-d_6 (a)	C 16
$(CH_3)_3N$			0.85	C_6D_{12}	P 2
			0.80	$CDCl_3$	P 2
			0.60	CD_3OD	P 2
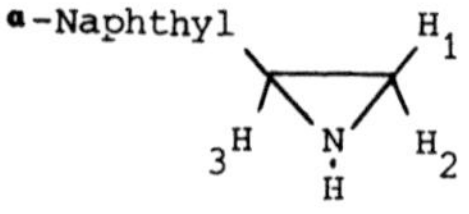			(-)3.6 to (-)4.9 (1) (-)0.4 to (-)1.4 (2) (-)0.7 to (-)1.7 (3)	Various solvents	O 3
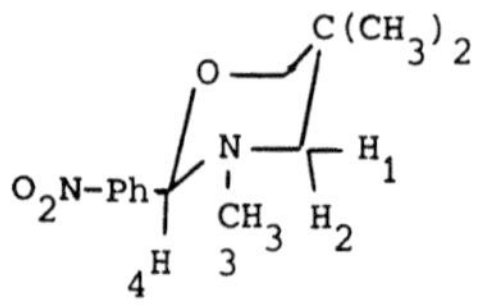			1.5 (1) 0 (2) 1.1 (3) 0 (4)	$CDCl_3$	R 9

Table 7.1 (continued)

Amines		^1J	^2J		Conditions	Ref.
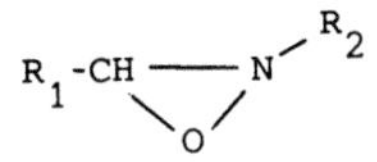						J 2 J 4
R_1	R_2		cis	trans		
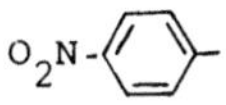	CH_3		-5.4	0		
	$C_6H_5-CH_2$		-5.3	0		
	$(CH_3)_2CH$		-5.0	0		
	$(CH_3)_3C$		/	0		
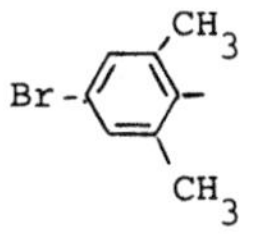	CH_3		-4.8	0		

- Obtained in ^{1}H-NMR except (a) measured using ^{15}N spectroscopy

(b) ^3J = 1.5 Hz (CH) with

R =

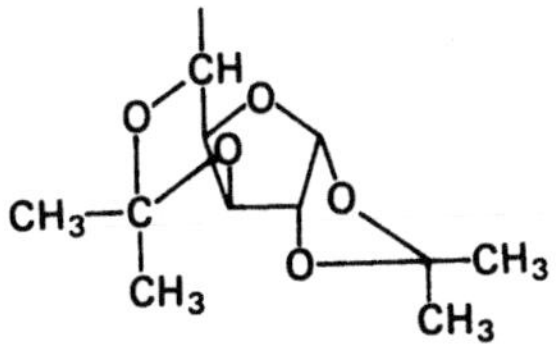

(c) R =

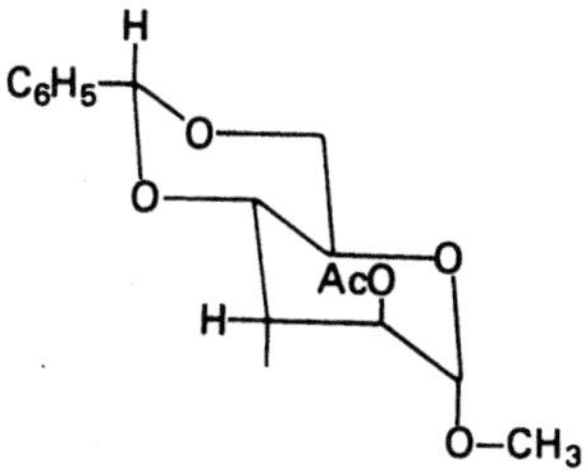

Table 7.2. $^{n}J_{^{15}N-^{1}H}$ Coupling constants of ammonium salts

Ammonium Salts		^{1}J	^{2}J	Conditions		Ref.
NH_4^+	Cl^-	73.2		H_2O		B 24
		73.3		5M H_2O, HCl M	(a)	A 23
		73		5M H_2O, H^+	(a)	B 57
NH_4^+	NO_3^-	73.7 $\pm$ 0.1		H_2O-HNO_3 saturated		B 3
$CH_3NH_3^+$	Cl^-	73		10M H_2O, H^+	(a)	B 57
		76.5	0.8	H_2O		B 24
		75.4		5M H_2O, HCl M	(a)	A 23
$CH_3NH_3^+$		-74.92	0.79	H_2O		R 14
$(CH_3)_2NH_2$	Cl^-	76.1		5M H_2O, HCl M	(a)	A 23
$(CH_3)_2NH_2^+$		-76.53	0.8	H_2O, H^+		R 1
$(CH_3)_3NH^+$	Cl^-	76.7		5M H_2O, HCl M	(a)	A 23
$(CH_3)_4N^+$	I^-		0.9	DMSO-d_6		B 5
$C_2H_5N{=}C{=}N{-}(CH_2)_3{-}\overset{+}{N}H(CH_3)_2$	Cl^-	88		DMSO	(a,b)	Y 14
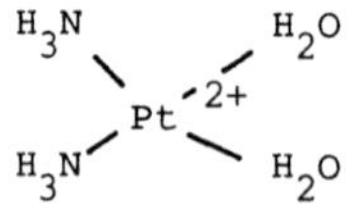		73.2		H_2O	(a)	A 22
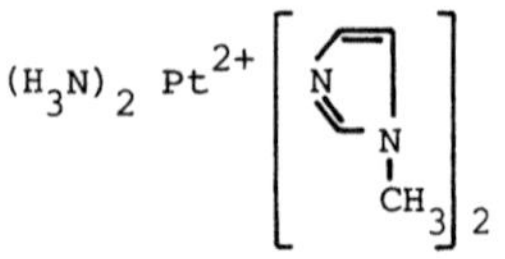	cis	72.0		H_2O	(a)	A 22
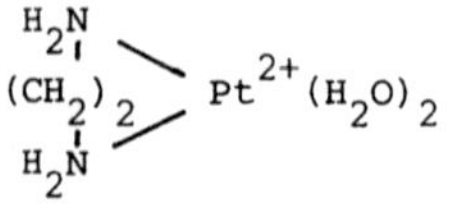		75.4		H_2O	(a)	A 22
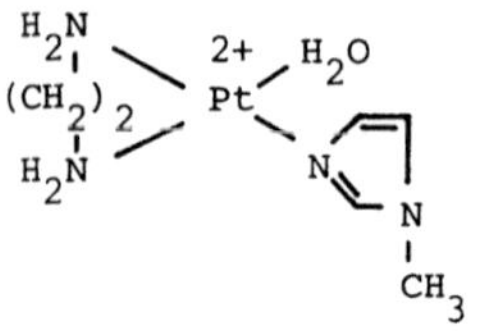		(c)		H_2O	(a)	A 22

Table 7.2 (continued)

Ammonium Salts	^{1}J	^{2}J	Conditions	Ref.
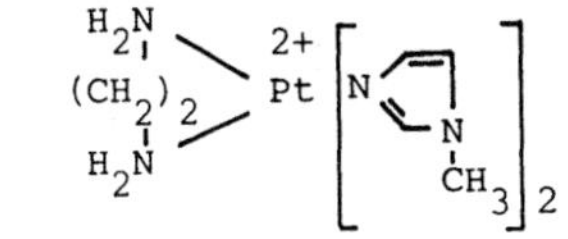	(c)		H_2O (a)	A 22
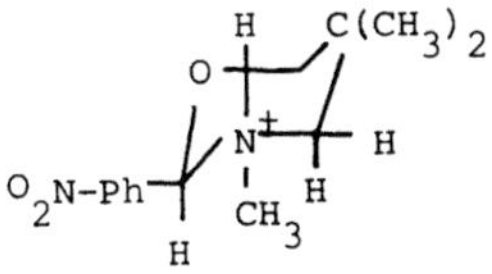		(d)	CF_3COOH	R 9

- Obtained in ^{1}H-NMR except (a) measured using ^{15}N spectroscopy

(b) Mixture with 3 other products (see Tab. 7.26)

(c) Not observed

(d) The ^{2}J couplings are unresolved

Table 7.3. ${}^{n}J_{{}^{15}N-{}^{1}H}$ Coupling constants of amino acids

Amino Acids	${}^{1}J$	${}^{2}J$	${}^{3}J$	Conditions	Ref.
H_2N-CH_2-COOH	74.7			H_2O	B 39
		0.5 ± 0.1		0.7M D_2O pH 6.33	L 17
		1.5 ± 0.1		pH 13.68	L 17
H_2N-CH_2-COOH, HCl	77.0			H_2O pH 1	B 24
d,l $H_2N-CH-COOH$ CH_3		<0.5		0.43M D_2O pH 6.37	L 17
		2.2 ± 0.1		pH 13.64	L 17
			3.1 ± 0.2	0.43M H_2O	L 1
			3.10	D_2O pH 1	S 5
d,l-$H_2N-CH-COOH$ $CH_2-C_6H_5$		<0.5		0.08M D_2O pH 5.6	L 17
		2.2 ± 0.1		pH 12.22	L 17
			2.9 ± 0.2	0.08M H_2O	L 1
			3.4 ± 0.2		L 1
			2.20-2.35	D_2O pH 1	S 5
$D_3N^+-CD-COO^-$ $CH_2-CD(CH_3)(CH_2D)$			-2.15	D_2O anion (a)	F 5
			-3.15	(a)	F 5
			-2.47	D_2O cation (a)	F 5
			-3.47	(a)	F 5
$H_2N-CH-COOH$ CH_2-COOH			3.0 ± 0.1	0.05M H_2O	L 1
			3.0 ± 0.3		L 1
			2.80-3.70	D_2O pH 10	S 5
$H_2N-CH-COOH$ $CH_2-C_6H_4-OH$			2.75-2.95	D_2O pH 10	S 5
$H_2N-CH-COOH$ $(CH_2)_4-NH_2$	73.7		2.95	H_2O	B 39

- Measured using ${}^{15}N$-NMR except (a) obtained from ${}^{1}H$ spectra

Table 7.4. $^{1}J_{^{15}N-^{1}H}$ Coupling constants of primary enamines (NH_2)

Enamines	^{1}J Free NH	^{1}J Associated NH	Conditions	Ref.
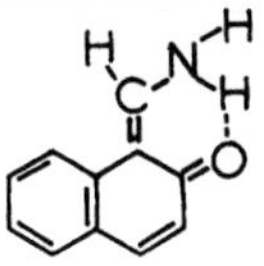	90.6	85.6	DMSO	D 20
	90	86.9	C_5D_5N	D 20
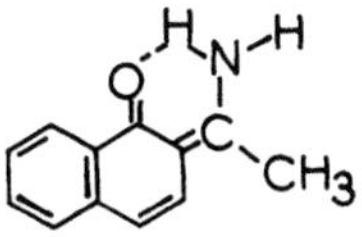	91.2	85.4	DMSO	D 20
	90.6	84.2	Acetone-d_6	D 20
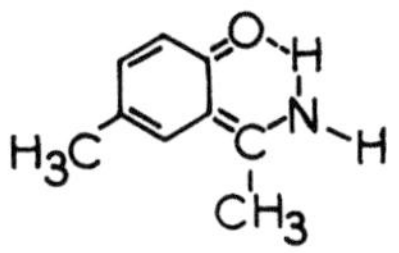	74.4	28.8	DMSO	D 20
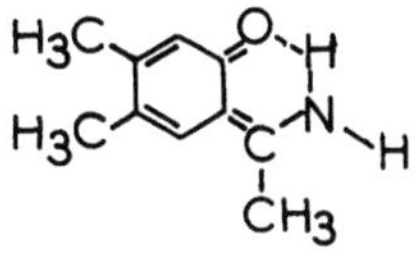	75.4	31.2	DMSO	D 20
	74.0	30.1	HMPT	D 20
	67.4	14.8	C_5D_5N	D 20
	62.8	/	$CDCl_3$	D 20
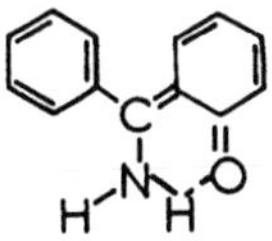	75.2	31.0	HMPT	D 20
	74.0	26.0	DMSO	D 20
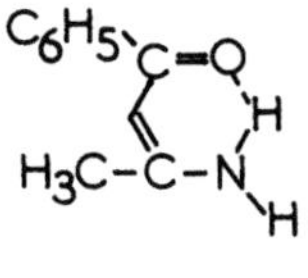	93.2	89.6	DMSO	D 20
	90.8	92.4 or 91.3	$CDCl_3$	D 20
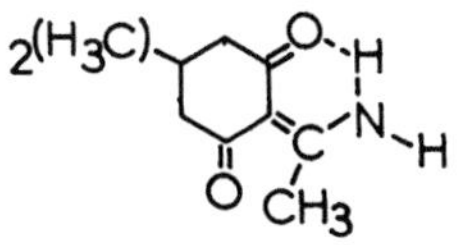	91.6	88.2	DMSO	D 20
	91.2	90.2	$CDCl_3$	D 20
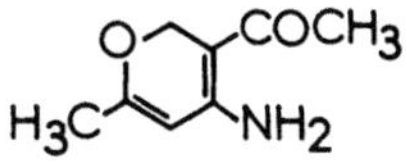	-92.3	-91.0	DMSO	N 2

Table 7.4 (continued)

Enamines	^1J Free NH	^1J Associated NH	Conditions	Ref.
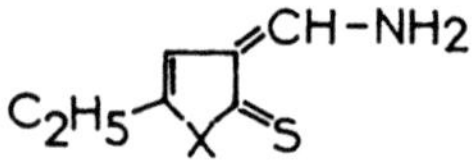 X = O, S		86	Various solvents	B 50

- Obtained in ^1H-NMR

Table 7.5. $^{n}J_{^{15}N-^{1}H}$ Coupling constants of secondary enamines (NHR)

Enamines	^{1}J	^{2}J	^{3}J	Conditions		Ref.
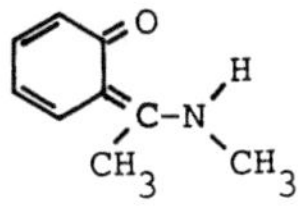	0			293 K	CH_2Cl_2	D 11
	0			293 K	$CDCl_3$	D 11
	21			273 K	$CHCl_3$	D 11
	32.4			223 K	$CHCl_3$	D 11
	68.2			227 K	CH_3OH	D 11
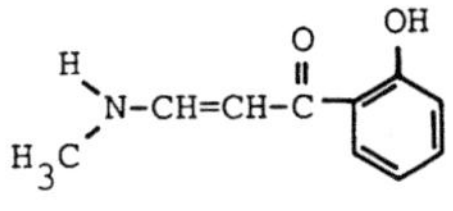	92.5	1.5 (CH_3)	4.2 (H)		$CDCl_3$	D 4
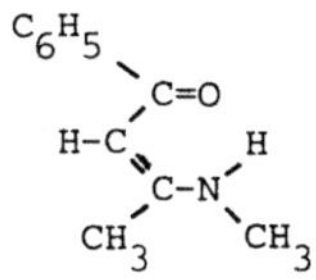	89.2				$CDCl_3$	D 20
	91.3		3.6 (H)	304 K	$CDCl_3$	D 3
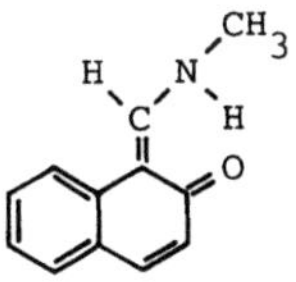	64.5	non - detected		304 K	$CDCl_3$	D 3
	81.4	1.3 (CH_3)		228 K	$CDCl_3$	D 3
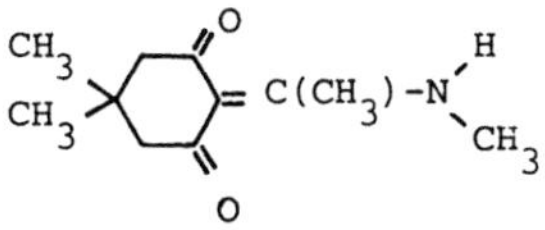	88.1				$CDCl_3$	D 20
	88.1	1.6	2.7		$CDCl_3$	D 3
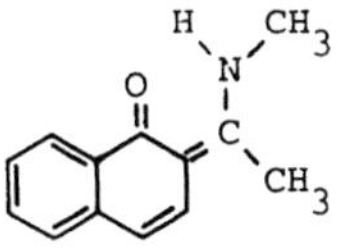	79.1				$CDCl_3$	D 3
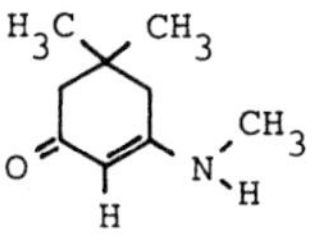	94.3	1.2	1.9 (H)		$CDCl_3$	D 3

Table 7.5 (continued)

Enamines	^1J	^2J	^3J	Conditions	Ref.
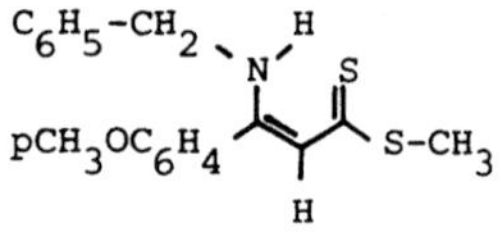	87	1.1	3.4	CCl_4	L 5

- Obtained in ^{1}H-NMR except (a) measured using ^{15}N spectroscopy

Table 7.6. $^{n}J_{^{15}N-^{1}H}$ Coupling constants of aromatic secondary enamines (NHC_6H_5)

Enamines	^1J	^2J	^3J	Conditions	Ref.
C_6H_5-NH-CH=C(CH_3)-C_6H_4-R					
R = OCH_3	91.1			DMSO-d_6	A 10
NO_2	91.2			DMSO-d_6	A 10
R-C_6H_4-NH-CH=C(CH_3)-C_6H_5				(a)	
R = H	90.8			DMSO-d_6	A 10 S 10
3-NO_2	92.3			DMSO-d_6	S 10
3-Cl	91.7			DMSO-d_6	S 10
3-OCH_3	90.6			DMSO-d_6	S 10
4-NO_2	90.4			DMSO-d_6	S 10
4-Cl	91.2			DMSO-d_6	S 10
4-OCH_3	92.9			DMSO-d_6	S 10
4-CH_3	90.2			DMSO-d_6	S 10
C_6H_5-NH-CH=C($COOC_2H_5$)$_2$	91.4 ± 0.2	≃0.2		$CDCl_3$	B 25
C_6H_5-NH-C(CH_3)=CH-CO-C_6H_5	89.2			303 K $CDCl_3$	D 5
	88.5			$CDCl_3$	D 20
C_6H_5-NH-C(CH_3)=CH-CO- 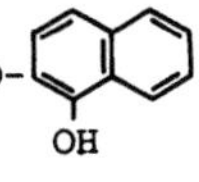	89.5		4.1 (H) 2.0 (CH_3)	$CDCl_3$	D 4
C_6H_5-NH-C(CH_2-C_6H_5)=CH-C_6H_5 Z E	89.6 ± 0.07 90.1 ± 0.07		5.17 (H) 2.50 (H)	DMSO-d_6	A 2
C_6H_5-NH-CH= 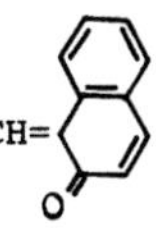	33			303 K $CDCl_3$	D 5
	49.4			233 K $CDCl_3$	D 5
	12.3			303 K CCl_4	D 5

Table 7.6 (continued)

Enamines	^1J	^2J	^3J	Conditions	Ref.
C_6H_5-NH-CH=CH-CO-C_6H_5	91.0		4.2	300 K $CDCl_3$	F 15
C_6H_5-NH-CH=CH-CS-C_6H_5	87.0		/	300 K $CDCl_3$	F 15
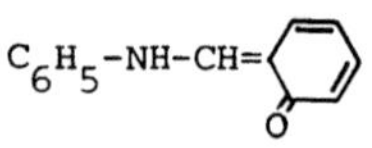	13.1			273 K C_2H_5OH	D 11
	13.2			303 K $CDCl_3$	D 5
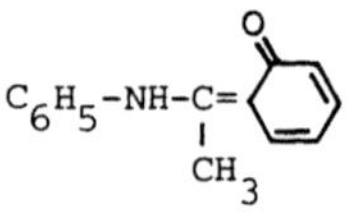	14.6			303 K $CDCl_3$	D 5
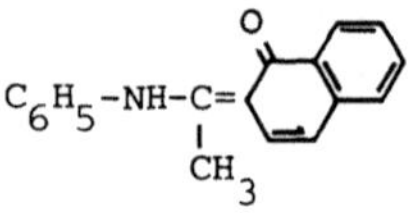	27.5			304 K $CDCl_3$	D 5
	36.5			243 K $CDCl_3$	D 5
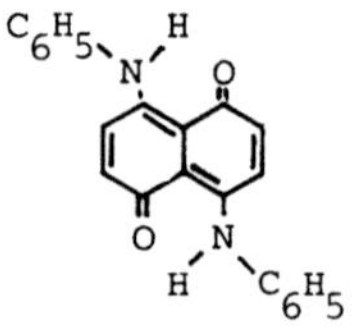	76.5			$CDCl_3$	B 52
	62.0			CCl_4	B 52
	61.2			C_5H_5N	B 52
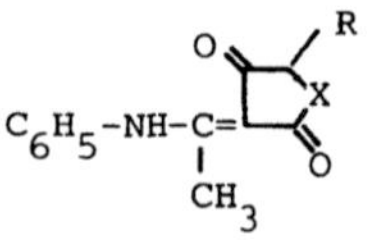					Y 13
X R					
NH H	90.0			307 K $CDCl_3$	
NH CH_2-C_6H_5	90.0			307 K $CDCl_3$	
O H	90.0			307 K $CDCl_3$	
S H	90.9			307 K $CDCl_3$	

Table 7.6 (continued)

Enamines		^1J	^2J	^3J	Conditions	Ref.
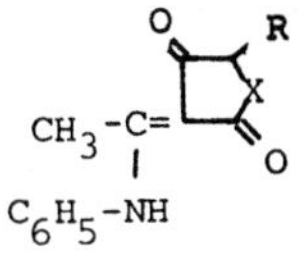					307 K $CDCl_3$	Y 13
X	R					
NH	H	90.0				
NH	$CH_2-C_6H_5$	90.0				
O	H	91.8				
S	H	86.4				
$C_6H_5-NH-C(C_6H_5)=CH-CH_3$	Z	86.9 ± 0.07		5.09	C_6D_6 DMSO-d_6	A 2
$C_6H_5-NH-C(C_6H_5)=CH-C_6H_5$	Z	87.5 ± 0.1		4.74	C_6D_6	A 2
$C_6H_5-NH-C(C_6H_4-NO_2)=CH-C_6H_4-NO_2$	Z	90.5 ± 0.1		4.58	DMSO-d_6	A 2
	E	89.9 ± 0.2		1.94	DMSO-d_6	A 2
$C_6H_5-NH-C(COOCH_3)=CH-CO-COO-i.C_3H_7$		89.5		4.0	CCl_4	K 6
$C_6H_5-NH-C(COOCH_3)=CH-CO-C_6H_5$		90.7		3.6	CCl_4	K 6
$C_6H_5-NH-C(COOCH_3)=CH-COOCH_3$	Z	91.2 ± 0.4		3.5 ± 0.2	CCl_4	B 25
	E	92.8 ± 0.3		1.7 ± 0.1	CCl_4	B 25
$C_6H_5-NH-\overset{+}{C}(O^-)=C(COCH_3)P(C_6H_5)_3$		95.5			0.03M CH_2Br_2	B 6
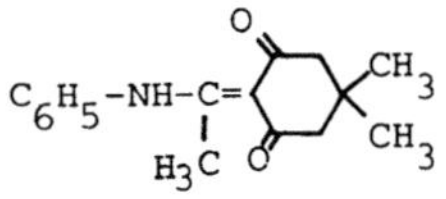		85.5			$CDCl_3$	D 5 D 20
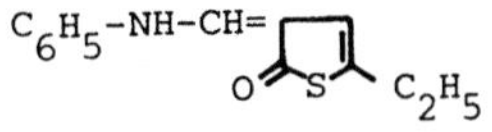		-93.0	≃0		Dioxane	N 2

- Obtained in ^{1}H-NMR except (a) measured using ^{15}N spectroscopy

Table 7.7. $^{n}J_{^{15}N-^{1}H}$ Coupling constants of aniline and 2-substituted anilines - Obtained in ^{1}H-NMR

Anilines	^{1}J	^{3}J	^{4}J	Conditions	Ref.
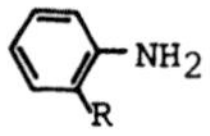					
H	-78.87	-1.93	-0.48	17%M C_6D_6 (a)	W 4
	78.6			$CDCl_3$	A 1 A 4 A 6
	81.9			Acetone	B 1
	78.0			$CDCl_3$	B 1
	82.6			DMSO	A 1 A 6
	78.0 - 82.3			Various solvents	P 2
	78.5			$(CH_3)_3SiNH-C_6H_5$	R 3
2-F	80.1			$CDCl_3$	A 6
	83.5			DMSO	A 6
2-Cl	-82.3	-1.98	-0.79 (3) -0.36 (5)	10%M C_6D_6 (b)	W 4 W 7
2-Br	81.4			$CDCl_3$	A 6
	84.3			DMSO	A 6
2-CF_3	83.6			$CDCl_3$	A 6
	86.5			DMSO	A 6
2-NO_2	90.3			$CDCl_3$	A 6
	91.0			DMSO	A 6
2-OCH_3	79.4			$CDCl_3$	A 6
	82.3			DMSO	A 6
2-$COCH_3$	-88.6	-2.46	-0.83 (3)	15%M $C_6D_5CD_3$	W 4
			-0.30 (5)	(c)	W 4
	-90.8 ± 0.2			15%M $(C_2H_5)_2O$	W 1
	-89.5 ± 0.2			183 K	
2-COC_6H_5	88.1			$CDCl_3$	A 6
	89.3			DMSO	A 6

(a) ^{5}J = 0.01 Hz (b) ^{5}J = 0.17 Hz (c) $^{5}J < \pm 0.1$ Hz

Table 7.8. $^{1}J_{^{15}N-^{1}H}$ Coupling constants of 3-substituted anilines - Obtained in ^{1}H-NMR (solvent $CDCl_3$)

Anilines	^{1}J	Ref.
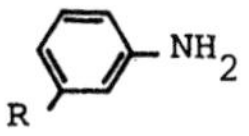		
3-F	80.1	A 4
3-Cl	80.9	A 1 A 4
3-Br	80.5	A 1 A 4
3-I	80.4	A 4
$3-CF_3$	81.0	A 4
$3-NO_2$	83.0	A 4 B 1
$3-CH_3O$	79.4	A 1 A 4
$3-CH_3$	78.2	A 1 A 4

Table 7.9. $^{n}J_{^{15}N-^{1}H}$ Coupling constants of 4-substituted anilines - Obtained in ^{1}H-NMR

Anilines	^{1}J	^{3}J	Ref.
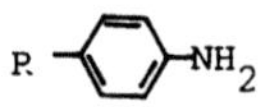			
4-F	77.8	1.8	A 1 A 4
4-Cl	78.9	1.9	A 1 A 4
4-Br	79.6	1.8	A 1 A 4
	83.6 (a)		B 1
4-I	79.7		A 1
	78.5		B 1
	79.7	1.8	A 4
4-NO_2	86.4	2.0	A 1 A 4
	78.7		B 1
4-OCH_3	75.6		A 1 A 4
4-CH_3	76.5	1.8	A 1 A 4
	76.9		B 1
4-NH_2	74.9		A 4
4-$N(CH_3)_2$	74.8		A 4

Solvent $CDCl_3$ except (a) acetone

Table 7.10. $^{n}J_{^{15}N-^{1}H}$ Coupling constants of di- and tri-substituted anilines - Obtained in ^{1}H-NMR

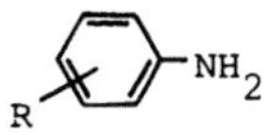

Anilines	^{1}J	^{3}J	^{4}J	Conditions	Ref.
2,4-Br_2	-83.11	-1.87	-0.67 (3)	2%M C_6D_6	W 4
			-0.30 (5)		W 4
	86.0			Acetone	B 1
	82.7	1.9		$CDCl_3$	A 4
2-Cl, 4-Br	-83.15	-1.94	-0.71 (3)	3%M C_6D_6	W 4
			-0.31 (5)		W 4
2-NO_2, 4-Cl	91.1			$CDCl_3$	A 6
2-Cl, 4-NO_2	89.2			$CDCl_3$	A 6
2,4-$(NO_2)_2$	92.6	2.2		DMSO-d_6	A 4
3,5-$(CH_3)_2$	77.5			$CDCl_3$	A 1 A 4
3,5-$(CH_3O)_2$	79.5			$CDCl_3$	A 4
2,4,6-Br_3	-85.46		-0.61	1%M C_6D_6	W 4
	87.8			Acetone	B 1
	85.5			$CDCl_3$	A 4 A 6
2-Cl, 4,6-Br_2	-85.60		-0.61 (3)	2%M C_6D_6	W 4
			-0.56 (5)		W 4
3,5-Br_2, 4-NO_2	83.2			$CDCl_3$	A 1
3,5-$(CH_3)_2$, 4-NO_2	83.2			$CDCl_3$	A 4
2-NO_2, 4,5-$(CH_3)_2$	96.0-98.7			Acetone-Glyme 198 K	A 18

Footnote to Table 7.11

(a) At 307 K the coupling constant is equal to 78 Hz

(b) Other couplings ^{2}J = -1.85 ^{3}J = -0.5 and ^{5}J = $\pm$0.21 Hz

(c) ^{3}J = 0.7 Hz (CH_3) (d) ^{4}J = 0.7 Hz (CH) (e) ^{4}J = 0.25 Hz (CH_3)

Table 7.11. $^{n}J_{^{15}N-^{1}H}$ Coupling constants of *N*-substituted anilines - Obtained in ^{1}H-NMR

Anilines	^{1}J	^{2}J	Conditions	Ref.
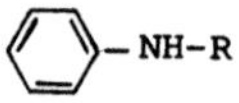				
$(CH_3)_3Si$	76.0		neat	R 3
OH	79.3		DMSO	A 13
trans-4-t-butylcyclohexyl	86.3		DMSO	A 13
CH_3	87.0		DMSO	A 13
	89.0		Acetone-d_6 193 K (a)	B 22
$CH_2C(CH_3)_3$	87.0		DMSO	A 13
cis-4-t-butylcyclohexyl	87.3		DMSO	A 13
CH_2CH_3	87.5		DMSO	A 13
CH_2COPh	88.2		DMSO	A 13
CH_2CN	88.5		DMSO	A 13
CH_2Ph	88.6		DMSO	A 13
NH_2	89.6		DMSO	A 13
NHPh	90.5		DMF	A 13
CH_2CF_3	90.2		DMSO	A 13
$COCH_3$	/		11%M Acetone-d_6 (b)	W 4
	-91.3		9% DMSO	W 4
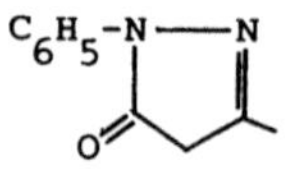	92 ± 1		303 K $CDCl_3$	L 24
	91.5 ± 0.5		323 K $CDCl_3$	L 24
	91.5 ± 0.5		311 K C_5D_5N	L 24
	91 ± 1		373 K C_5D_5N	L 24
	92 ± 0.5		311 K DMSO-d_6	L 24
C_6H_5-N(CH_2-CH_3)(CH_2-CH_3) (1)		0 (1)	neat (c)	B 24
C_6H_5-N(CH_3 (1))(CH_2-C≡CH (2))		0.2 (1) 0.2 (2)	50% CD_2Cl_2 (d)	B 22
C_6H_5-N(CH_3 (1))(C≡C-CH_3)		±0.8 (1)	Acetone-d_6 50% (e)	B 22

Table 7.12. $^{1}J_{^{15}N-^{1}H}$ Coupling constants of anilinium salts - Obtained in ^{1}H-NMR

Anilinium Salts	^{1}J	Conditions	Ref.
$C_6H_5-NH_3^+$	76.9	HSO_3F	A 13
	75.0 ± 0.3	CF_3COOH	B 1
	76.0	H_2SO_4	A 7
p $O_2N-C_6H_4-NH_3^+$	76.0	H_2SO_4	A 7

<u>Table 7.13.</u> $^{n}J_{^{15}N-^{1}H}$ Coupling constants of five-membered nitrogen heterocyclic compounds - Obtained in ^{1}H-NMR. See also K 43 for substituted pyrroles.

Five-Membered Heterocyclic Compounds	^{1}J	^{2}J	^{3}J	Conditions	Ref.
	-96.40	-5.36	-4.55	85% C_6D_6 (a)	J 9
	-96.53	-4.52	-5.39	neat (a)	R 5
	96.56	4.36	5.35	neat	R 8
	-96.48	-4.55	-5.35	neat (b)	B 9 B 10
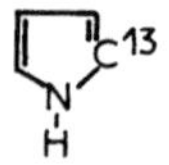	-96.49	-4.49	-5.35	neat (b)	B 9 B 10
	-96.49	-4.55	-5.38	neat (b)	B 9 B 10
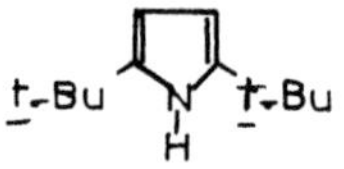	91.5		5.2	CCl_4	G 2
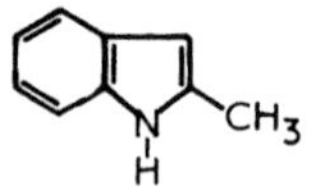	96.5			DMSO-d_6	A 7
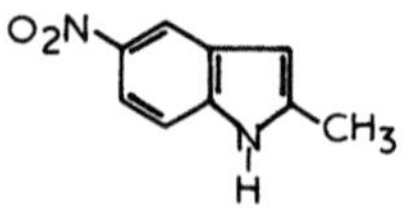	96.9			DMSO-d_6	A 7
	98.0			80% H_2SO_4	A 7
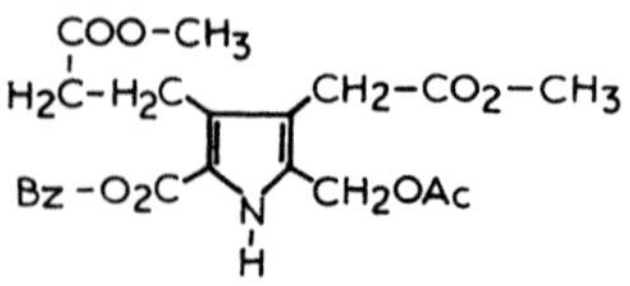	98.0		2.8		B 55
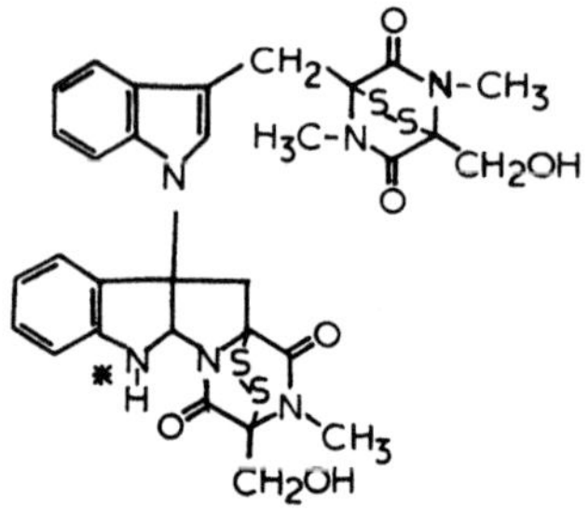	87.8			$CDCl_3$ (a)	B 54

Table 7.13 (continued)

Five-Membered Heteroxyclic Compounds	^1J	^2J	^3J	Conditions	Ref.
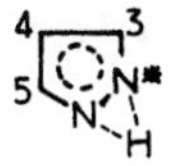		-8.94	-3.37 (4) -4.28 (5)	4% CCl_4	J 3
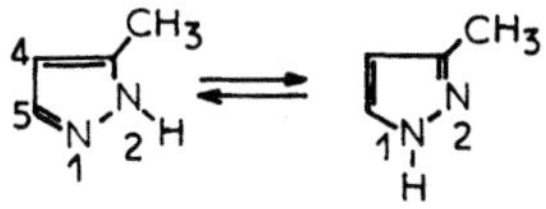		14 (1-5)	6 (1-4)	(a)	S 21
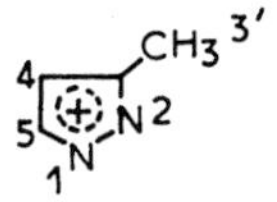		6.6 (1-5)	4.6 (1-4) 3.4 (2-4) 3.4 (2-5) 3.4 (2-3')	(a)	S 21
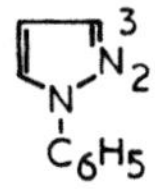		4.4 (1-5) 14.2 (2-3)	7.4 (1-3) 6.0 (1-4) 1.0 (2-4)	0.9M $CDCl_3$	H 4
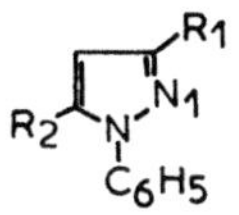				1-1.5M $CDCl_3$	A 15
R_1 / R_2			$J_{N_1-R_1}$		A 15
CH_3 / C_6H_5			2.8		
CH_3 / i C_4H_9			2.9		
i C_4H_9 / CH_3			/		
$C_6H_5-CH_2$ / C_6H_5			2.2		
CH_3 / C_2H_5			2.8		
C_2H_5 / CH_3			2.4		
CH_3 / $C_6H_5-CH_2$			2.9		
$C_6H_5-CH_2$ / CH_3			2.7		
C_2H_5 / $C_6H_5-CH_2$			2.4		
$C_6H_5-CH_2$ / C_2H_5			2.6		
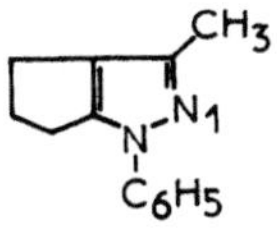			3.0 (1)		A 15
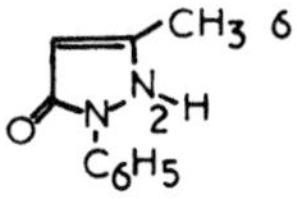			3.5 (2-6)	0.7M $CDCl_3$	H 4

Table 7.13 (continued)

Five-Membered Heterocyclic Compounds	^1J	^2J	^3J	Conditions	Ref.
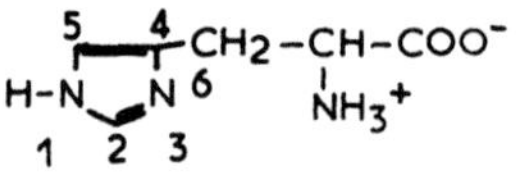		-4.6 (1-2) -4.8 (1-5) -6.1 (3-2)	 -3.0 (3-5)	H_2O pH 1.3	B 35
		-8.2 (1-2) -5.9 (1-5) -10.2 (3-2)	 -1.8 (3-5)	H_2O pH 7.6	B 35
		-8.8 (1-2) -6.6 (1-5) -9.6 (3-2)	 -2.2 (3-5)	H_2O pH 10.9	B 35
			2.6 (3-6)	H_2O pH 6.14	B 35
3-(π)-^{15}N Histidyl- α-Lytic Protease	(c)			H_2O pH 8.3 (a)	B 44
1,3-(τ-π)-^{15}N Histidyl- α-Lytic Protease	(c)			H_2O pH 8.3 (a)	B 44
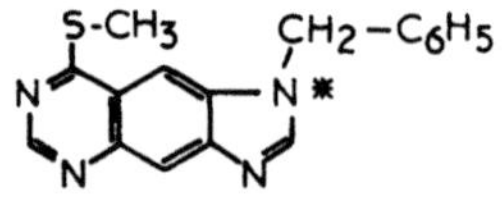		1.2 (CH_2)			W 5
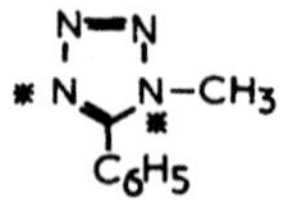		1.85(CH_3)		$CDCl_3$	F 10
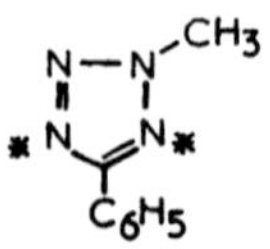			1.7 (CH_3)	$CDCl_3$	F 10
		14.9 14.4 14.25	1.7 1.8 1.8	C_5H_{12} $CDCl_3$ H_2O	K 8 K 8 K 8
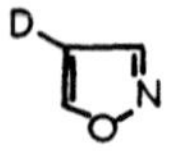		15		DMSO	K 7
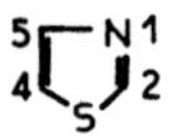		-10.56 (2) -10.60 (5)	-1.97	10mg in 300 1 CS_2	B 33

Table 7.13 (continued)

Five-Membered Heterocyclic Compounds	^1J	^2J	^3J	Conditions	Ref.
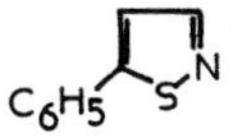		14.2		$CDCl_3$	K 7 K 8
		4.2		H_2SO_4	K 7 K 8
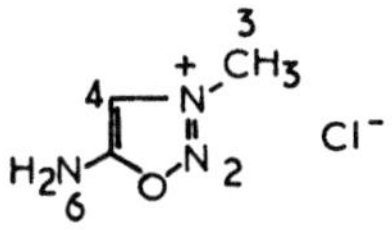					
2-Labelled			2.9 (3') 0 (4)	CD_3OD or CF_3COOH	S 17
3-Labelled		2.3 (3') 4.2 (4)		CD_3OD	S 17
		2.5 (3') 4.3 (4)		CF_3COOH	S 17
6-Labelled	(d)		0 (4)	CD_3OD	S 17
	96.8±0.2		0 (4)	CF_3COOH	S 17
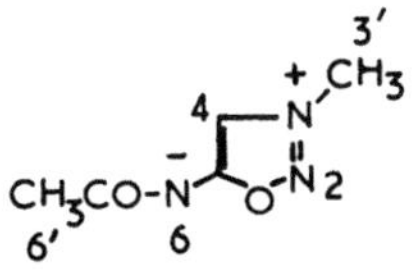		2.1(3-3') 5.5(3-4)	2.6(2-3') 0 (2-4) 1.9(6-6') 0 (6-4)	$CDCl_3$	S 17
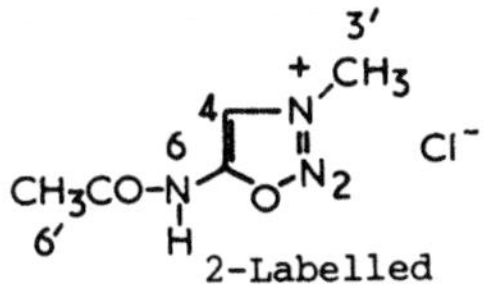					
2-Labelled			3.0 (3') 0 (4)	CD_3OD or CF_3COOH	S 17
3-Labelled		2.4 (3') 4.6 (4)		CD_3OD	S 17
		2.3 (3') 4.8 (4)		CF_3COOH	S 17
6-Labelled			1.4 (6') 0 (4)	CD_3OD	S 17
			1.6 (6') 0 (4)	CF_3COOH	S 17

(a) Measured using ^{15}N spectroscopy (b) Measured using ^1H and ^{15}N spectroscopy

(c) Linewidth = 55 Hz (proton decoupled) (d) Undetectable

= 70 Hz (no proton decoupling)

Table 7.14. ${}^{n}J_{{}^{15}N-{}^{1}H}$ Coupling constants of six-membered nitrogen heterocyclic compounds

Six-Membered Heterocyclic Compounds	^{1}J	^{2}J	^{3}J	^{4}J	Conditions	Ref.
		11			neat (a)	B 57
		-10.76	-1.53	±0.21	neat	L 8
		-10.06	-1.56	±0.18	30% CH_3OH	L 8
		-10.93	-1.48	+0.27	2M Acetone-d_6	B 23
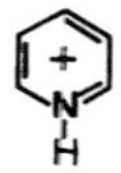		-3.01	-3.98	±0.69	/	L 8
	90.5 ± 1				H_2SO_4	B 24
		-3.01	-3.98	+0.69	2M CD_3OH	B 23
		+0.47	-5.32	+1.11	1M $CDCl_3$	W 8 B 23
		0.35	-5.17	+1.03	0.2M CS_2	W 8
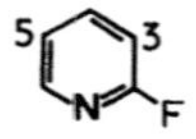		-11.35	-0.69 (3)	+0.69	85% Acetone-d_6 (a)	J 1
			-1.94 (5)			
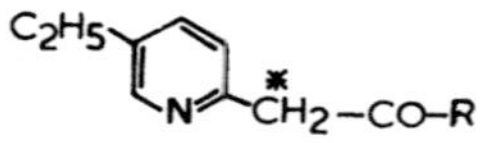 R = C_6H_5			2.18 ± 0.25			K 16
i-C_3H_7			2.77 ± 0.15			K 16

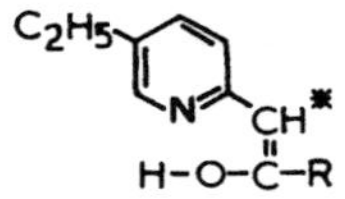	R = C_6H_5	(d)		3.91 ± 0.25		K 16
	i-C_3H_7	(d)		4.27 ± 0.15		K 16
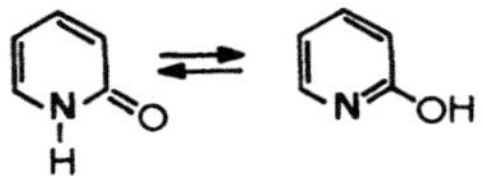		(d)			DMSO	C 22
		(d)			293 K C_5H_5N	C 22
		(d)			Acetone-d_6	C 22
		90			217 K $CDCl_3$	C 22
		90			189 K CH_2Cl_2	C 22
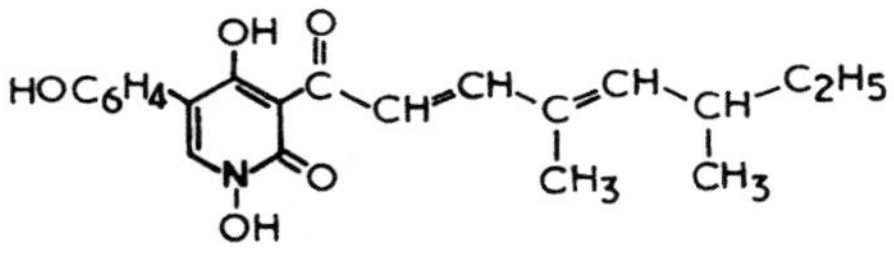			1.0		DMSO-d_6	M 47
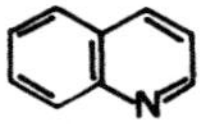			-11.0	-1.33	CCl_4	C 9 C 10
			11.1	1.4	CCl_4	K 8 C 9
			-11.0	-1.33	CH_3CN	C 9 C 10
			-11.1	-1.4	Aprotic solvents	T 2
			11.8 ± 0.5	≃1.5 ± 0.5	(a)	P 3

Table 7.14 (continued)

Six-Membered Heterocyclic Compounds	^1J	^2J	^3J	^4J	Conditions	Ref.
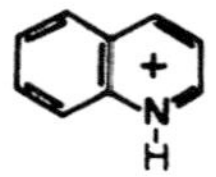		-2.0	-4.5		CF_3COOD - D_2SO_4	C 9 C 10 K 8 T 2
	96.0				HSO_3F	A 7
					D_2O-Dioxane-DCl	
		3.0			pH 2.10	K 22
		10.4			pH 6.30	K 22
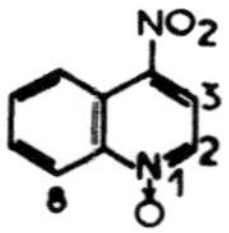		<1 (N_1-2)	5.1 (N_1-3)			
			2.5 (NO_2-3)		Dioxane	K 5
			(c) (N_1-8)			
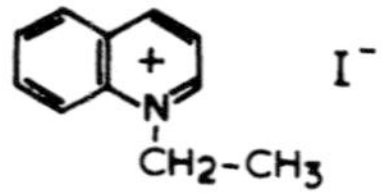		-1.6	-4.3		D_2O	T 2
		/	-4.4		D_2SO_4 or CF_3COOD	T 2
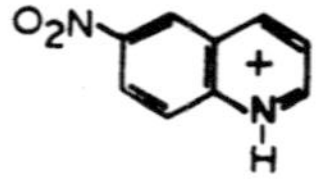	96.5				HSO_3F	A 7
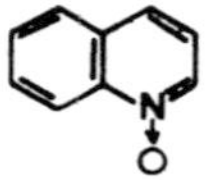		≃0	-5.0		Aprotic solvents	T 2
		+1.2	-6.0		D_2SO_4 CF_3COOD	T 2

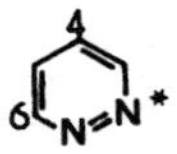		-12.04	-1.12 (4) -3.70 (6)	-0.36	25% CD_2Cl_2	J 3
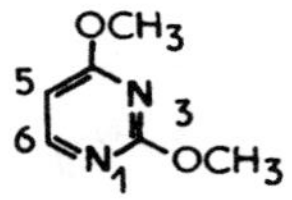		12.2 (1-6)	1.4 (5) 0.6 (5)	0.6 (3-6)	/	R 4
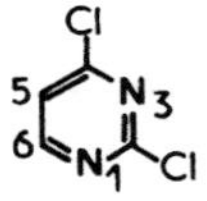		12.5 (1-6)	1.4 (5) 0.8 (5)	0.9 (3-6)	CS_2	R 4
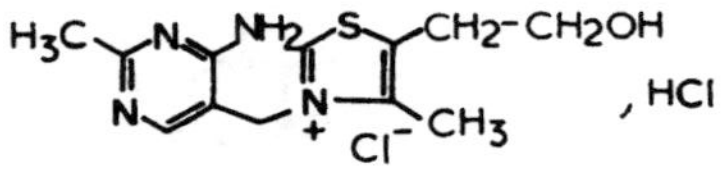	91 (NH_2)				H_2O (a)	C 20
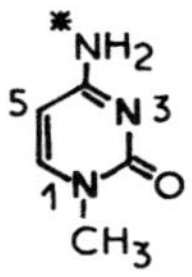	90 90.0				DMF-d_7 DMSO	S 2 B 41
id, HCl	91.4 - 92.6 94.2		0.7 (5)		DMSO-d_6 DMSO	B 41 R 4

Table 7.14 (continued)

Six-Membered Heterocyclic Compounds	^{1}J	^{2}J	^{3}J	^{4}J	Conditions	Ref.
	92 (b)		0.7 (5)		DMSO	M 19
	94.0				243 K SO_2	R 4
	94 (b)				213 K SO_2	M 14
	93.8 - 96.8		0.7 (5)		213 K SO_2	B 41
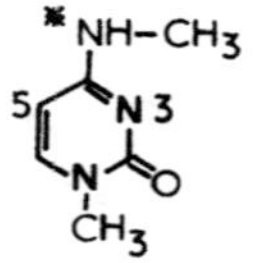	94	1.2 (CH_3)			DMSO-d_6	B 41
id, HCl	94	1.5 (CH_3)			DMSO-d_6	B 41
	93.5	0.7 (CH_3)	1.5 (5)		213 K SO_2	B 41
id, HI	94				DMSO-d_6	B 41
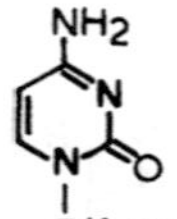 Ribose-3'Phosphate	86.0				0.08M H_2O pH 7	B 36
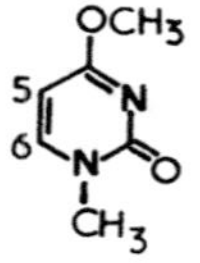		1.2 (1) 1.8 (6)	3.9 (5)		CH_3COOH	R 4

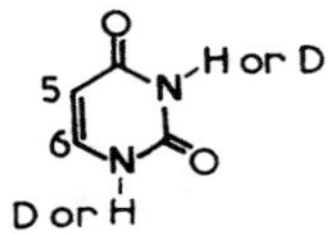	91 (1) 97 (3)	3.5 (1-6)	4.4 (5) 2.5 (5)		DMSO saturated	R 4
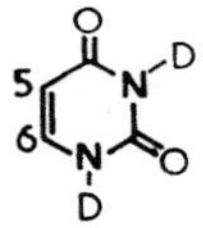		3.3 (1-6)	4.46 (1-5) 2.63 (3-5)	0.25 (3-6)	3mg in 0.5ml D_2O	L 18
mono anionic species		6.54 (1-6)	2.81 (1-5) 1.73 (3-5)	≃0 (3-6)	Idem pD 12	L 18
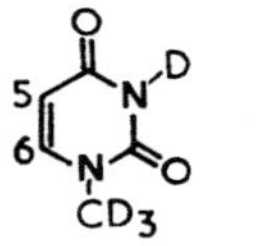		2.48 (1-6)	4.73 (1-5) 2.69 (3-5)	≃0 (3-6)	3mg in 0.5ml D_2O	L 18
mono anionic species		2.32 (1-6)	3.97 (1-5) 0.70 (3-5)	≃0 (3-6)	Idem pD 12	L 18
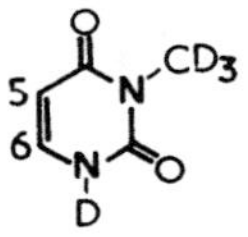		3.36 (1-6)	4.49 (1-5) 2.75 (3-5)	≃0 (3-6)	3mg in 0.5ml D_2O	L 18
mono anionic species		10.46 (1-6)	1.68 (1-5) 2.72 (3-5)	≃0 (3-6)	Idem pD 12	L 18

Table 7.14 (continued)

Six-Membered Heterocyclic Compounds	^1J	^2J	^3J	^4J	Conditions	Ref.
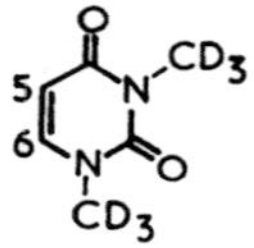		2.54 (1-6)	4.82 (1-5) 2.96 (3-5)	≃0 (3-6)	3mg in 0.5ml D_2O	L 18
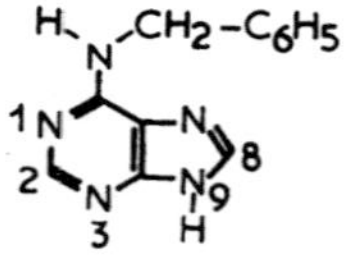		16 (1-2) 9 (9-8)			DMSO-d_6	L 31
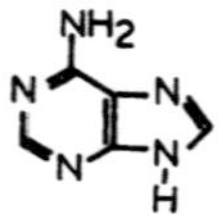		10 (9-8)			DMSO-d_6	L 31
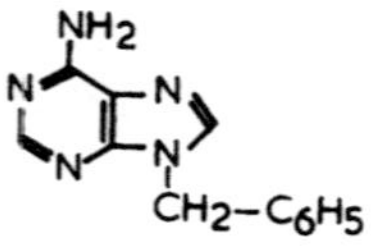		8 (9-8)			DMSO-d_6	L 31
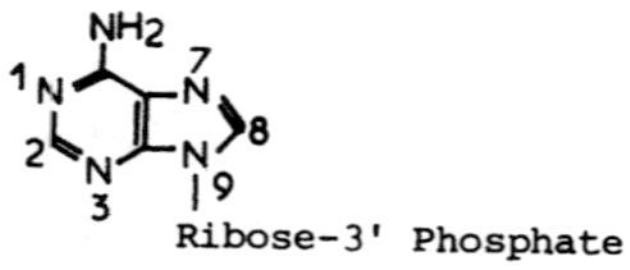		12.5 (1-2) 14.0 (3-2) 10.5 (7-8) 9.5 (9-8)			0.08M in H_2O pH 3	B 36

	88.2	14.0	(1-2)				pH 7	B 36
		15.5	(3-2)					
		10.5	(7-8)					
Adenosine Triphosphate		16	(1-2)			D_2O (a)		H 11
		16	(3-2)					
		10	(7-8)					
		0	(9-)					
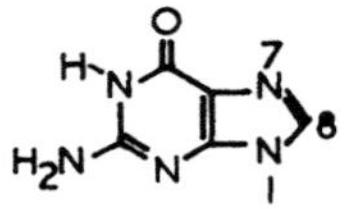 Ribose-3' Phosphate	90.7	10.0	(7-8)			0.08M in H_2O	pH 7	B 36
		10.0	(7-8)				pH 10	
2',3',5'-tri-O-benzoyl uridine [3-^{15}N]	91.3					0.5M in $CDCl_3$		P 6
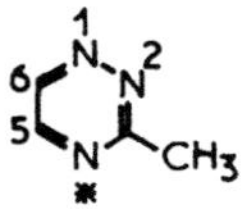		10	(5)	0.2	(6)	CCl_4 or $CDCl_3$		N 7
				2.6	(CH_3)			

- Obtained in ^{1}H-NMR except (a) measured using ^{15}N spectroscopy (b) For each proton of NH_2

(c) A few Hz are indicated for $^3J_{N_1-H_8}$ (d) Not detected

Table 7.15. $^{n}J_{^{15}N-^{1}H}$ Coupling constants of porphyrins

Porphyrins	^{1}J	^{3}J (b)	Conditions		Ref.
Meso tetraphenyl porphyrin	(c)		0.011M $CHCl_3$ + Acetone-d_6 18%	(a)	G 8
Meso tetraphenyl porphyrin, $2H^+$	95.3		0.016M $CHCl_3$ + CF_3COOH excess	(a)	G 8
Octaethyl porphyrin	98	4.7	0.019M $CDCl_3$	(a)	K 14a
	24.5		0.019M $CDCl_3$ 301 K	(a)	K 14b
	98	4.7	0.019M $CDCl_3$ 223 K	(a)	K 14b
	23 $\pm$ 3		0.019M C_5D_5N 343 K	(a)	K 14b
Octaethyl porphyrin (CH_3)	100		$CDCl_3$	(a)	K 14b
Octaethyl porphyrin, $2H^+$	93		CF_3COOH	(a)	K 14b
Octaethyl porphyrin (CH_3), $2H^+$	92		CF_3COOH	(a)	K 14b
Octaethyl porphyrin $(CH_3)_2$, H^+	48		$CDCl_3$	(a)	K 14b
1,3,5,7-tetraethyl 2,4,6,8-tetramethyl porphyrin	97.5	4.55 $^{4}J_{CH_3-N}$ 0.8	$CDCl_3$ 307 K		M 1
Etioporphyrin, 2HCl	92.5	4.75 $^{5}J_{CH_3-N}$ 1.5	$CDCl_3$		M 1
Protoporphyrin IX methyl ester	103 $\pm$ 4		10-30mg/cm^3 $CDCl_3$	(a)	I 5
	90 $\pm$ 3		CF_3COOH	(a)	I 5
CoproporphyrinIII methyl ester	100 $\pm$ 3		10-30mg/cm^3 $CDCl_3$	(a)	I 5
	85 $\pm$ 3		CF_3COOH	(a)	I 5
Pheophytin a	98	3	0.22M $CDCl_3$		B 32

- Obtained in ^{1}H-NMR except (a) measured using ^{15}N spectroscopy

(b) Each meso proton has a coupling ^{3}J with two nitrogen atoms

(c) Not detected because of exchange

Table 7.16. ${}^{n}J_{{}^{15}N-{}^{1}H}$ Coupling constants of amides

Amides	^1J	^2J	^3J	Conditions	Ref.
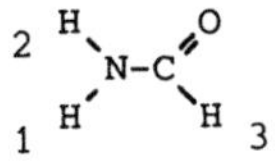	89			neat (a)	B 57
	90.7 (1)	14.6		neat	M 4
	88.3 (2)				
	92.0 (1)	19.0		neat	S 4
	88.0 (2)				
	91.9 (1)	23.3		8%M H_2O	S 4
	95.4 (2)				
	-91.30(1)	-14.47		15% H_2O	C 8
	-86.93(2)				
	-91.3 (1)	-14.4		H_2O	R 10
	-86.9 (2)				
	91.0 (1)	16.4		10%M Acetone	S 4
	88.0 (2)				
	89.7 (1)	16.4		29.6% Acetone	B 26
	86.4 (2)				
	90.0 (1)	15.5		9.4% $CH_3COC_3H_7$	D 6
	87.7 (2)				
	89.3 (1)	15.6		14%M Diglyme	D 6
	87.1 (2)				
	90.1 (1)	16.3		0.1M CH_3CN	G 1
	87.5 (2)				
	89.0 (1)	15.9		0.1M CH_3CN +	G 1
	86.6 (2)			0.5M $(Bu)_4N^+I^-$	
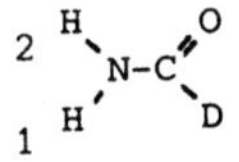	91.6 (1)			neat	S 4
	88.0 (2)				
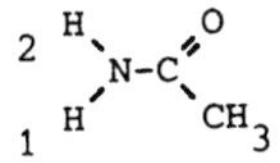	89 ± 2		1.3	H_2O	B 24
	90.9 (1)		1.3	1M H_2O	L 7
	88.4 (2)				
	89 (1)			1M DMSO	L 7
	87 (2)				
	-88.5 (1)		-1	0.6M DMSO	D 9
	-87.2 (2)				
$H_2\overset{*}{N}$-CO-$(CH_2)_2$-CH(NH_2)-COOH	89.7			H_2O (a)	B 39

Table 7.16 (continued)

Amides	^1J	^2J	^3J	Conditions	Ref.
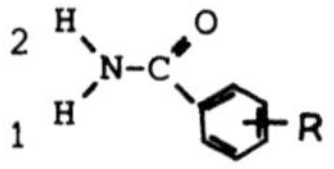					
H	89 (1)				L 29b
	89 (2)				
2-F	-90.5 (1)			5% $CDCl_3$	F 2
	-89.0 (2)				
	-89.8 (1)			5% DMSO-d_6	F 2
	-88.5 (2)				
2-Br	90.0 (1)			DMSO	A 13 A 18
	88.2 (2)				
2-CF_3	89.7 (1)			DMSO	A 13 A 18
	88.7 (2)				
2-NO_2	89.6 (1)			DMSO	A 13 A 18
	88.4 (2)				
3-Br	88.8 (1)			DMSO	A 13 A 18
	88.2 (2)				
3-CF_3	89.1 (1)			DMSO	A 13 A 18
	88.7 (2)				
4-F	88.6 (1)			DMSO	A 13 A 18
	88.2 (2)				
4-I	89.2 (1)			DMSO	A 13 A 18
	88.1 (2)				
3,5-$(CH_3)_2$	88.8 (1)			DMSO	A 13 A 18
	88.0 (2)				

- Obtained in ^{1}H-NMR except (a) measured using ^{15}N spectroscopy

Table 7.17. $^{n}J_{^{15}N-^{1}H}$ Coupling constants of N-substituted amides

N-substituted amides	^{1}J	^{2}J	^{3}J	Conditions	Ref.
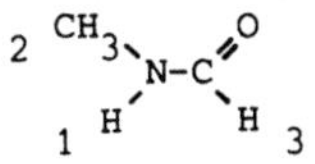	92.6	1.2 (2)		neat	B 26
		15.6 (3)			
	93.9	1.4 (2)		49% H_2O	B 26
		14.4 (3)			
CH_3-NH-CO-CH_3	93.0			neat	M 4
	93			H_2O (a)	Y 4
	93			DMSO (a)	Y 4
CH_3-NH-CO-C_2H_5	92.9			neat	M 4
CH_3-NH-CO-i.C_3H_7	92.7			neat	M 4
CH_3-NH-CO-t.C_4H_9	92.3			CCl_4	M 4
C_4H_9-NH-CHO (2 1 3)	Z (-)92.2	(-)15 (3)		neat	R 6
	E (-)89.8	(-)14.3 (3)		neat	R 6
C_4H_9-NH-CO-CH_3	Z (-)92.0			neat	R 6
C_6H_5-NH-CHO (2 1 3)	Z 91.2	16.3 (3)		38.6%M $CDCl_3$	B 27
	E 88.0	15.0 (3)		38.6%M $CDCl_3$	B 27
C_6H_5-NH-CO-CH_3	90.9		0.7	Acetone	B 24
	-91.3			9%M DMSO	W 4
	89.9			$CDCl_3$	B 25
HO-C(=O)-CH_2-NH-CO-CH_3	94.5			DMSO	B 24
CH_3O-C(=O)-CH_2-NH-CO-CH_3	93.5			Acetone	M 4
CH_3O-C(=O)-CH(C_2H_5)-NH-CO-CH_3	92.6			Acetone	M 4
CH_3O-C(=O)-CH(i.C_3H_7)-NH-CO-CH_3	92.9			Acetone	M 4

Table 7.17 (continued)

N-substituted amides		^1J	^2J	^3J	Conditions	Ref.
OHC-NH-CO-CH_3		-90.2	-13.4		8% CF_3COOH	K 9
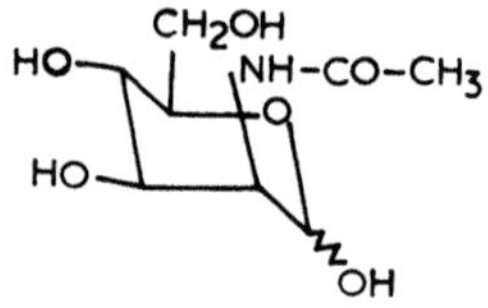		93 93 (b)			9 : 1 (a) H_2O : DMSO-d_6	C 19 C 19
CH_3-CO-NH-CH_2-R 1 2	(c)	91.3	0.5 (1)	1.5 (2) 1.2 (CH_3)	C_5D_5N	C 6
CH_3-CO-ND-CH_2-R 1 1' 2	(c)		0.7 (1) 0.6(1')	1.7 (2) 1.3 (CH_3)	C_5D_5N : D_2O 5 : 2	C 6
CF_3-CO-NH-CH_2-R 2	(c)	92.3 92.5		1.4 (2)	C_5D_5N $CDCl_3$	C 6
CF_3-CO-ND-CH_2-R 1 1' 2	(c)		0.8 (1) 0.6(1')	1.3 (2)	C_5D_5N : D_2O 9 : 1	C 6

- Obtained in ^1H-NMR except (a) measured using ^{15}N spectroscopy

(b) Two doublets due to anomers α and β

(c) R =

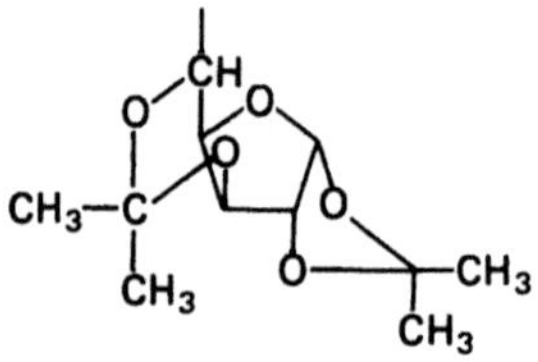

(d) For neat dimethylformamide, the coupling constants $^2J_{N-H}$ have been measured (B 4)

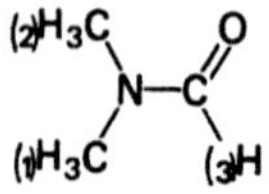

$^2J_{N-H}$ = 1.1 Hz(1), 1.2 Hz(2) and 15.6 Hz(3)

(e) The coupling constant $^1J_{N-H}$ has been measured for two polyamides in

CF_3COOH (K13) nylon-3 ^1J = 93.5 ± 0.5 Hz

nylon-4 ^1J = 94.0 ± 0.5 Hz

Table 7.18. $^{n}J_{^{15}N-^{1}H}$ Coupling constants of lactams and imides

Lactams		^{1}J	^{2}J	^{3}J	Conditions	Ref.
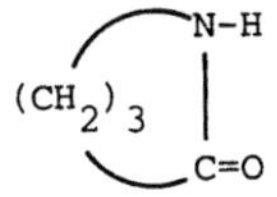		92.5 ± 0.5			CF_3COOH (a)	K 13
		93			H_2O (a)	Y 4
		93			DMSO (a)	Y 4
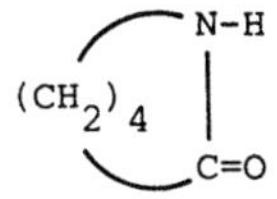		91			H_2O (a)	Y 4
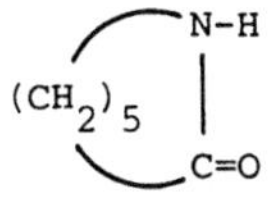		91			7.7M H_2O (a)	B 57
		92.5 ± 0.5			CF_3COOH (a)	K 13
		90			H_2O (a)	Y 4
		90			DMSO (a)	Y 4
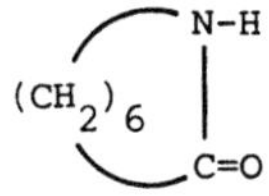		89			H_2O (a)	Y 4
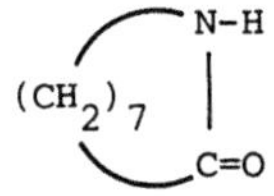	cis	89			2.5M DMSO (a)	Y 3 Y 4
	trans	91				
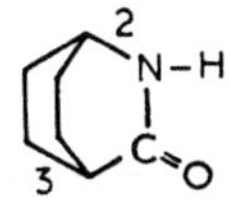		92.8	1.5 (2)	1.3 (3)	0.1M D_2O	K 1
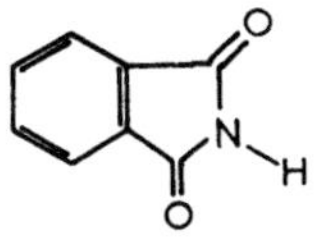		93.0 ± 0.8			DMSO	B 24
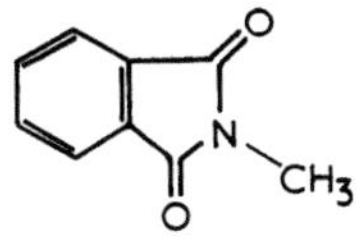			1.4			B 24
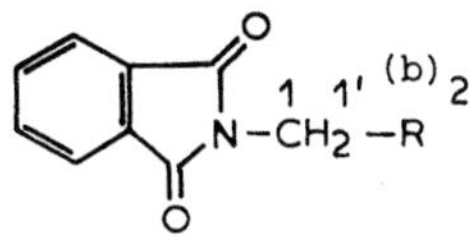			1.0 (1)	<0.9 (2)	C_6D_6	C 6
			1.1 (1')			

- Obtained in ^{1}H-NMR except (a) measured using ^{15}N spectroscopy

(b) R = sugar residue, see (c) in the caption of Table 5.17

Table 7.19. $^{n}J_{^{15}N-^{1}H}$ Coupling constants of ureas

Ureas	^1J	^2J	^3J	Conditions	Ref.
NH_2-CO-NH_2	89 ± 1			DMSO	B 24
	90.3			0.1M D_2O (a)	L 6
	86.8		1.7	1:1:1 20mg/mole Acetone:DMSO:TMU	S 6
NH_2-CO-NH-CH_3	87.9 (1)			H_2O - C_2H_5OH (a)	M 9
1 2	91.6 (2)			(a)	M 9
CH_3-NH-CO-NH-CH_3	88.7	0.7	0.1	40mg/mole DMSO-d_6 (c)	S 6
$(CH_3)_2$N-CO-NH-CH_3 1 2	91.3 (2)			H_2O - C_2H_5OH (a)	M 9
C_6H_5-NH-CO-NH-C_6H_5	89.7 ± 0.2			DMSO-d_6	B 25
NH_2-CO-NH-CO-NH_2 1 2	88.4 ± 1 (1)			DMSO (b)	B 24

- Obtained in ^{1}H-NMR except (a) measured using ^{15}N spectroscopy
(b) From the ^{15}N satellites of the proton spectrum
(c) Other coupling measured ^4J < 0.05 Hz

Footnote to Tabe 7.20

- Obtained in ^{1}H-NMR except (a) measured using ^{15}N spectroscopy
(b) Signal at low frequency

Table 7.20. $^{n}J_{^{15}N-^{1}H}$ Coupling constants of complexed and protonated amides

Amides Salts	^{1}J	^{2}J	^{3}J	Conditions	Ref.
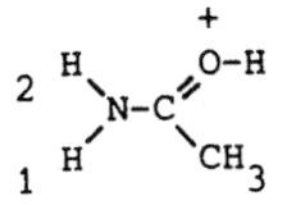	96.6 (1)		1.8	1M HSO_3F	L 13
	93.2 (2)				
	96.5			HSO_3F	L 29a
	92.5				
	96.0 (1)		1.3	1M H_2SO_4 100%	L 13
	92.8 (2)				
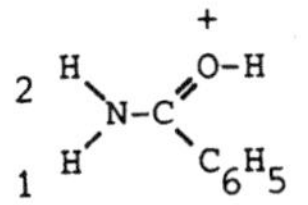	95.5 (1)			HSO_3F	L 29b
	94.3 (2)				
	98 (1)			H_2SO_4 100%	L 29b
	95 (2)				
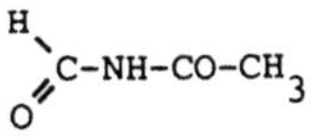	-90.2	-13.4		8% CF_3COOH	K 9
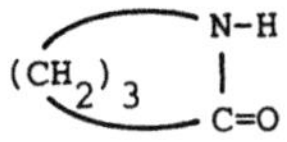	92.5 ± 0.5			CF_3COOH (a)	K 13
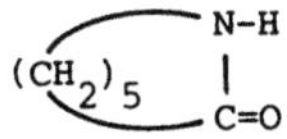	92.5 ± 0.5			CF_3COOH (a)	K 13
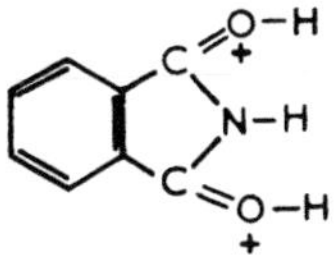	105			$HF-SbF_5-SO_2$ 193 K (a)	O 6
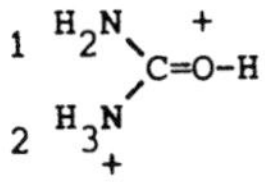	96.8 (1)			193 K HSO_3F-SbF_5	O 5
	76.6 (2)				
$CH_3-NH-C(=O\cdot SbCl_5)-NH-CH_3$	92.0 ± 0.5 (b)			0.4M CD_2Cl_2	S 18
	96.5 ± 0.5			253 K	
$CH_3-NH-C(=O\cdot BF_3)-NH-CH_3$	92.0 ± 1.5 (b)			0.25M CD_2Cl_2	S 18
	92.0 ± 1.5			238 K	
Nylon-3	93.5 ± 0.5			CF_3COOH	K 13
Nylon-4	94.0 ± 0.5			CF_3COOH	K 13

Table 7.21. $^{n}J_{^{15}N-^{1}H}$ Coupling constants of peptides

Peptides	^{1}J	^{2}J	$^{3}J_{1}(a)$	$^{3}J_{2}(a)$	Conditions	Ref.
Ac-Gly^{15}N			(-)1.5			K 11
	94.5				DMSO	B 24
Ac-Gly-Me ester	93.5				Acetone	M 4
Ac-DL-Ala^{15}N	93.30			3.04	H_2O pH 6	S 5
Ac-Ala(^{15}N)-NH-CH_3	93.4	1.1		1.2 2.9	H_2O	K 1
Ac-Val-Me ester	92.9				Acetone	M 4
Ac-L-Phe(^{15}N)-NH-CH_3	91.9	<1		1.8 3.6	DMSO-d_6	G 7
	91.8	1.3		2.0 3.2	CD_3OH	G 7
Gly-Gly^{15}N	93	1.1	<1		H_2O pH 1.5	S 8
Gly^{15}N-Gly^{15}N	94				1M H_2O pH 6	I 2 I 4
L-Phe-Gly^{15}N	93.15				H_2O pH 2	S 5
	93.1				H_2O pH 5.3	K 15
L-Val-Gly^{15}N	94.45		<1		H_2O pH 2	S 5
L-Ala-L-Ala^{15}N	93.24		1.20	3.05	H_2O pH 2	S 5
L-Ala-L-Phe^{15}N	92.15		<1	2.30 2.40	H_2O pH 2	S 5
Gly-Leu-Met-NH_2 I Tyr-Phe(^{15}N)-Lys	92				DMSO-d_6	B 16
cyclo(L-Phe-Gly^{15}N)	90.50		<1.0		DMSO	S 5
cyclo(L-Val-Gly^{15}N)	91.07		1.02		H_2O pH 2	S 5
	90.00		<1		DMSO	S 5
cyclo(L-Ala-L-Ala^{15}N)	91.13				H_2O pH 2	S 5
cyclo(L-Ala-L-Phe^{15}N)	89.30		1.85	1.90 3.20	DMSO	S 5

Table 7.21 (continued)

Peptides		^1J	^2J	$^3J_1$(a)	$^3J_2$(a)	Conditions	Ref.
Oxytocin(L-Tyr^{2} ^{15}N)		93				H_2O pH 3	B 38
(Tyr^{2} ^{15}N)		94				H_2O pH 4	L 22
(L-Ile^{3} ^{15}N)		93				H_2O pH 3	B 38
(Asn^{5} ^{15}N)		92				H_2O pH 4	L 22
(Cys^{6} ^{15}N)		93				H_2O pH 4	L 22
(Leu^{8} ^{15}N)		93				H_2O pH 4	L 22
(Gly^{9} ^{15}N)		95				H_2O pH 4	L 22
Valinomycin(L-Val^{15}N)		93.55		1.20		CCl_4	S 5
		94				CCl_4	B 16
(^{15}N-Gly)hemoglobin (CO-Hb-Gly^{15}N)		95.2				pH 5 Phosphate Buffer	L 53 L 19
(β-Ala-Sar-Gly)$_n$ (a')	Gly	93				H_2O	K 25
	β-Ala	93					
(β-Ala-Sar-D,L-Ala)$_n$ (a')	β-Ala	93				H_2O	K 25
	D,L-Ala	93					

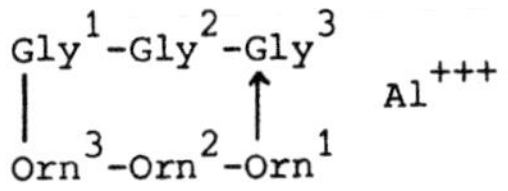

Orn = NH-CH(α)-CO-
CH_2-CH_2-CH_2-N(OH)-CO-CH_3
β_2 γ_2 δ_2
β_3 γ_3 δ_3

Alumichrome (b)

	^1J	^2J			^3J		Ref.
Orn^1	90 ± 0.1			1.5 ± 0.2			D 8 L 16
				α 0.1 ± 0.2 (c)			
		δ_2	0.2 ± 0.2		β_2	0.5 ± 0.5	
		δ_3	1.3 ± 0.4		β_3	5.0 ± 0.5	
					γ_2	5.4 ± 0.4	
					γ_3	(d)	
					CH_3	2.0 ± 0.2	

Table 7.21 (continued)

Peptides	^1J	^2J	^3J		Ref.
Orn[2]	91.3 ± 0.2		α 1.0 ± 0.2 (c)		D 8 L 16
			α 1.0 ± 0.2 (c)		
		δ_2 0.2 ± 0.2		β_2 0.2 ± 0.2	
		δ_3 1.6 ± 0.4		β_3 5.8 ± 0.5	
				γ_2 0.5 ± 0.5	
				γ_3 (d)	
				CH_3 2.0 ± 0.2	
Orn[3]	93.2 ± 0.3		α 1.7 ± 0.2 (c)		D 8 L 16
			α 0.5 ± 0.2 (c)		
		δ_2 1.0 ± 0.4		β_2 2.5 ± 0.4	
		δ_3 (d)		β_3 0.4 ± 0.2	
				γ_2 5.8 ± 0.4	
				γ_3 0.3 ± 0.2	
				CH_3 2.0 ± 0.2	
Gly[1]	94.1 ± 0.3		α_2 0.2 ± 0.2 (c)		D 8 L 16
			α_2 0.2 ± 0.2 (c)		
			α_3 1.1 ± 0.2 (c)		
			α_3 1.1 ± 0.2 (c)		
Gly[2]	94.8 ± 0.3		α_2 1.4 ± 0.2 (c)		D 8 L 16
			α_2 0.0 ± 0.2 (c)		
			α_3 0.9 ± 0.2 (c)		
			α_3 0.9 ± 0.2 (c)		
Gly[3]	92.5 ± 0.2		α_2 1.1 ± 0.2 (c)		D 8 L 16
			α_2 0.0 ± 0.2 (c)		
			α_3 1.1 ± 0.2 (c)		
			α_3 0.4 ± 0.2 (c)		

(a) $^3J_1$ and $^3J_2$ refer to the structure $-\overset{1}{C}H-CO-NH-CH-\overset{2}{C}H_2-$

(a') Sar denotes $N(CH_3)-CH_2-CO-$. No differences in J_{N-H} between cis and trans forms were detected

(b) Obtained with 0.07 M solutions in DMSO-d_6 at 336 K
α_2 and α_3 refer to the two protons of the glycyl residue

(c) The values have not been assigned to particular N atoms

(d) Not observed

Table 7.22. ${}^{n}J_{{}^{15}N-{}^{1}H}$ Coupling constants of thioamides and thioureas

Thioamides and Thioureas	^{1}J	^{2}J	^{3}J	Conditions	Ref.
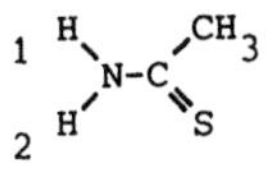	91 (1) 94 (2)		2	$CDCl_3$ saturated	W 6
	91 (1) 92 (2)		2	$CDCl_3$: CD_3NO_2 1 : 1	W 20
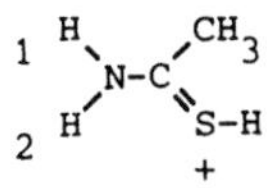	94.6 (1) 94.6 (2)		2.9 (CH_3) 2.8 (SH) (b)	$CDCl_3$: CD_3NO_2 1 : 1 + CF_3SO_3H	W 20
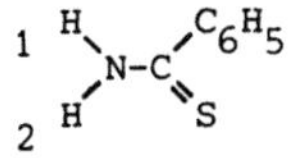	90.5 (1) 93.0 (2)			$CDCl_3$: CD_3NO_2 1 : 1	W 20
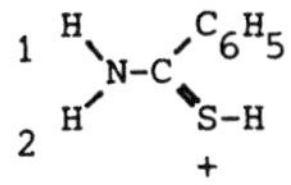	95.5 (1) 94.5 (2)		2.0 (SH)(b)	$CDCl_3$: CD_3NO_2 1 : 1 + CF_3SO_3H	W 20
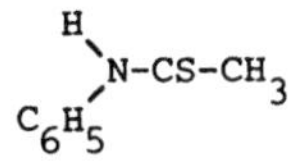 Z E	90 92		3 (CH_3) 2 (CH_3)	$CDCl_3$ saturated	W 6
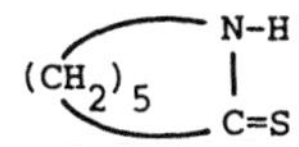	92			DMSO 0.03M in 20 cc (a)	Y 6
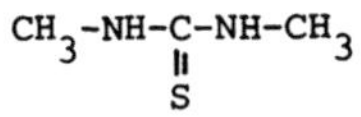	94.8			C_2H_5OH (a)	M 9
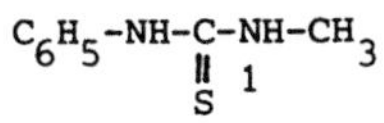	91.2 (1)	0.9 (1)		C_2H_5OH	B 24
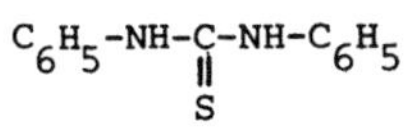	89.8 ± 0.2			DMSO-d_6	B 25
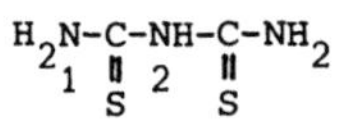	92 ± 1.8 (1) 105.0 (2)			DMSO saturated (a)	W 14

- Obtained in ^{1}H-NMR except (a) measured using ^{15}N spectroscopy

(b) $^{3}J_{N-SH}$ is observable on cooling to 258 K

Table 7.23. ${}^{n}J_{{}^{15}N-{}^{1}H}$ Coupling constants of imines

Imines		^{1}J	^{2}J	^{3}J	Conditions	Ref.
$(C_6H_5)_2$-C=NH		51.2 ± 0.4			pentane	B 24
		51.2			233 K 5% pentane	L 9
sec.Bu(C$_6$H$_5$)C=NH		50.6			213 K 5% pentane	L 9
		50.9				
C_6H_5-CH=N-CH_3 2 1			3.9 (2)		neat	B 24
			0.6 (1)			
	E		3.7 (2)		233 K CD_3OD	Y 5
			0.8 (1)			
	Z		-10.0 (2)		233 K CD_3OD	Y 5
			0.7 (1)			
Naphthyl-CH=N-CH_3	E		3.6 (2)		233 K CD_3OD	Y 5
			0.7 (1)			
	Z		-10.4 (2)		233 K CD_3OD	Y 5
			0.6 (1)			
9-Anthryl-CH=N-CH_3 2 1	E		3.4 (2)		333 K CD_3OD	Y 5
			0.8 (1)			
	Z		-9.8 (2)		233 K CD_3OD	Y 5
			0.6 (1)			
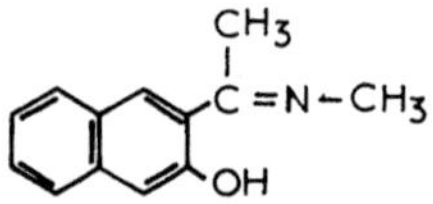			<1	1.9	$CDCl_3$	D 3
9-Anthryl-CH=N-t.Bu	E		3.8	2.0	283 K C_6D_6	Y 5
	Z		-10.0	2.2	283 K C_6D_6	Y 5
C_6H_5-CH=N-C_6H_5			3.8 ± 1		$CDCl_3$	B 25
$(CH_3)_2$-C=N-C_6H_5				1.8 (b)	(a)	A 13
				3.4 (c)	(a)	A 13
CH_3(t.Bu)C=N-C_6H_5	E			1.8	(a)	A 13
C_6H_5-C(C_2H_5)=N-C_6H_5	E			1.20	C_6D_6	A 2
$(C_6H_5-CH_2)_2$-C=N-C_6H_5				1.15(b) 3.45(c)	C_6D_6	A 2
C_6H_5-CH_2-C(C_6H_5)=N-C_6H_5				1.37	C_6D_6	A 2

Table 7.23 (continued)

Imines	^1J	^2J	^3J	Conditions	Ref.
$(CH_3)_2N$-CH=N-C_6H_5		2.4±0.1		CCl_4	B 25
$(CH_3)_2N$-C(CH_3)=N-C_6H_5			1.1±0.1	CCl_4	B 25

- Obtained in ^{1}H-NMR except (a) measured using ^{15}N spectroscopy

(b) Coupling with CH_2 or CH_3 cis with respect to C_6H_5

(c) Coupling with CH_2 or CH_3 trans with respect to C_6H_5

Values for salicylidene-aniline in $CDCl_3$ are also given (K 42)

Footnote to Table 7.24

- Obtained in ^{1}H-NMR except (a) measured using ^{15}N spectroscopy

(b) Refers to protons Z or E with respect to the lone pair

Table 7.24. ${}^{n}J_{{}^{15}N-{}^{1}H}$ Coupling constants of oximes

Oximes		^2J		^3J		Conditions	Ref.
$R_1 \searrow C=N-OH$, $R_2 \nearrow$							
R_1	R_2	Z	E	Z	E		
H	H	-13.88 (b)	2.68			H_2O	C 10
H	CH_3	-15.93	2.93	-4.58	-4.16	D_2O	C 10
				-2.58	-4.16	0.1M $CDCl_3$	B 27
H	C_2H_5	16.20	2.95			CH_2Cl_2	C 9
		16.25	2.95			CH_2Cl_2	K 7
		15.45	3.0			Pentane	K 7
		14.64	3.13			CH_3COOH	C 9
H	$C_6H_5-CH_2$	15.8	2.9			$CDCl_3$	K 7
		16.0	2.9			DMSO	K 7
H	t.C_4H_9		2.8			CH_2Cl_2	K 7 C 9
H	C_6H_5		2.6			$CDCl_3$	K 7
H	mesityl		2.4±0.2			DMSO (a)	C 14
H	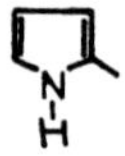	15.2				Acetone	C 2
H		14.8	2-3			Ether	C 2
H		14.1	2.35			CH_3OH-D_2O	C 2
H	2-pyridyl		2.7			CH_3OH	C 2
H	3-pyridyl		3.2			CH_3OH	C 2
H	4-pyridyl		2.7			CH_3OH	C 2
H	3-indolyl	15.6				DMSO	C 2
H	I^- 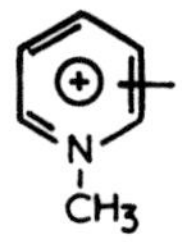 2	13.8	2.75			$DMSO-D_2O$	C 2
H	3	14.5	2.5			D_2O	C 2
H	4	14.5	2.8			D_2O	C 2

Table 7.24 (continued)

Oximes		^2J		^3J		Conditions	Ref.
H	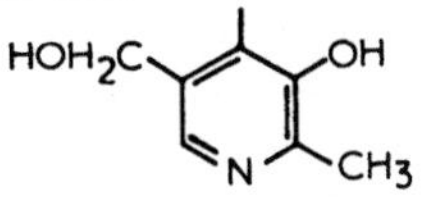		3.2			D_2O	C 2
H	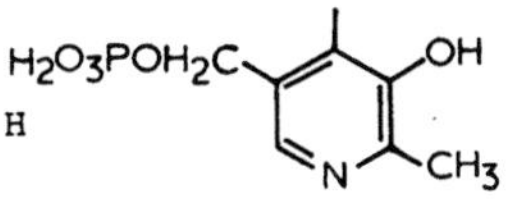		2.7			D_2O	C 2
CH_3	CH_3			-4.0 (b)	-2.2	H_2O	C 10
				-4.0 (b)	-1.8	C_6H_6	C 10
				-3.7 (b)	-2.0	CS_2	C 10
				-4.5 (b)	-3.2	CF_3COOD	C 10
CH_3	t-C_4H_9				-1.8	C_6H_6	C 10
					-1.9	CS_2	C 10
					-3.2	HCOOH	C 10
					-3.3	CF_3COOD	C 10
					-3.4	D_2SO_4	C 10
CH_3	C_6H_5			-4.20	-2.0	0.1M $CDCl_3$	B 17
			1.77	4.12	1.90	DMSO-d_6	O 1
CH_3	2-quinoxalinyl				1.84	DMSO-d_6	O 1
C_2H_5	C_6H_5		1.86		1.29	DMSO-d_6	O 1
				3.72		$CDCl_3$	O 1
t.Bu-CH_2	C_6H_5	1.75	1.92		1.38	DMSO-d_6	O 1
				4.60		$CDCl_3$	O 1
$(CH_3)_2N$-CH_2	C_6H_5			1.23	3.71	DMSO-d_6	O 1
CH_3O-CH_2	C_6H_5	2.01	1.63	1.39	3.75	DMSO-d_6	O 1
Cl-CH_2	C_6H_5			1.22		DMSO-d_6	O 1
					4.10	$CDCl_3$	O 1
		1.93				Ether	O 1
Br-CH_2	C_6H_5			1.19		DMSO-d_6	O 1
					4.33	$CDCl_3$	O 1
		1.88				Ether	O 1
I-CH_2	C_6H_5	1.91		1.15		DMSO-d_6	O 1
$(C_2H_5)_2N$-CH_2	2-quinoxalinyl			1.57	3.34	DMSO-d_6	O 1
Cl-CH_2	2-quinoxalinyl			1.07		DMSO-d_6	O 1
Mesityl 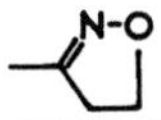					2.3 ± 0.2	CH_2Cl_2 (a)	C 14

Table 7.25. $^{n}J_{^{15}N-^{1}H}$ Coupling constants of miscellaneous compounds with C=N bond

Miscellaneous C=N- Derivatives		^{2}J	^{3}J	Conditions	Ref.
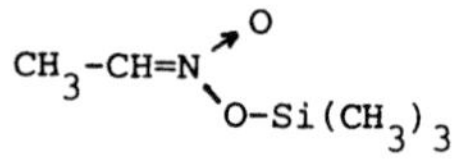	(E)	0	1.5	/	I 6
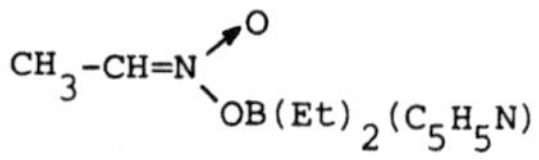	(E)	0	4.4	/	I 6
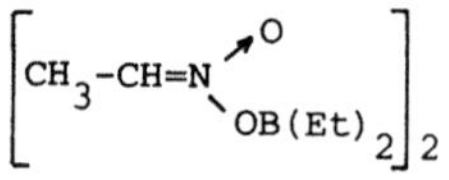	(Z)	2.9	4.8	/	I 6
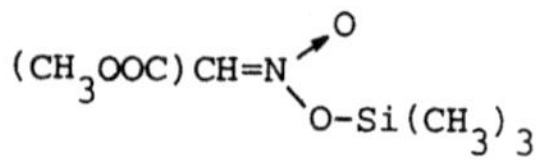	(E)	0	/	/	I 6
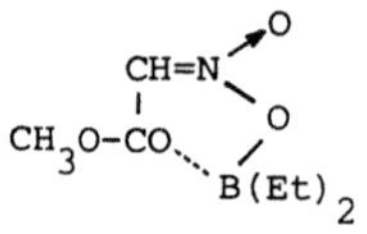	(Z)	1.3	/	/	I 6
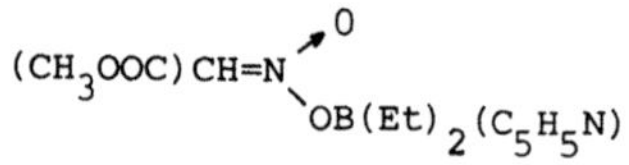	(E)	0	/	/	I 6
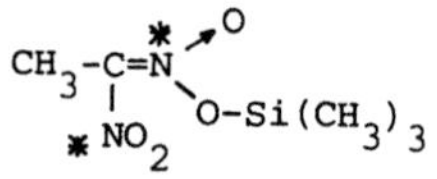			2.1 and 2.3	/	I 6
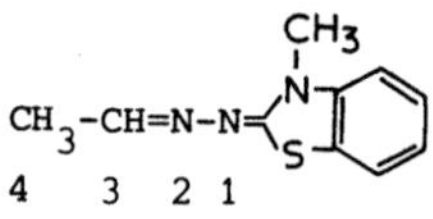	(E)		5.4 (1-3) 3.7 (2-4)	(a)	N 4
	(Z)		10 (1-3) 2.3 (2-4)	(a)	N 4
CH_3-N=C=S		3.3		neat	B 24
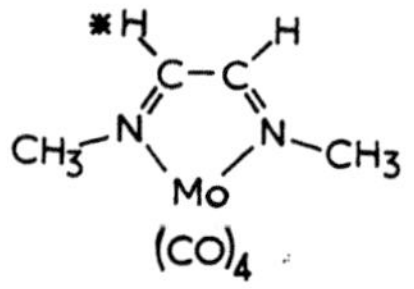		(b)		$CDCl_3$	S 13

- Obtained in ^{1}H-NMR except (a) measured using ^{15}N spectroscopy

(b) Not detected

Table 7.26. $^{n}J_{^{15}N-^{1}H}$ Coupling constants of compounds with N-C=N bonds

Miscellaneous N-C=N- Derivatives	^{1}J	^{2}J	^{3}J	Conditions	Ref.
$(CH_3)_2$-N*-CH*=N-C_6H_5		8.4 ± 0.1	(c)	CCl_4 (a)	B 25
$(CH_3)_2$-N*-CH*=N-$C_6H_4(OCH_3)_p$		8.3 ± 0.1	(c)	CCl_4 (a)	B 25
$(CH_3)_2$-N*-CH*=N-$C_6H_4(NO_2)_p$		7.5 ± 0.1	(c)	CCl_4 (a)	B 25
$H_2N-C_6H_4-SO_2-N=C(NH_2)_2$	85.0 (1) 91.0 (2)			8gr in 18ml DMSO	S 11
$CH_3O-C(=O)-CH(NH_3^+)-(CH_2)_3-N^{1}H-C(=N-NO_2)-{}^{2}NH_2$ Cl^-	97.0 (1) 94.0 (2)			58% H_2O	G 9
$^-OOC-CH(H_2N)-(CH_2)_3-NH-C(^{2}NH_2^+)(^{1}NH_2)$	91.7 (1)			H_2O pH 9.9	L 6
			4 (d)	H_2O pH 4.6	L 6
	92 (2)			H_2O pH 6	Y 8
	91 (1)				
	91.5 (1)			H_2O	B 39
	80 (1)			H_2O	S 19
	80 (2)				
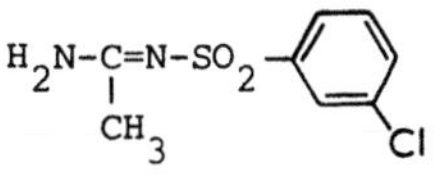	93			$CDCl_3$ (a)	D 19
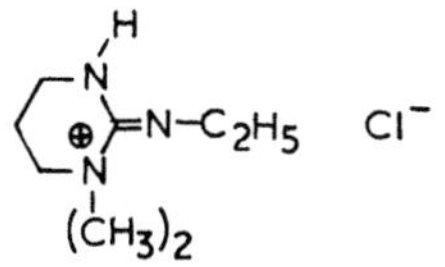	88 (b)			DMSO	Y 10
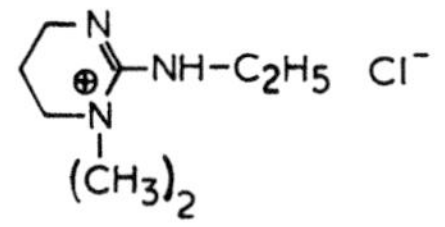	88			DMSO	Y 10

- Obtained in ^{15}N-NMR except (a) measured using ^{1}H spectroscopy

(b) For Z and E isomers

(c) $^{2}J_{N-CH_3} \simeq 0$ Hz

(d) Refers to $J_{N_1-H_2}$

(e) The $^{1}J_{N-H}$ coupling constant of the cyanamide $H_2N-C\equiv N$ has been measured in DMSO from the ^{15}N satellites of the proton spectrum (B24) $^{1}J_{N-H} = 89 \pm 1$ Hz

Table 7.27. $^{n}J_{^{15}N-^{1}H}$ Coupling constants of iminium salts

Iminium Salts		^{1}J	^{2}J	^{3}J	Conditions	Ref.
$(C_6H_5)_2-C{=}NH_2^+\ Cl^-$		92.6			SO_2	B 24
$(t.C_4H_9)CH{=}\overset{+}{N}(H)-OH$	(E)	106	3.5		H_2SO_4	C 9 K 7
1 CH_3, 2 CH_3 $>C{=}\overset{+}{N}(D)OH$				-4.5 (1) -3.2 (2)	CF_3COOD	C 10
$t.C_4H_9$, CH_3 $>C{=}\overset{+}{N}$(H or D)OH				-3.2	HCOOH	C 10
				-3.3	CF_3COOD	C 10
				-3.4	D_2SO_4	C 10
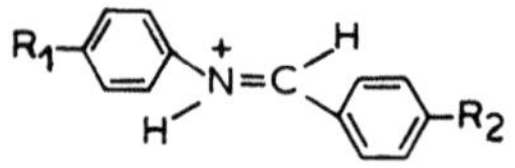						
R^1	R^2					
H	CH_3	91.5			HSO_3F	A 13
CH_3	H	91.5			HSO_3F	A 13
H	H	91.5			HSO_3F	A 13
H	Cl	91.5			HSO_3F	A 13
NO_2	H	92.0			HSO_3F	A 13
H	NO_2	91.9			HSO_3F	A 13
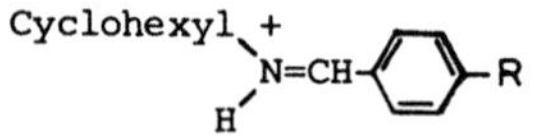					20% CF_3COOH (a)	B 53
R = OCH_3		91.1	(b)			
CH_3		92.0	(b)			
H		90.3	(b)			
Cl		88.5	(b)			
NO_2		92.0	(b)			

- Obtained in ^{1}H-NMR except (a) measured using ^{15}N spectroscopy

(b) Not observed ^{2}J < 1 Hz

Table 7.28. $^{n}J_{^{15}N-^{1}H}$ Coupling constants of nitriles, isonitriles and protonated species - Obtained in ^{1}H-NMR

Nitriles and Isonitriles	^{1}J	^{2}J	^{3}J	Conditions	Ref.
$H-C\equiv N$		8.7 ± 0.1		neat	B 21
$CH_3-C\equiv N$			1.8	SO_2	O 4
			1.7	/	B 24
			1.8 ± 0.1	/	M 28
$C_6H_5-CH_2-\overset{+}{N}\equiv C^-$		3.2		$CDCl_3$	B 5
$H-C\equiv\overset{+}{N}-H$	134	19		$HSO_3F-SbF_5-SO_2$	O 4
$CH_3-C\equiv\overset{+}{N}-H$	136		2.8	$HSO_3F-SbF_5-SO_2$	O 4
$CH_3-CH_2-C\equiv\overset{+}{N}-H$	130		1.5 - 2	193 K $HF-BF_3$	H 3
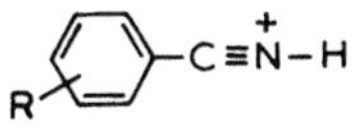				$HSO_3F-SbF_5-SO_2$	A 13 A 18
R					
$4-CH_3$	136.2				
H	136.2				
2-Br	135.1				
$4-CH_3O$ (a)	134.7				
$4-NO_2$	132.6				
$3,5-(CH_3O)_2$ (a)	131.8				
$3,5-(CH_3O)_2$ (b)	131.6				
$4-NO_2$, 2-Cl	132.5				

(a) O-protonated species

(b) Ring-protonated species

Table 7.29. $^{n}J_{^{15}N-^{1}H}$ Coupling constants of nitro derivatives

Nitro Derivatives	^{2}J	^{3}J	^{4}J	Conditions	Ref.
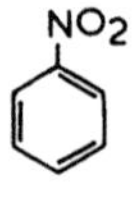		-1.94	-0.83	70% Acetone-d_6 (b)	E 1
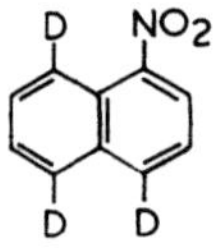		2.5		10% $CHCl_3$	K 10
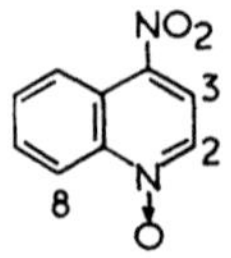	1	5.1 ± 0.15 (1-3)		Dioxane	K 5
		a few (1-8)			K 5
		2.2 ± 0.15 (4-3)			K 5
		2.7 (4-3)		10% $CHCl_3$	K 10
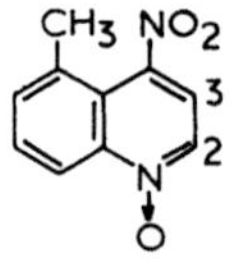		2.8 (4-3)		10% $CHCl_3$	K 10
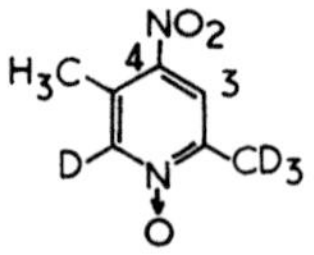		2.0 (4-3)		10% $CHCl_3$	K 10
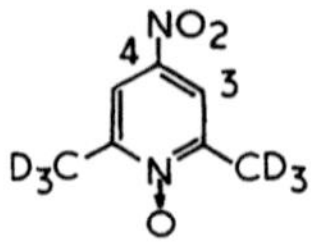		2.0 (4-3)		10% $CHCl_3$	K 10
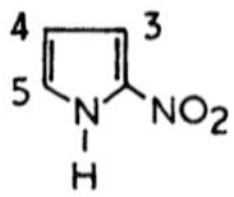		-0.30 (2-3)	-0.80 (2-4)		N 3
			-0.85 (2-6)		N 3
id, H^+		0.35 (2-3)	-0.60 (2-4)		N 3
			-0.50 (2-5)		N 3

Table 7.29 (continued)

Nitro Derivatives	^2J	^3J	^4J	Conditions	Ref.
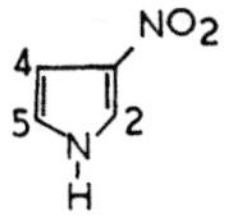		-0.2 (2-3)			N 3
		-0.0 (3-4)	0.80 (3-5)		N 3
id, H^+		0.45 (2-3)			N 3
		-0.0 (3-4)	-0.70 (3-5)		N 3
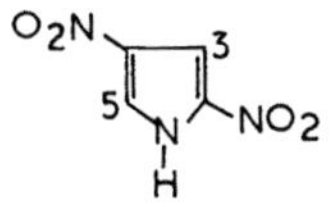		-0.40 (2-3)			N 3
		-0.45 (3-4)	-0.95 (2-5)		N 3
		-0.30 (4-5)			N 3
id, H^+		-0 (2-3)			N 3
		-0 (3-4)	-0.4 (2-5)		N 3
		-0 (4-5)			N 3
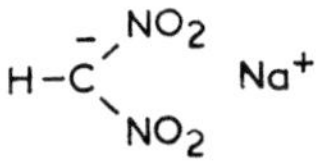	6			DMF (a)	E 3
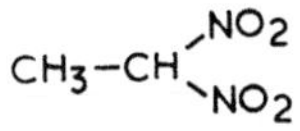	1.1	3.4			I 6
$O_2N-O-CH_3$		4.2		Acetone-d_6	B 5

- Obtained in ^{1}H-NMR except (a) measured using ^{15}N spectroscopy

(b) ^5J = -0.29 Hz

Table 7.30. $^{n}J_{^{15}N-^{1}H}$ Coupling constants of hydrazines, hydrazones and azines

Hydrazines, Hydrazones, Azines		^{1}J	^{2}J	^{3}J	Conditions	Ref.
$(CH_3)_2$-N-NH_2 (1 2)				3.2 (2)	neat	L 8
C_6H_5-NH-NH_2 (1 2)		88 (1)			neat (a)	Y 7
		88 (2)				
		89.6 (1)			DMSO	A 13
p.NO_2-C_6H_4-$\overset{*}{N}$H-NH_2		99.2			DMSO	A 13
C_6H_5-$(NH)_2$-C_6H_5		90.5			DMF	A 13 B 24
C_6H_5-NH-NH-CO-CH_3 (1 2)	E	92.3 (1)	1.2 (1)		DMSO	E 2
		97.6 (2)	1.1 (2)			
	Z	90.1 (1)	5.5 (1)		DMSO	E 2
		101.6 (2)	$<$0.4 (2)			
CH_3-CH_2-NH-$\overset{*}{N}$=CH-CH_3 (5 1 2 3 4)		(H anti)	(-)14.5 (3)	(-)5.1 (4)	/	L 8
		(H syn)	(+)4.5 (3)	(-)2.3 (4)		
				2.5 (5)		
C_6H_5-$\overset{*}{N}$H-N=CH-CH_2-CH_3		90.5 89.5 (b)			DMSO	A 13
C_6H_5-$\overset{*}{N}$H-N=CH-CH_2-C_6H_5	E			2.2	(a)	A 13
	Z			3.6	(a)	A 13
C_6H_5-$\overset{*}{N}$H-N=C$(CH_3)_2$				1.8 (c)	(a)	A 13
				3.6 (d)	(a)	A 13
C_6H_5-$\overset{*}{N}$H-N=C(t.Bu)CH_3	E			1.8	(a)	A 13
C_6H_5-$\overset{*}{N}$H-N=CH-C_6H_5 (1 2 3)		92.7		6.6 (3)	DMSO-d_6	B 25
C_6H_5-$\overset{*}{N}$H-N=CH-$C_6H_4NO_2$m		93.7			DMSO	A 13
C_6H_5-$\overset{*}{N}$H-N=CH-$C_6H_4NO_2$p		93.8			DMSO	A 13
C_6H_5-$\overset{*}{N}$H-N=C$(CHO)_2$		96			CCl_4	A 13
		96 $\pm$ 1			CCl_4	R 15b

Table 7.30 (continued)

Hydrazines, Hydrazones, Azines		^1J	^2J	^3J	Conditions	Ref.
$C_6H_5-NH-N=C(CHO)COCH_3$	E	96 ± 1			$CDCl_3$	R 15a
	Z	96 ± 1			$CDCl_3$	R 15a
$C_6H_5-NH-N=C(COOC_2H_5)-COCH_3$		94.7 96.1 (b)			CCl_4	B 25
$C_6H_5-\overset{*}{N}H-N=S=O$		99			DMSO	A 13
$pNO_2C_6H_4-\overset{*}{N}H-N=C(CH_2)_2$		94.3			DMSO	A 13
$pNO_2C_6H_4-\overset{*}{N}H-N=CH-C_6H_4NO_2m$		95.1			DMSO	A 13
$pNO_2C_6H_4-\overset{*}{N}H-N=CH-C_6H_4OCH_3p$		94.9			DMSO	A 13
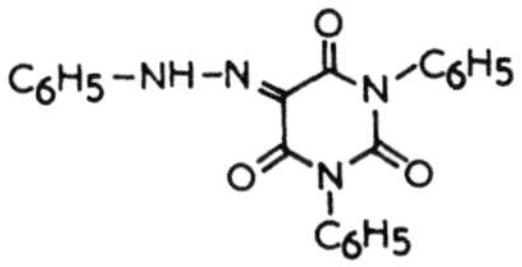					5% DMSO-d_6	Y 1
		95			308 K	
		92			350 K	
		83			366 K	
		79			376 K	
		very broad singlet			406 K	
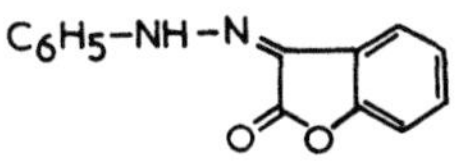		97 97 (b)			5% DMSO-d_6	Y 1
		97			5% C_6D_5N	Y 1
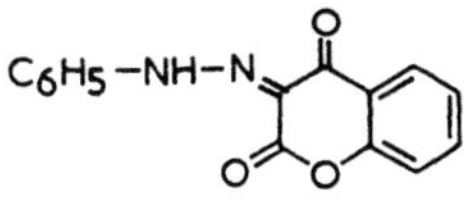		95 93 (b)			311 K 5% $CDCl_3$	Y 1
		97			5% DMSO-d_6	Y 1
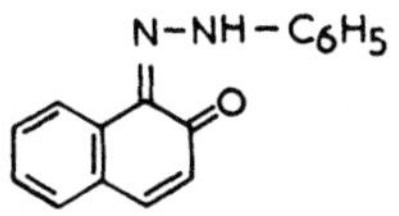		58.5			302 K CH_2Cl_2	B 43
		72.4			221 K	
		83.6			154 K	
		65.0			293 K $CDCl_3$	
		51.0			293 K CCl_4	
		60.0			293 K DMSO	

Table 7.30 (continued)

Hydrazines, Hydrazones, Azines	^1J	^2J	^3J	Conditions	Ref.
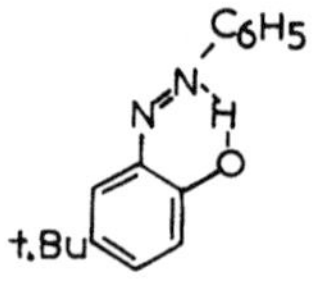	2			293 K CH_2Cl_2	B 43
	2			193 K CH_2Cl_2	
	0			293 K DMSO	
	2			267 K CH_2OH 95%	
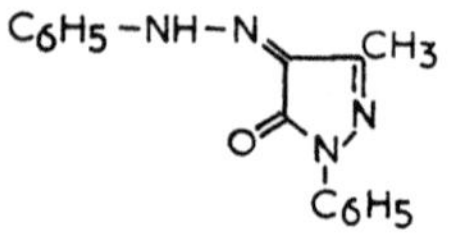	96.5			from 213 K to 273 K $CDCl_3$ CCl_4 CH_2Cl_2	B 51
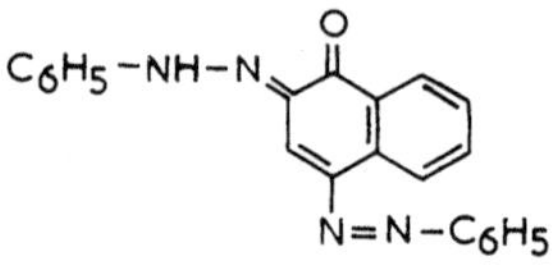	72			304 K CH_2Cl_2	B 51
	77.3			275 K	
	81			267 K	
	87			198 K	
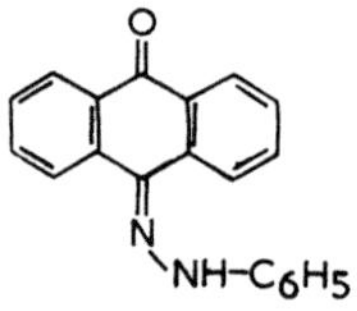	90.5			CH_2Cl_2	B 51
	92.0			Acetone	B 51
	93.0			DMF	B 51
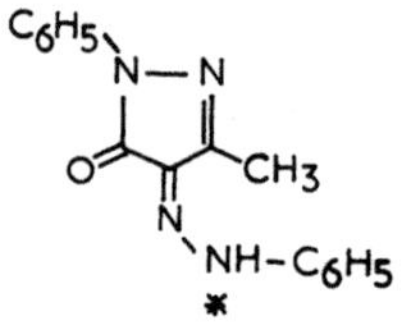	96 ± 1			$CDCl_3$	L 23
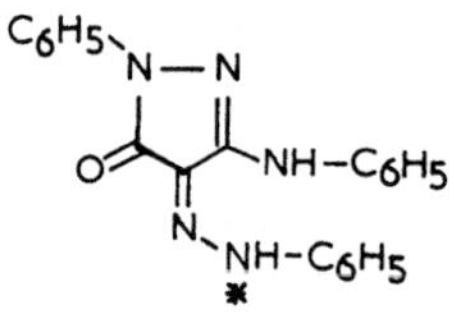	97 ± 1			311 K $CDCl_3$	L 23
	96 ± 1			311 K DMSO-d_6	L 23
	97 ± 1			233 K C_5D_5N	L 23
$CH_2=\overset{+}{N}=\overset{-}{N}$		±0.14	(-)1.10	/	J 8

Table 7.30 (continued)

Hydrazines, Hydrazones, Azines	^1J	^2J	^3J	Conditions	Ref.
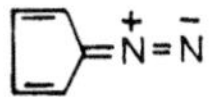			-0.34	/ (e)	J 8
$\overset{-}{N}=\overset{+}{N}=CH-C\underline{OO}C_2H_5$		2.8	1.0	293 K 1M HCl (a)	L 2
$H-N=\overset{+}{N}=\overset{-}{N}$	70.18 69	2.26	2.25	10% Ether (a)	M 11
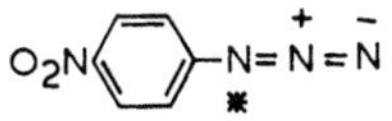			1.8	/	M 24
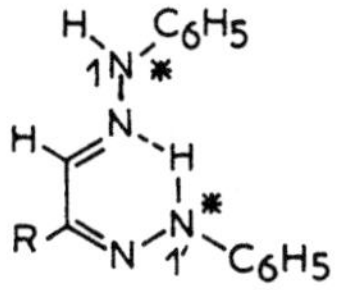 R= Sugar Residue	93 (1) 94.5 (1')			Dioxane	M 15
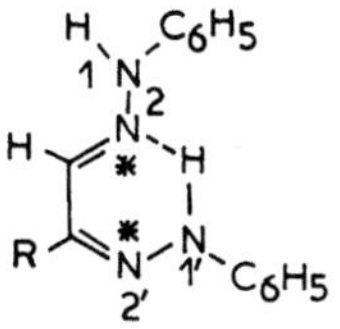 R= Sugar Residue		-5.0 (f) -2.0 (f)		Dioxane or C_5H_5N	M 15
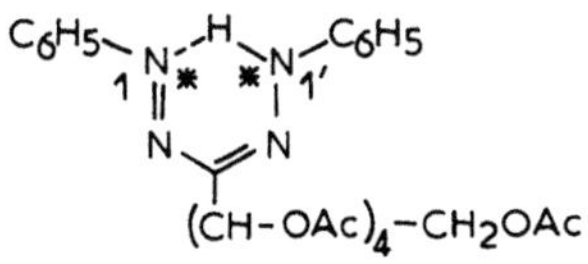	46.6 (1) 46.6 (1')			$CDCl_3$	M 16
id, 1-labelled only	46.5 (1)			$CDCl_3$	M 16

Table 7.30 (continued)

Hydrazines, Hydrazones, Azines		^1J	^2J	^3J	Conditions	Ref.
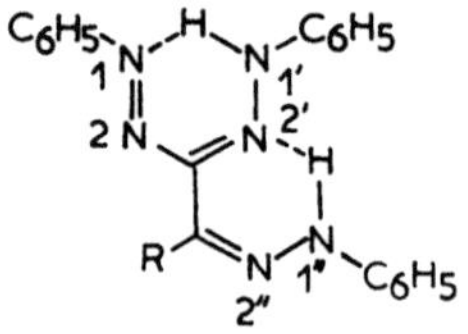		46.0 (1)			$CDCl_3$	M 17
		94.5(1")				
		2.5 (g)				
R= $(CH\text{-}OAc)_3\text{-}CH_2OAc$						
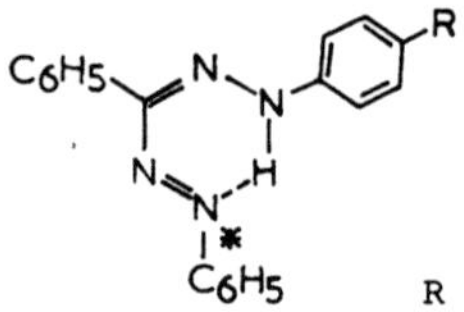	R = H	45			$CDCl_3$	F 9
	NO_2	0 (h)			$CDCl_3$	F 9
	OCH_3	86.5			$CDCl_3$	F 9
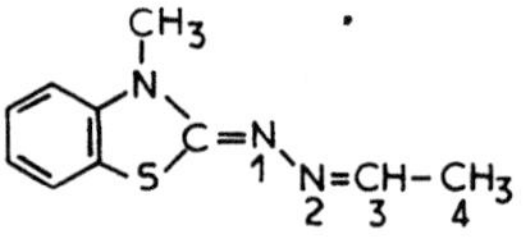	Z			10 (1-3)	(a)	N 4
				2.3(2-4)		
	E			5.4(1-3)		
				3.7(2-4)		

- Obtained in ^{1}H-NMR except (a) measured using ^{15}N spectroscopy

(b) Two doublets: syn and anti form

(c) Refers to CH_3 anti with respect to the lone pair

(d) Refers to CH_3 syn with respect to the lone pair

(e) $^4J = -1.11$ Hz

(f) ^1J or ^2J coupling of H_1 with N_2 and $N_{2'}$

(g) Coupling of $H_{1''}$ with N_2 or $N_{2'}$

(h) Not detected

Table 7.31. $^{n}J_{^{15}N-^{1}H}$ Coupling constants of nitrosamines, nitramines and salts - Obtained in ^{1}H-NMR

Nitrosamines Nitramines and Salts		^{1}J	^{2}J	^{3}J	Conditions	Ref.
$R_1R_2N\text{-}N{=}O$ (1 2)	cis			0 - 0.8 (2)	$CDCl_3$	A 9b
	trans			2.1 - 3.1 (2)	$CDCl_3$	A 9b
$R_1(R_2R_3N)N\text{-}N{=}O$ (1 2)	cis			0 (2)	$CDCl_3$	A 9a
	trans			1.9 - 2.2 (2)	$CDCl_3$	A 9a
$R_1(R_2O)N\text{-}N{=}O$	cis			0 - 0.7 (2)	$CDCl_3$	A 8a
	trans			1.6 - 2.6 (2)	$CDCl_3$	A 8a
$CH_3(N{\equiv}C\text{-}CH_2)N\text{-}N{=}O$ (4: CH_3; 3: C≡N; 5: CH_2; 1 2)	E		1.5 (1-4)	0 (2-4)	$CDCl_3$ (a)	S 3
			1.6 (1-5)	1.7 (2-5)		
				1.5 (3-5)		
	Z		1.7 (1-4)	0 (2-5)	$CDCl_3$ (a)	S 3
			1.4 (1-5)	1.8 (2-4)		
				1.5 (3-5)		
$CH_3\text{-}NH\text{-}NO_2$		100 ± 5	1 ± 0.2	3.0 ± 0.2	CH_2Cl_2	I 1
$(CH_3)_2N\text{-}NO_2$				2.3	$CDCl_3$	B 5
			1 ± 0.3	2.5 ± 0.5	CH_2Cl_2	I 1
$CH_3(CH_3OOC)N\text{-}NO_2$			0.6 ± 0.3	2.5 ± 0.4	CH_2Cl_2	I 1
$CH_3(C_2H_5OOC)N\text{-}NO_2$			0.4 ± 0.2	2.0 ± 0.4	CH_2Cl_2	I 1
$CH_3((CH_3)_3Si)N\text{-}NO_2$ (1 2)				0.6 (1)	243 K CH_2Cl_2	I 1
				3.0 ± 0.2 (2)		
$(CH_3)_3Si(C_2H_5OOC)N\text{-}NO_2$				0.9 ± 0.1	CH_2Cl_2	I 1

Table 7.31 (continued)

Nitrosamines Nitramines and Salts		^1J	^2J	^3J	Conditions	Ref.
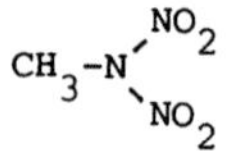			1.85 ± 0.3	3.6 ± 0.3	CH_2Cl_2	I 1
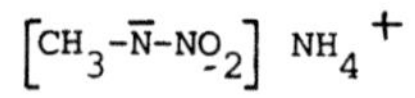				5.0 ± 0.3	H_2O	I 1
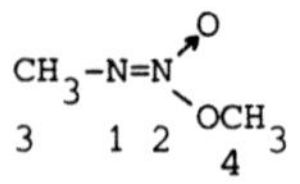	cis			5.5 ± 0.3 (3) 3.5 ± 0.3 (4)	CH_2Cl_2	I 1
	trans			5.3 ± 0.3 (3) 3.8 ± 0.3 (4)	CH_2Cl_2	I 1
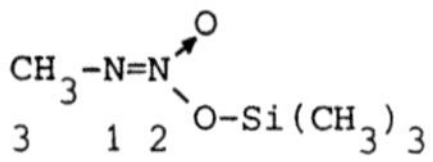	trans			5.4 ± 0.3 (3)	CH_2Cl_2	I 1
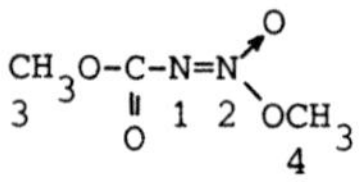				3.15 ± 0.3 (4)	CH_2Cl_2	I 1
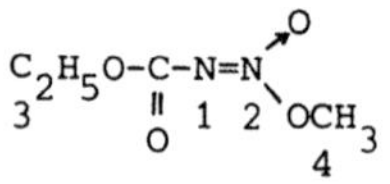				3.8 ± 0.3 (4)	CH_2Cl_2	I 1
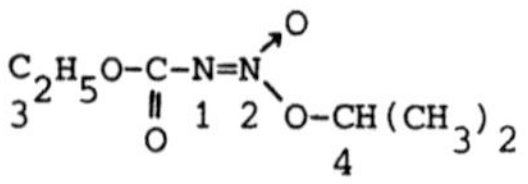				2.5 ± 0.3 (4)	CH_2Cl_2	I 1
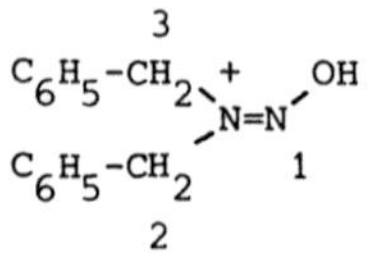			2.6 (1)	3.6 (2) 0 (3)	243 K HSO_3F	A 3

(a) $^5J_{3-4} \simeq 0$ Hz

Table 7.32. $^{n}J_{^{15}N-^{1}H}$ Coupling constants of nitrites and other O-N-O derivatives - Obtained in ^{1}H-NMR

Nitrites and Other O-N-O Derivatives		^{2}J	^{3}J (a)		Conditions	Ref.
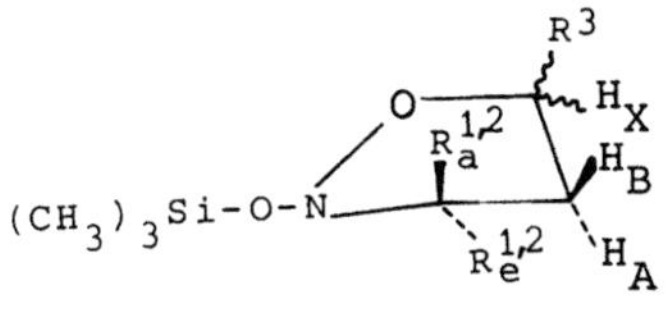			H_B	CH_3		S 20
$R^1=R^2=R^3=COOCH_3$	55%		1.1			
	45%		1.0	(b)		
$R^1=R^2=COOCH_3$ $R^3=C_6H_5$	70%		0.5			
	30%		1.3	(c)		
$R^1=R^2=CH_3$ $R^3=COOCH_3$	85%		1.2	2.0		
				3.6 (e)		
	15%			2.1 (d)		
				3.4 (e)		
$R^1=R^2=H$ $R^3=COOCH_3$	100%	/	(a)			
$R^1=R^3=COOCH_3$ $R^2=H$	50%	4.6 (H_e)	(a)			
$R^1=COOCH_3$ $R^2=H$ $R^3=C_6H_5$	60%	<0.1 (H_a)	(a)			
	40%	3.6 (H_e)	(a)			
$R^1=CH_3$ $R^2=H$ $R^3=COOCH_3$	73%			2.6		
	15%			3.9 (e)		
	12%			2.7		
			trans	cis		
$C_6H_5-CH_2-O-N=O$			2.4	0	198 K $CDCl_3$	A 8b
			1.7		310 K	
$C_6H_5-CH_2-CH_2-O-N=O$			2.3	0	198 K $CDCl_3$	A 8b
			1.8		310 K	
$C_6H_5(CH_3)-CH-O-N=O$			2.4	0	198 K $CDCl_3$	A 8b
			1.9		310 K	

Table 7.32 (continued)

Nitrites and Other O-N-O Derivatives	^2J	^3J (a)			Conditions	Ref.
$C_6H_5(C_2H_5)$-CH-O-N=O		2.1		0	198 K $CDCl_3$	A 8b
			2.1		310 K	
t.C_4H_9-CH_2-CH_2-O-N=O		2.4		0	198 K $CDCl_3$	A 8b
			1.7		310 K	

(a) $^3J_{N-H_A}$ < 0.1 Hz; $^3J_{N-H_X}$ < 0.1 Hz except (b) 2.4 Hz

(c) 1.3 Hz

(d) 2.3 Hz

(e) Axial CH_3 group

Table 7.33. $^{n}J_{^{15}N-^{1}H}$ Coupling constants of compounds with N-S bonds

Amino-Sulfur Derivatives	^{1}J	^{3}J	Conditions	Ref.
CF_3-S-NH_2	80.6		/ (a)	C 11 C 13
$(CF_3S)_2-NH$	99.1		/ (a)	C 11 C 13
$C_6H_5-SO_2-NH_2$	80.8		9% DMSO	S 12
$H_2N-C_6H_4-SO_2-\overset{*}{N}H_2$	78.7		9% DMSO	S 12
$C_6H_5-SO_2-NH-C_2H_5$	84.2		9% DMSO	S 12
$C_6H_5-SO_2-NH-CH_2-C_6H_5$	85.1		9% DMSO	S 12
$C_6H_5-SO_2-NH-CH(CH_3)(C_3H_7)$	86.8		9% DMSO	S 12
$C_6H_5-SO_2-NH-CH\langle CH_2, CH_2 \rangle (CH_2)_3$	86.0		9% DMSO	S 12
$C_6H_5-SO_2-NH-C_6H_4-R$			9% DMSO	S 12
R = OCH_3	84.2			
F	84.2			
CH_3	84.2			
Br	82.5			
H	82.5			
$\{Rh\ Cl\ [P(iPr)_3]_2(C_6H_5N{=}SO)\}$		7		M 3

- Measured using ^{15}N-NMR except (a) obtained from ^{1}H spectroscopy

Table 7.34. $^{n}J_{^{15}N-^{1}H}$ Coupling constants of compounds with N-P bonds

Amino-Phosphorus Derivatives	^{1}J	^{3}J (c)	Conditions	Ref.
H_2N-PF_2	-80.4		$(CH_3)_4Si:CFCl_3$	A 17
	-80.4		$(CH_3)_4Si:C_6D_6$ 308 K 1:1	A 16
	82.7		223 K	C 13
$H_2N-P(CF_3)_2$	-85.64			C 12
	85.6			C 11 C 13
$H_2N-P(CH_3)(CF_3)$	78.9			C 11
	79.9			C 13
$H_2N-P(C_2F_5)_2$	85.8			C 13
$H-N(SiH_3)-PF_2$	73.1	^{2}J= -4.1	$(CH_3)_4Si:CFCl_3$	A 17
$H-N-(PF_2)_2$	-74.7		$(CH_3)_4Si:C_6D_6$ 1:1	A 16
$H-N-[P(CF_3)_2]_2$	81.9			C 13
$H-N-[P(CH_3)(CF_3)]_2$	77.4 78.9 (b)			C 11
	77.5 78.7 (b)			C 13
$H-N(C_6H_5)-P(CH_3)_2$	-81.0	-2.7	C_6H_6	M 26
$H-N(C_6H_5)-P(t.Bu)_2$	-80.5		Mesitylene	M 26
$H-N(C_6H_5)-P[N(CH_3)_2]_2$	-79.0		C_6H_6	M 26
$[H-N(C_6H_5)]_2-P-N(CH_3)_2$	-79.5	-3.5±0.5 (d)	C_6H_6	M 26
$H-N(C_6H_5)-P[N(CH_3)CH_2]_2$	-78.5		C_6H_6	M 26
$CH_3-N-[PCl_2]_2$		^{2}J= +0.6	(a)	C 3
$C_6H_5-N[Sn(CH_3)_3]-P(CH_3)_2$		-2.1	C_6H_6	M 26
$H_2N-P(CCl_3)_2Cl_2$	88		0.4M $POCl_3$	K 11
H_2N-PF_4	90.3		228 K	C 13

Table 7.34 (continued)

Amino-Phosphorus Derivatives	^1J	^3J (c)	Conditions	Ref.
$(H_2N)_2$-PF_3	87.5			C 11 C 13
H-N(C_6H_5)-PO$(CH_3)_2$	-83.0	-1.2	DMSO-d_6	M 26
H-N(C_6H_5)-PO(t.Bu)$_2$	-80.5		CH_2Cl_2+Mesitylene	M 26
H-N(C_6H_5)-PS$(CH_3)_2$	-82.0	-1.2	Dioxane	M 26
H-N(C_6H_5)-PS(t.Bu)$_2$	-78.5		$CHCl_3$+ Mesitylene	M 26
H-N(C_6H_5)-PS$[N(CH_3)CH_2]_2$	-83.0		C_6H_6 + $CHCl_3$	M 26
H-N(C_6H_5)-PSe$(CH_3)_2$	-83.0	-1.0	CH_2Cl_2	M 26
H-N(C_6H_5)-PSe(t.Bu)$_2$	-78.5		$CHCl_3$+ Mesitylene	M 26
H-N(C_6H_5)-PSe$[N(CH_3)CH_2]_2$	-85.8		C_6H_6	M 26
H-N(C_6H_5)-PTe$(CH_3)_2$	-79.5	-1.7	C_6H_6 + CH_2Cl_2	M 26
C_6H_5-N$[Sn(CH_3)]_3$-PS$(CH_3)_2$		$\pm$0.2	C_6H_6	M 26
HN=P$(CCl_3)_2$Cl	exchange		298 K CCl_4	K 21
	80		(e) CCl_4	K 21
H-N(C_6H_5)-P$(CH_3)_3$ I	-83.3	-1.4	CH_2Cl_2	M 26
H-N(C_6H_5)-P $[(CH_3)(t.Bu)_2]$ I	-77.0	-1.2	DMSO-d_6	M 26
H-N(C_6H_5)-P $[(SCH_3)(CH_3)_2]$ I	-84.5	-1.3	$CHCl_3$	M 26
H-N(C_6H_5)-P $[(SeCH_3)(CH_3)_2]$ I	-82.5	-1.2	CH_2Cl_2	M 26
H-N(C_6H_5)-P $[(SeCH_3)(t.Bu)_2]$ I	-81.0		DMSO-d_6	M 26
H-N(C_6H_5)-P $[(CH_3)\{N(CH_3)CH_2\}_2]$ I	-79.8	-1.2	CH_2Cl_2	M 26

- Obtained in ^{1}H-NMR except (a) measured using ^{15}N spectroscopy

(b) The observation of two J_{N-H} values is due to the phosphorus chiral centres

(c) $^3J_{N-P-C-H}$

(d) $^3J_{N-P-N-H}$

(e) At low temperature

Table 7.35. $^{n}J_{^{15}N-^{1}H}$ Coupling constants of other amino-heteroelement derivatives

Other Amino-Heteroelement Derivatives	^{1}J	^{2}J	^{3}J	Conditions	Ref.
$H_2N-As(CF_3)_2$	73.4				C 11 C 13
$H-N[As(CF_3)_2]_2$	79.0				C 11 C 13
$(HN-BH)_3$	78.0				B 37 C 13
$(HN-B-NH_2)_3$	79.1				F 7
$H-N(C_6H_5)-Ge(CH_3)_3$	77.1		1.1(CH_3)		R 7
$CH_3-N(C_6H_5)-Pb(CH_3)_3$		<0.3	0.48	20% C_6H_6 (a)	K 12
$H-N(SiH_3)_2$	71.4			223 K	C 13
	-71.3	-4.3		C_6H_6 (a)	A 17
$H-N-[Si(CH_3)_3]_2$	66.5				C 13 R 7
$H-N(PF_2)-SiH_3$	-73.1	-4.1		$(CH_3)_4Si:CFCl_3$ (a)	A 17
$H-N(C_6H_5)-Si(CH_3)_3$	76.0		1.1(CH_3)		R 7
	76.0		1.0(CH_3)		R 3
$N(SiH_3)_3$		-4.2		$(CH_3)_4Si$ (a)	A 17
$H-N(C_6H_5)-Sn(CH_3)_3$	73.8		(d)(CH_3)		R 7
$(C_6H_5)N[P(\overset{*}{C}H_3)_2]-Sn(CH_3)_3$			-2.1	$C_6H_6 + CH_2Cl_2$	M 26
$(C_6H_5)N[P(\overset{*}{C}H_3)_2S]-Sn(CH_3)_3$			±0.2	C_6H_6	M 26
$[MoF(N-NH_2)dppe_2]BF_4$ trans (b)	86±3			CH_2Cl_2 (a)	C 21
$[MoCl(N-NH_2)(C_5H_5N)(PMe_2Ph)_3]Cl$ (c)	83±1			CH_2Cl_2 (a)	C 21
$\{RhCl[P(iPr)_3]_2(C_6H_5NSO)\}$			7	(a)	C 21
$[WCl(N-NH_2)(C_5H_5N)(PMe_2Ph)_3]Cl$ (c)	83±3			CH_2Cl_2 (a)	C 21

Table 7.35 (continued)

Other Amino-Heteroelement Derivatives		^1J	^2J	^3J	Conditions	Ref.
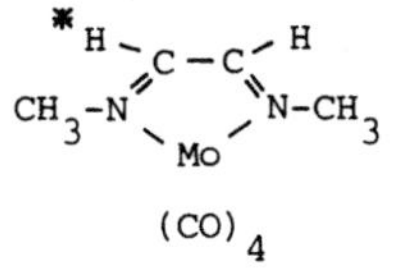			(d)		$CDCl_3$	S 13
$(NH_3)_2\,Pt^{2+}\,(H_2O)_2$	cis	73.2			H_2O (a)	A 22
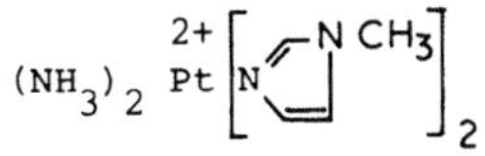	cis	72.0			H_2O (a)	A 22
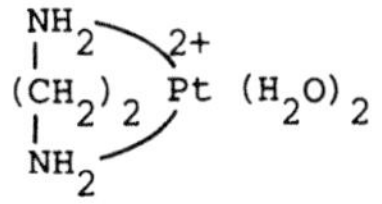		75.4			H_2O (a)	A 22
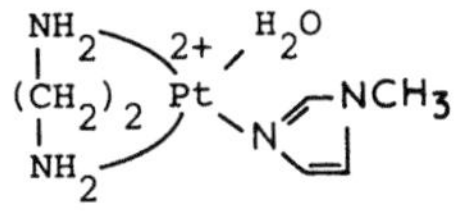		(d)			H_2O (a)	A 22
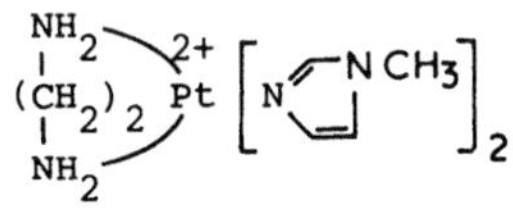		(d)			H_2O (a)	A 22
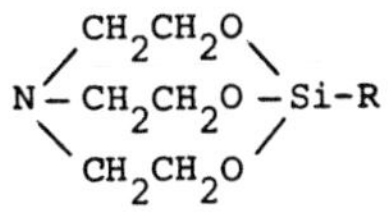	R=CH_3		0	2.4	$CDCl_3$ (a)	L 42
	CH_2=CH		0.1	2.3	Acetone-d_6 (a)	L 42
	C_6H_5		0.1	2.3	Acetone-d_6 (a)	L 42
	$ClCH_2$		0.1	2.3	$CDCl_3$ (a)	L 42
	CH_3O		0	2.2	Acetone-d_6 (a)	L 42
	C_2H_5O		0	2.3	Acetone-d_6 (a)	L 42

- Obtained in ^{1}H-NMR except (a) measured using ^{15}N spectroscopy

(b) dppe Ph_2-P-CH_2-CH_2-P-Ph_2

(c) Cl is trans with respect to NNH_2

(d) Not detected

A coupling constant ^1J = 72 ± 2 Hz was measured for the three NH_2 groups of *trans* $Co(Gly)_3$ (F 17)

Table 7.36. $^{1}J_{^{15}N-D}$ Coupling constants - Obtained in ^{15}N-NMR

$^{15}N-^{2}H$	^{1}J	Conditions	Ref.
NH_2D	9.45 $\pm$ 0.2	neat	L 10
NHD_2	9.45 $\pm$ 0.2	neat	L 10
ND_3	9.45 $\pm$ 0.2	neat	L 10
	9.8 $\pm$ 1	vapor	L 57

Table 7.37. $^{1}J_{^{15}N-^{11}B}$ Coupling constants

$^{15}N-^{11}B$	^{1}J	Conditions	Ref.
$(CH_3)_3N-BF_3$	14 - 15	Acetone or CH_3CN	C 5
$(CH_3)_3N-BCl_3$	14	Acetone or CH_3CN	C 5
$(CH_3)_3N-BBr_3$	13	Acetone or CH_3CN	C 5
$(C_2H_5)_3N-BH_3$	(b)	neat (a)	G 12
$C_6H_5NH-B(CH_3)_2$	<10	/	F 6
$[(C_2H_5)_2N]_3B$	(c)	neat (a)	G 12

- Obtained in ^{11}B-NMR except (a) measured using ^{15}N spectroscopy

(b) Half-Height width $\simeq$ 10 Hz

(c) Half-Height width $\simeq$ 30 Hz

See also the review concerning ^{11}B-NMR spectroscopy (N9)

Table 7.38. $^{n}J_{^{15}N-^{13}C}$ Coupling constants of amines, amine hydrochlorides and amino acids

Amines, Aminoacids and Salts	^{1}J	^{2}J	^{3}J	Conditions	Ref.
CH_3-NH_2	7 ± 1 -4.5±0.1			neat (a) gaz (b)	B 24 P 1
CH_3-NH_2, HCl	<8			H_2O (a)	B 24
$CH_3-CH_2-CH_2-NH_2$	3.9	1.2	1.5	$CDCl_3$	B 11
$CH_3-CH_2-CH_2-NH_2$, HCl	4.4	<0.2	1.3	$CDCl_3$	B 11
$(CH_3)_3-C-CH_2-NH-CH(CH_3)_2$ 3 2 1 4 5	3.5 (1) 4.2 (4)	1.6 (2) 2.6 (5)	1.0 (3)	$CDCl_3$	B 19
id, HCl	4.3 (1) 3.6 (4)	/ /	0.8 (3)	$CDCl_3$	B 19
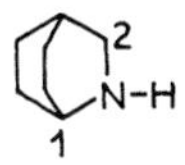	2.6 (1) 2.5 (2)			$CDCl_3$	B 19
id, HCl	(c)				B 19
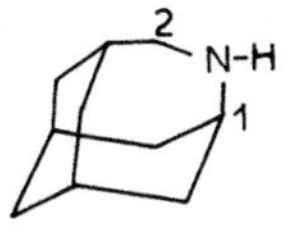	3.1 (1) 3.4 (2)			$CDCl_3$	B 19
id, HCl	3.5 (1) 3.7 (2)			$CDCl_3$	B 19
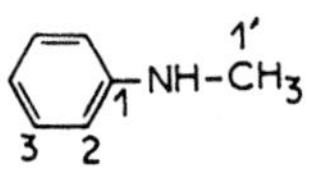	13 (1) 10.3 (1')	2.5	1.4	50% Acetone-d_6 (d)	B 22
1 CH_3, 2 CH_3 >N-C(3)(=Cr(CO)$_5$)-CH_3 (4)	12.4 (1) 8.3 (2) 4.0 (3)	0.5 (4)	<0.25 <0.25	C_6D_6 (e)	K 3
1 CH_3, 2 CH_3 >N-C(3)(=W(CO)$_5$)-CH_3 (4)	12.5 (1) 8.1 (2) 4.7 (3)	1.0 (4)	<0.25 <0.25	C_6D_6 (e)	K 3

Table 7.38 (continued)

Amines, Aminoacids and Salts	^1J	^2J	^3J	Conditions	Ref.
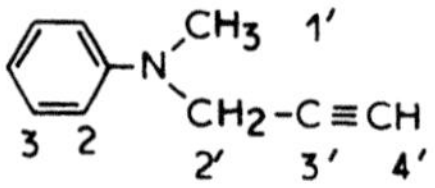	12.5 (1)	-2.1 (2)	-1.16(3)	50% CD_2Cl_2 (d)	B 22
	9.6 (1')				
	9.6 (2')	0.9(3')	0.9(4')		
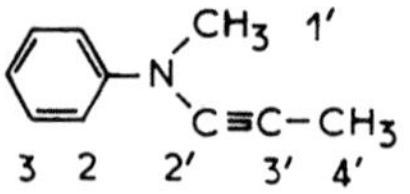	36.2 (2')	5.5(3')	-1.8 (3)	50% Acetone-d_6 (d)	B 22
	16.2 (1)				
	∓12.2 (1')	-2.3 (2)	<0.5(4')		
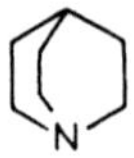	2.1	<0.2	2.8	$CDCl_3$	B 11
id, HCl	4.8	<0.2	6.7	$CDCl_3$	B 11
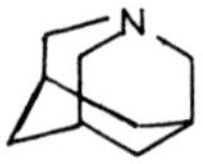	2.5	<0.3	<0.3	$CDCl_3$	B 56
id, HCl	4.0	<0.3	0.6	$CDCl_3$	B 56
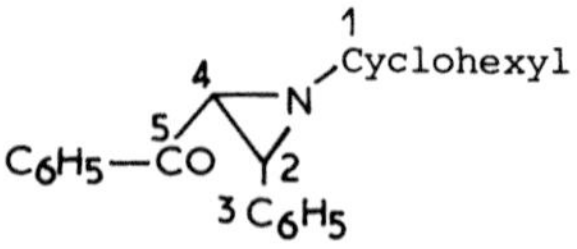	≤3.4 (1)	-4.0 (3)			T 3
	+7.8 (2)	-3.4 (5)			
	+7.3 (4)				
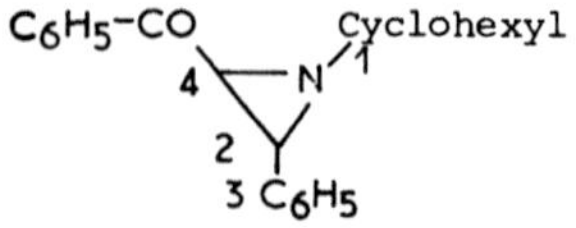	<2.0 (1)	-3.1 (3)	-5.7		T 3
	+5.2 (2)				
	+8.2 (4)				
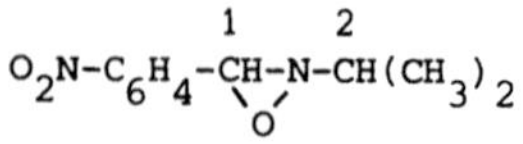 Z	+4.9 (1)			CH_2Cl_2	J 2
	+4.9 (1)				T 3
	-5.9 (2)				T 3
E	3.1 (1)				J 2
	-3.1 (1)				T 3
	-5.5 (2)				T 3

Table 7.38 (continued)

Amines, Aminoacids and Salts	^1J	^2J	^3J	Conditions	Ref.
H_2N-CH_2-COOH	6.2	0		H_2O pH 6.33	L 8
	4.8	0		H_2O pH 13.68	L 8
$H_2N-CH(CH_3)-COOH$	5.6	0		H_2O pH 6.37	L 8
	4.4	0		H_2O pH 13.64	L 8
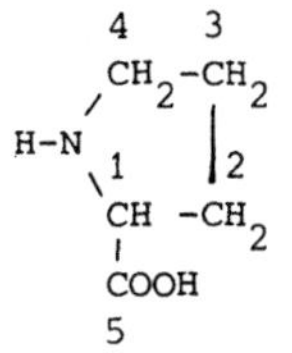	6.6 (1)	1.7 (2)		H_2O pH 0.4	B 20
	4.6 (4)	0.8 (3)		(a)	
		<0.3 (5)			
	5.5 (1)	1.7 (2)		H_2O pH 6.5	B 20
	4.9 (4)	1.0 (3)			
		<0.2 (5)			
	3.4 (1)	0.7 (2)		H_2O pH 12	B 20
	3.5 (4)	0.4 (3)			
		0.3 (5)			
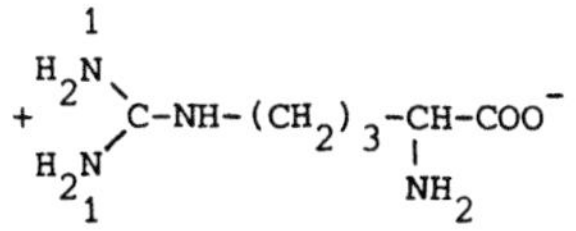	21.3 (1)			H_2O pH 9.9	L 6
	18.5 (1)			H_2O pH 13.4	L 6
L[Guanido $^{15}N_2$ ^{13}C] arginine,HCl	21.3			0.30mg/ml H_2O	H 9
Geldanamycine	26.5				H 9

- Obtained in ^{13}C-NMR except (a) measured using ^{15}N spectroscopy

(b) Obtained from the ^{13}C satellites of the proton spectrum

(c) Not detected

(d) ^4J < 0.5 Hz

(e) Refers to *Z* and *E* forms

Table 7.39. $^{n}J_{^{15}N-^{13}C}$ Coupling constants of aromatic compounds - Obtained in ^{13}C-NMR

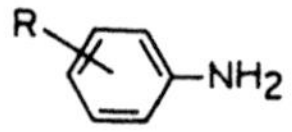

Aromatic Compounds	^{1}J	^{2}J	^{3}J	^{4}J	Conditions	Ref.
H	10.9				/	E 2
	12.1				1M DMSO-d_6	A 14
	-11.47	-2.68	-1.29	0.27	50% Acetone-d_6	H 5
	-10.9±0.3	-2.8±0.3			none or $CDCl_3$	W 10
2-OCH_3	11.4	0.9 (2)	1.3(5)	0	$CDCl_3$	N 1
		2.6 (6)	<0.5(3)			
2-CH_3	-10.5±0.3				$CDCl_3$	W 10
2-COOH	-10.7±0.5		≤0.5 (COOH)		NaOD pD 12	W 10
3-CH_3	11.8				1M DMSO-d_6	A 14
3-OCH_3	12.5				1M DMSO-d_6	A 14
3-CF_3	12.5				1M DMSO-d_6	A 14
3-Cl	13.2				1M DMSO-d_6	A 14
3-Br	13.2				1M DMSO-d_6	A 14
3-I	12.5				1M DMSO-d_6	A 14
4-CH_3	11.8				1M DMSO-d_6	A 14
	-10.5±0.3				$CDCl_3$	W 10
4-OCH_3	11.0				1M DMSO-d_6	A 14
4-Cl	12.5				1M DMSO-d_6	A 14
4-Br	12.5				1M DMSO-d_6	A 14
4-NO_2	14.7				1M DMSO-d_6	A 14
	-15.0±0.3				DMSO	W 10
	-14.9±0.3				Acetone-d_6	W 10
2,4-$(NO_2)_2$	16.9				1M DMSO-d_6	A 14
2,4,6-$(Br)_3$	17.6				1M DMSO-d_6	A 14

Table 7.39 (continued)

Aromatic Compounds	^1J	^2J	^3J	^4J	Conditions	Ref.
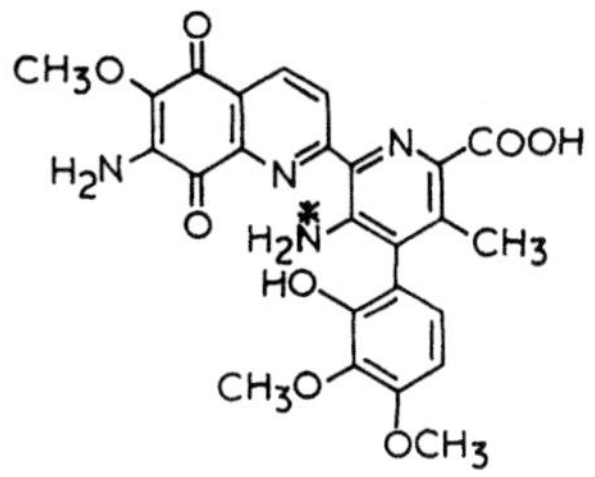	14.6				50mg in 0.5ml DMSO-d_6	G 3
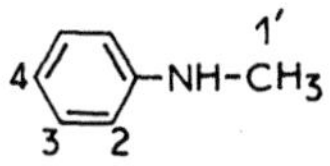	13 (1) 10.3(1')	2.5	1.4	<0.5	50% Acetone-d_6	B 22
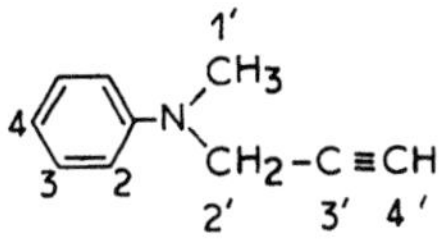	12.5 (1) 9.6(1') 9.6(2')	-2.1 (2) 0.9(3')	-1.16(3) 0.9(4')	<0.5	50% CD_2Cl_2	B 22
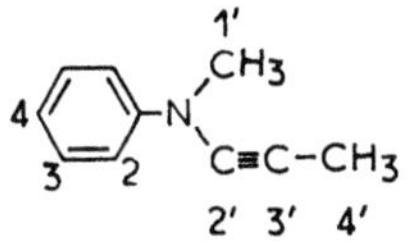	36.2(2') 16.2 (1) ∓12.2(1')	5.5(3') -2.3 (2)	-1.8 (3) <0.5(4')	<0.5	50% Acetone-d_6	B 22
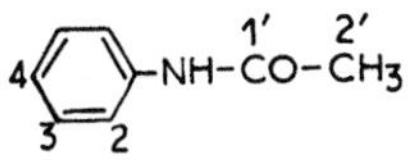	-14.3 (1) -14.7 (1)	9±0.3(2')			$CDCl_3$ Acetone-d_6	W 10 W 10
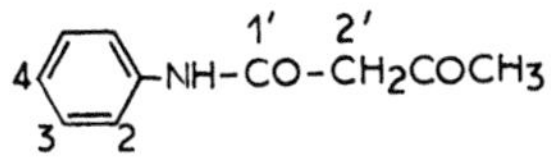	14.4 (1) 15.5(1')	<0.5 (2) 7.2(2')	1.4 (3)		$CDCl_3$	N 1
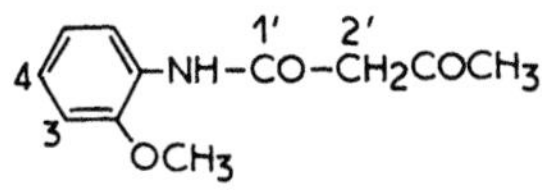	14.9 (1) 15.4(1')	 7.7(2')	<0.5 (3) 1.0 (5)		$CDCl_3$	N 1
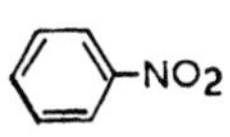	14.5 -14.57	 -1.67	 -2.32	 ±0.6	/ 70% Acetone-d_6	N 1 W 13 E 1

Similar values were obtained with aniline in $CDCl_3$ (K 42)

Table 7.40. $^{n}J_{^{15}N-^{13}C}$ Coupling constants of anilinium and benzenonium ions - Obtained in ^{13}C-NMR

Aromatic Salts	^{1}J	Conditions	Ref.
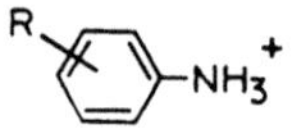			
H	-8.9 ± 0.3	CF_3COOD (a)	W 10
	8.6	1M HSO_3F	A 14
2-Cl	10.9	1M HSO_3F	A 14
2-Br	11.0	1M HSO_3F	A 14
2-NO_2	10.4	1M HSO_3F	A 14
3-Br	7.9	1M HSO_3F	A 14
4-F	8.6	1M HSO_3F	A 14
4-Cl	9.1	1M HSO_3F	A 14
4-Br	9.2	1M HSO_3F	A 14
4-NO_2	9.2	1M HSO_3F	A 14
	-9.2 ± 0.3	DCl	W 10
2-NO_2, 4-Cl	11.0	1M HSO_3F	A 14
2,4,6-$(Br)_3$	13.5	1M HSO_3F	A 14
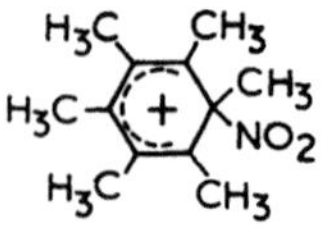	8.3 ± 0.3	203 K HNO_3 - HSO_3F	D 2
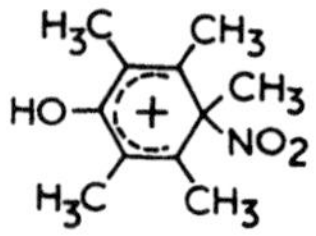	8.3 ± 0.3	203 K HNO_3 - HSO_3F	D 2

(a) Other coupling constants obtained $^{2}J = 1.5 \pm 0.3$ Hz and $^{3}J = 2.1 \pm 0.3$ Hz
A similar value was obtained for aniline in CD_3COOH (K 42)

Table 7.41. $^{n}J_{^{15}N-^{13}C}$ Coupling constants of five-membered nitrogen heterocyclic compounds - Obtained in ^{13}C-NMR

Five-Membered Heterocyclic Compounds	^{1}J	^{2}J	^{3}J	Conditions	Ref.
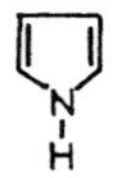	-12.98			/	W 10
	-12.97	-3.92		40% Acetone-d_6	B 12 B 23
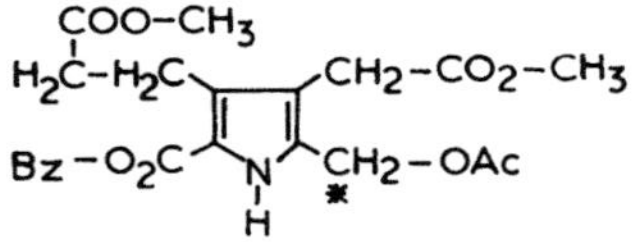		1.9		/	B 55
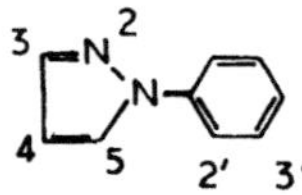	12.1 $\pm$ 0.3 (1-5)	1.6 $\pm$ 0.3 (1-2') 6.2 $\pm$ 0.3 (4) 2.1 $\pm$ 0.3 (4)	1.6 $\pm$ 0.3 (2-2') 2.0 $\pm$ 0.3 (1-3')	0.9M $CDCl_3$ (a)	H 4
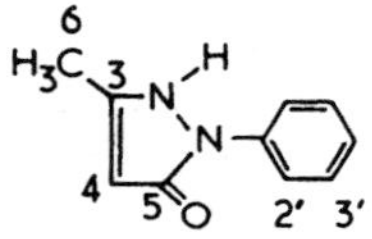	11.0 $\pm$ 0.6 (1-5) 12.2 $\pm$ 1.2 (1-1')	13.3 $\pm$ 0.6 (1-4) 1.5 $\pm$ 0.6 (2-4) $<$1.2 (2-5) 9.8 $\pm$ 2.4 (2-6) 1.2 $\pm$ 0.3 (1-2')	1.7 $\pm$ 0.3 (1-3') 1.2 $\pm$ 0.3 (2-2')	0.7M $CDCl_3$ (b)	H 4

Table 7.41 (continued)

Five-Membered Heterocyclic Compounds	^1J	^2J	^3J	Conditions	Ref.
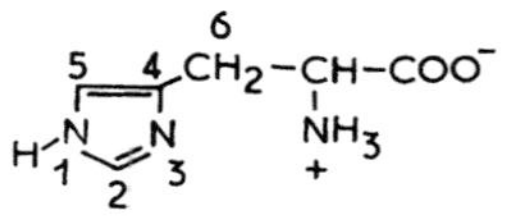	-16.1 (1-2) -11.6 (1-5) -16.0 (2-3) -9.9 (3-4)	/ (3-6)		H_2O pH 1.3 cation	B 35
	-10.1 (1-2) -10.4 (1-5) -2.7 (2-3) +4.1 (3-4)	-4.6 (3-6)		H_2O pH 7.6 amphion	B 35
	-6.4 (1-2) -7.3 (1-5) -6.9 (2-3) -4.7 (3-4)	-3.8 (3-6)		H_2O pH 10.9 anion	B 35
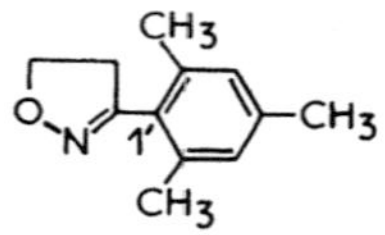		7 ± 1 (1')		Dioxane	C 14
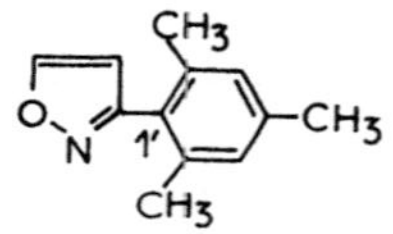	4.0 ± 0.5	7.0 ± 0.5 (1')		Dioxane	C 14

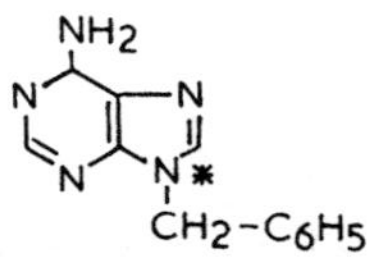	9.3 (CH_2)	50% DMSO-d_6 50% CF_3COOH	W 5
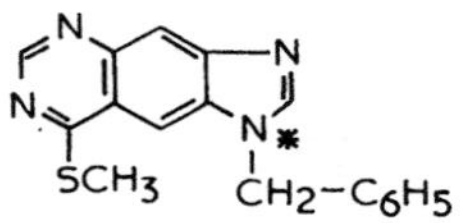	8.6 (CH_2)	50% DMSO-d_6 50% CF_3COOH	W 5

(a) $^{n}J_{C_3} = 1.2 \pm 0.3$ Hz
(b) $^{n}J_{C_3} = 3.1 \pm 0.6$ Hz

Table 7.42. $^{n}J_{^{15}N-^{13}C}$ Coupling constants of six-membered nitrogen heterocyclic compounds - Obtained in ^{13}C-NMR

Six-Membered Heterocyclic Compounds	^{1}J	^{2}J	^{3}J	Conditions	Ref.
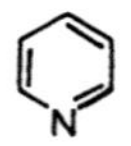	0.45	2.4	3.6	neat	L 8
	+0.62	+2.53	-3.85	/	B 12 B 23
	0.7	2.6	3.8	30% CH_3OH	L 8
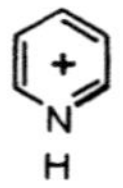	12.0	2.1	5.3	30% CH_3OH	L 8
	-11.85	+2.01	-5.3	/	B 12 B 23
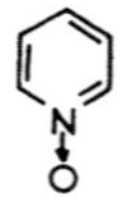	-15.23	+1.43	-5.17	/	B 12 B 23
	$\pm$15.24	1.32	-5.13	1M $CDCl_3$	W 8
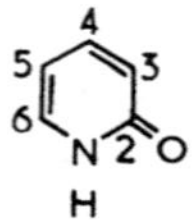	11.2 (2)	<0.5 (3)	10.5 (4)	/	B 28 N 1
	2.5 (6)	5.2 (5)			
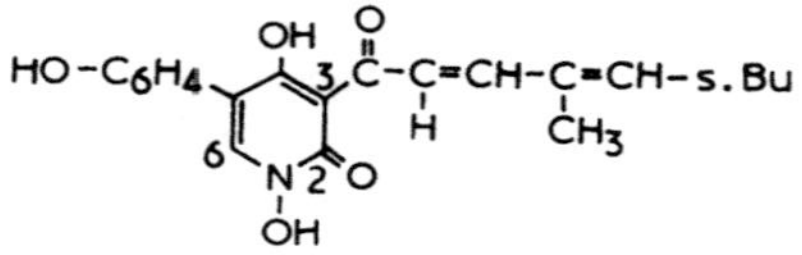	11.0 (2)	9.2 (3)		DMSO-d_6	M 47
	15.0 (6)				

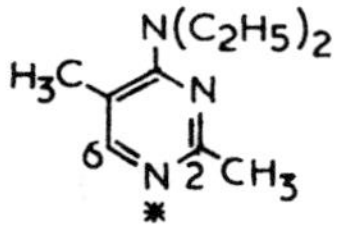	0.8 (2) 0.0 (6)	10.3 (2') 2.7 (5)	0.4 (5') 2.8 (4)	$CDCl_3$	B 14
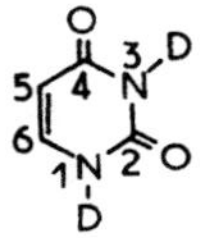	16.9 (1-2) 11.2 (1-6) 16.9 (2-3) 8.5 (3-4)	6.8 (3-5)		DMSO-d_6	L 18
monoanionic species	6.7 (1-6)	(a) (3-5)		30mg/ml D_2O pD 12	L 18
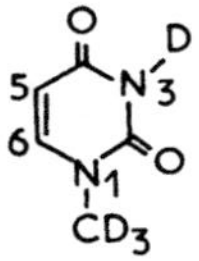	12.8 (1-6)	5.5 (3-5)		30mg/ml D_2O	L 18
monoanionic species	12.8 (1-6)	$\leqslant$1.0 (3-5)		30mg/ml D_2O pD 12	L 18
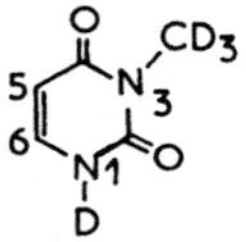	11.6 (1-6)	4.9 (3-5)		30mg/ml D_2O	L 18
monoanionic species	$\leqslant$1.0 (1-6)	(a) (3-5)		30mg/ml D_2O pD 12	L 18

Table 7.42 (continued)

Six-Membered Heterocyclic Compounds	^1J	^2J	^3J	Conditions	Ref.
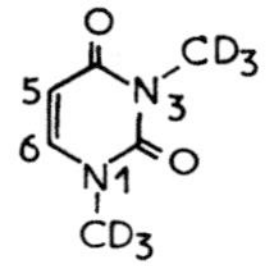	12.2 (1-6)	5.5 (3-5)		30mg/ml D_2O	L 18
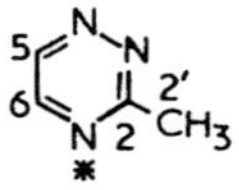	3.6 (2) 1.0 (6)	9.3 (2') 0.9 (5)		$CDCl_3$	B 14
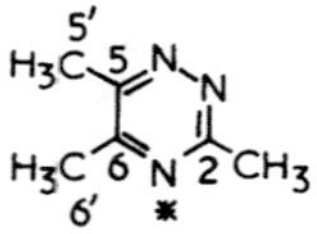	2.6 (2) 0.7 (6)	9.2 (2') 8.9 (6') 1.0 (5)	0.5 (5')	$CDCl_3$	B 14
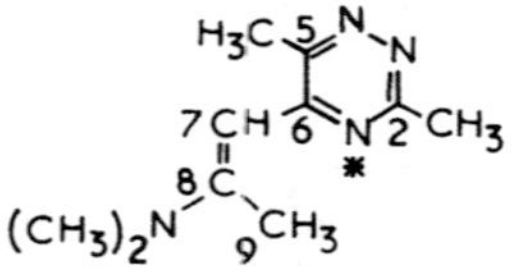	1.2 (2) 0.3 (6)	10.5 (2') 5.5 (7) 0.0 (5)	0.0 (5') 0.0 (8) ^4J = 3.9 (9)	$CDCl_3$	B 14
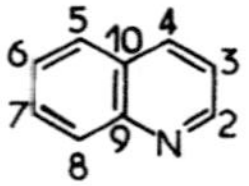	1.4 (2) 0.6 (9)	2.7 (3) 2.1 (10) 9.3 (8)	3.5 (4) 0.0 (5) 3.9 (7) ^4J = 0.9 (9)	50% CCl_4	P 3

	15.9 (2) 13.8 (9)	1 (3) 1 (8) <1 (10)	4.6 (4) 0 (5) 2.7 (7) ^4J = 0 (9)	30% H_2SO_4	P 3
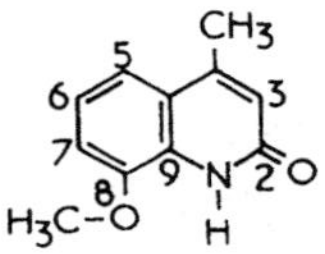	13.3 (2) 14.3 (9)	7.6 (3)	1.0 (5) 0.5 (7)	$CDCl_3$	N 1
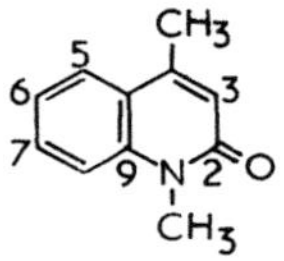	12.2 (2) 14.2 (9) 11.5 (1')	6.5 (3)	1.0 (5) 2.0 (7)	$CDCl_3$	N 1
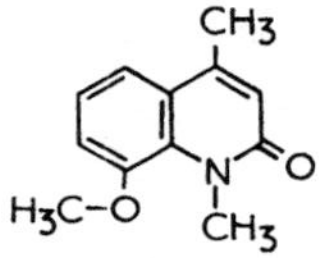	11.6 (2) 13.9 (9) 10.2 (1')	6.5 (3) 0.9 (8)	0.9 (5)	$CDCl_3$	N 1
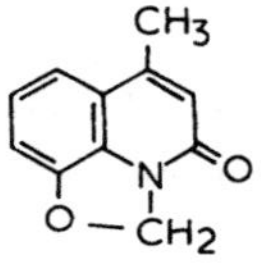	13.5 (2) 13.9 (9) 8.8 (1')	7.9 (3) 3.2 (8)	0.5 (5) 0.5 (7)	$CDCl_3$	N 1
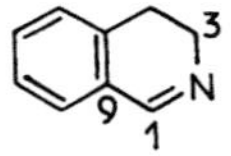	2.9 (1) 3.4 (3)	2 (9)	<1	$CDCl_3$	C 15

Table 7.42 (continued)

Six-Membered Heteroxyclic Compounds	^{1}J	^{2}J	^{3}J	Conditions	Ref.
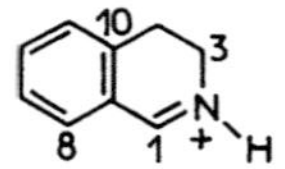	15.6 (1) 5.9 (3)	< 1	1.5 (8) 2 (10)	$CDCl_3$ + CF_3COOH	C 15
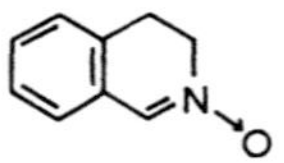	21.5 (1) 7.8 (3)	< 1	3.4 (8) 2.4 (10)	$CDCl_3$	C 15
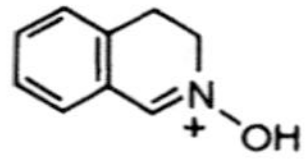	20.5 (1) 6.8 (3)	< 1	3.9 (8) 2.4 (10)	$CDCl_3$ + CF_3COOH	C 15
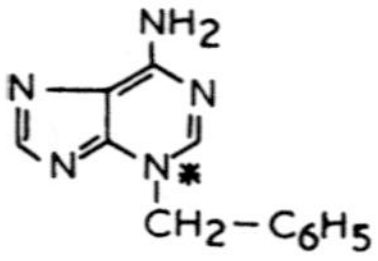	7.2 (CH_2)			50% DMSO-d_6 50% CF_3COOH	W 5

(a) Unresolved

Table 7.43. $^{n}J_{^{15}N-^{13}C}$ Coupling constants of polyheterocyclic nitrogen compounds - Obtained in ^{13}C-NMR

Polyheterocyclic Compounds	^{1}J	^{2}J	Conditions	Ref.
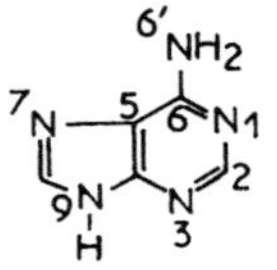	20.5 (6-6') 7.3 (5-7) 9.5 (3-4)		DMSO-d_6	Y 2
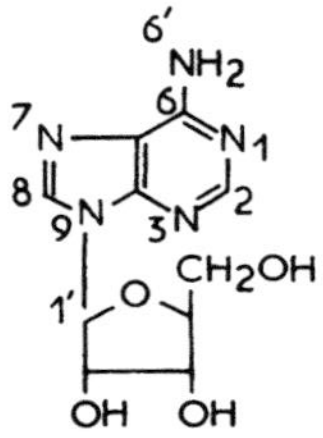	20.5 (6-6') 19.3 (4-9) 11.1 (9-1') 10.4 (7-8) 8.5 (5-7) 4.4 (3-4)		100mg in 1.5ml DMSO-d_6	K 4
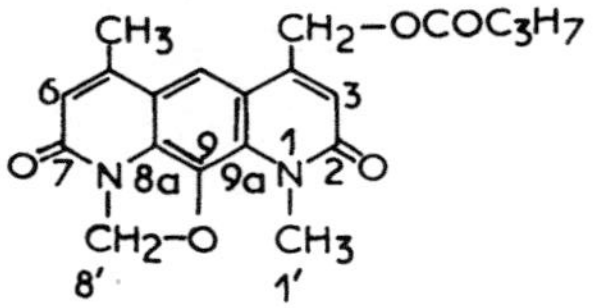	12.2 (2) 15.8 (9a) 10.1 (1') 13.6 (7) 12.2 (8a) 9.3 (8')	6.5 (3) 3.6 (9) 7.2 (6)	$CDCl_3$	N 1
$[^{15}N_4]$ octaethyl porphyrin H_2^{2+}	13.4		CF_3COOH	K 14b

Table 7.43 (continued)

Polyheterocyclic Compounds	^{1}J	^{2}J	Conditions	Ref.
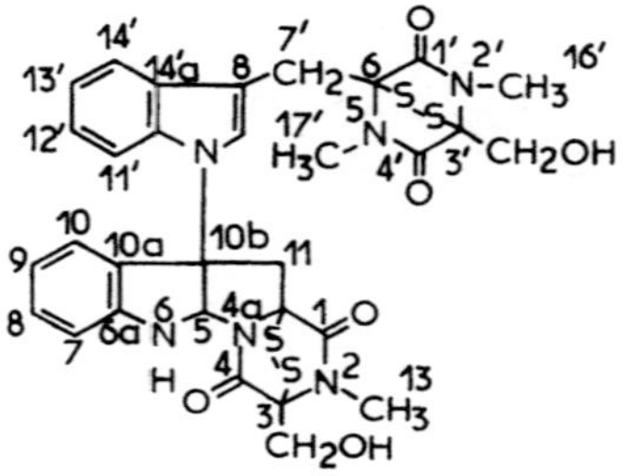	13.7 (2-1)	7.3 (2-11a)	$CDCl_3$	B 54
	7.3 (2-3)			M 48
	9.3 (2-13)			
	14.6 (4a-4)	6.2 (4a-3)		
	5 ± 1 (4a-5)	4.2 (4a-10b)		
	5.0 (4a-11a)	< 1.0 (4a-11)		
	13.7 (2'-1')	7.3 (2'-6')		
	7.3 (2'-3')			
	9.3 (2'-16')			
	13.7 (5'-4')	6.7 (5'-3')		
	6.3 (5'-6')			
	9.3 (5'-17')			
	8.1 (6-5)	3.7 (6-10a)	^{3}J = 1.3	
	11.8 (6-6a)	1.1 (6-10b)	(6-11a)	
	13.8 (10'-9')	4.3 (10'-8')		
	14.5 (10'-10'a)	4.7 (10'-14'a)		
	11.5 (10'-10b)	1.5 (10'-5)		

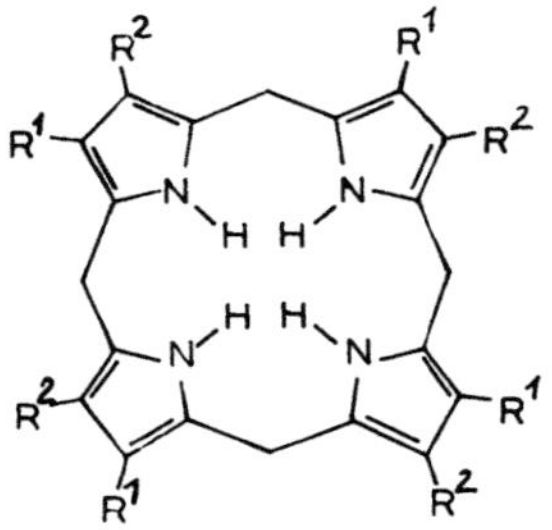	(a)	6		/	B 55
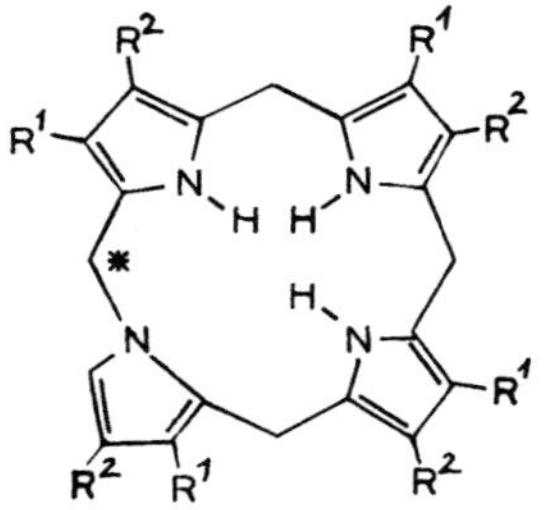	(a)	6		/	B 55
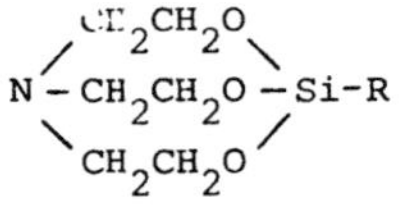	R=CH_3	7.8	1.5	$CDCl_3$	L 42
	CH_2=CH	7.3	1.2	Acetone-d_6	L 42
	C_6H_5	7.0	1.2	Acetone-d_6	L 42
	$ClCH_2$	7.0	1.2	$CDCl_3$	L 42
	CH_3O	6.7	0.8	Acetone-d_6	L 42
	C_2H_5O	7.0	0.9	Acetone-d_6	L 42

(a) $R^1 = CH_2COOH$ $R^2 = CH_2CH_2COOH$

Table 7.44. $^{n}J_{^{15}N-^{13}C}$ Coupling constants of oximes - Obtained in ^{13}C-NMR

Oximes			^{1}J	^{2}J	^{3}J	^{5}J	Conditions	Ref.
R^1R^2-C=N-OH								
R^1	R^2							
H	H		2.96				H_2O	L 12
H	CH_3	E	4.0	9.0			H_2O	L 12
		Z	2.3	1.8			H_2O	L 12
H	C_2H_5	E	2.4	7.3	0		H_2O	L 12
		Z	1.6	1.4	0		H_2O	L 12
CH_3	C_2H_5	E	3.6	0 (CH_3) 10.5 (CH_2)	0		$CDCl_3$	L 12
		Z	3.4	11.1 (CH_3) 1.8 (CH_2)	0		$CDCl_3$	L 12
C_2H_5	C_2H_5		2.7	10.2 (a) 2 (b)	2.1 (a) 0 (b)		$CDCl_3$	L 12
CH_3	CH_2=CH	E	4.3	1.8 (CH_3) 11.6 (=CH)	6.1		C_6D_6	L 12
		Z	3.5	11.6 (CH_3) 1.8 (=CH)	2.4		C_6D_6	L 12
H	C_6H_5-CH=CH (Z)	E	5.2	9.1	5.6		C_6D_6	L 12
		Z	(c)	(c)	3.9		C_6D_6	L 12
H	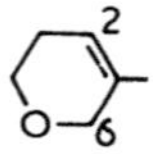	E	4.4	7.1	4.6 (2) 2.5 (6)		$CDCl_3$	D 1
		E	5.0	11.0 (2) 1.4 (6)	6.1 (3)		C_6D_6	L 12
		Z	4.0	1.8 (2) 2.4 (6)	3.1 (3)		C_6D_6	L 12

	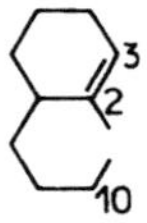	E	3.7	8.2 (2)	1.5 (10)		3.3 (3)		$CDCl_3$	D 1
	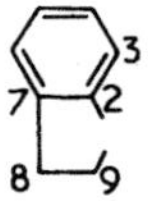	E	3.8	9.9 (2)	+0.9 (9)	0.7 (3)	2.4 (7) (d) (8)		0.1M $CDCl_3$	B 18
	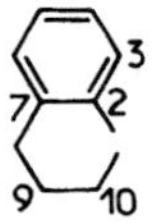	E	4.2	8.6 (2)	+1.4 (10)	3.8 (3)	2.9 (7) (d) (9)		0.1M $CDCl_3$	B 18
						C_2	C_6			
H	C_6H_5	E	4.8		7.3	2.8	2.8	0.8	$CDCl_3$	L 12
H	C_6H_4-2Cl	E	5.0		7.9	2.9	3.4	0.7	0.5M $CDCl_3$	D 1
H	C_6H_4-2OCH_3	E	5.1		7.3	2.1	2.9	0.7	0.5M $CDCl_3$	D 1
H	C_6H_4-2NO_2	E	4.2		8.7	2.4	3.8	0.4	0.5M DMSO	D 1
H	C_6H_4-3Cl	E	4.7		7.6	3.0	3.0	0.9	0.5M $CDCl_3$	D 1
H	C_6H_4-3OCH_3	E	4.9		7.3	3.1	2.9	0.8	0.5M $CDCl_3$	D 1
H	C_6H_4-3NO_2	E	3.8		8.4	3.5	3.0		0.5M DMSO	D 1
H	C_6H_4-3CN	E	4.0		8.2	3.0	2.9		0.5M DMSO	D 1
H	C_6H_4-4Cl	E	4.5		7.6	3.0	3.0		0.5M $CDCl_3$	D 1
H	C_6H_4-4OCH_3	E	4.9		7.3	3.7	3.7		0.5M $CDCl_3$	D 1

Table 7.44 (continued)

	Oximes		^{1}J	^{2}J		^{3}J		^{5}J	Conditions	Ref.
H	C_6H_4-4N$(CH_3)_2$	E	4.9		7.3	2.9	2.9		0.5M $CDCl_3$	D 1
H	C_6H_4-4NO_2	E	3.7		8.5	3.7	3.7		0.5M DMSO	D 1
H	C_6H_4-4CN	E	4.9		7.3	3.7	3.7	1.0	0.5M $CDCl_3$	D 1
H	C_6H_4-4CF_3	E	5.1		9.2	3.2			0.5M $CDCl_3$	D 1
H	Mesityl	E	2.8 ± 0.1		6.2 ± 0.1	1.2 ± 0.1			DMSO	C 14
				C_1	$C_{(CH_3)}$					
CH_3	C_6H_5	Z	-3.1	+1.8	-11.6	(d)			0.1M $CDCl_3$	B 17
		E	-3.9	-9.1	+1.0	(-)2.9			0.1M $CDCl_3$	B 17
		E	3.9	9.3	1.0	2.9			0.5M $CDCl_3$	B 13
		E	2.4	9.7	1.5	2.7			0.5M Acetone-d_6	B 13
CH_3	C_6H_4-2CH_3	E	3.8	8.8	<0.6	1.1 (2)	1.2 (6)		0.1M $CDCl_3$	B 18
CH_3	C_6H_4-4CH_3	E	3.9	9.2	0.7	2.8			0.5M $CDCl_3$	B 13
CH_3	C_6H_4-4CL	E	4.0	9.5	0.8	2.9			0.5M $CDCl_3$	B 13
CH_3	C_6H_4-4Br	E	3.8	9.5	0.6	2.7			0.5M $CDCl_3$	B 13
CH_3	C_6H_4-4NO_2	E	2.7	10.1	<0.6	3.4			0.5M Acetone-d_6	B 13
CH_3	Mesityl	E	3.3	8.0	<0.6	1.2			0.1M $CDCl_3$	B 18

(a) Anti with respect to OH
(b) Syn with respect to OH
(c) Not observed
(d) Non-resolvable (i.e. < 0.6 Hz)

Table 7.45. $^{n}J_{^{15}N-^{13}C}$ Coupling constants of compounds with C=N bonds

Compounds with -C=N- Bonds	^{1}J	^{2}J		^{3}J	Conditions	Ref.
C_6H_5-CH=N-CH_3	7.1 ± 0.3 (CH) < 3 (CH_3)				neat (a)	B 24
S=C=N-CH_3	13.4 ± 0.2 (CH_3)				neat (a)	B 24
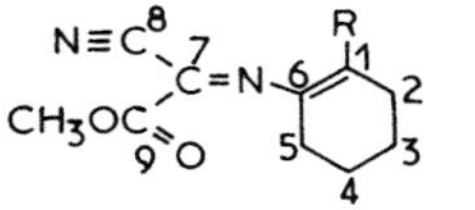 (b)	5.8 (6) 9.0 (7)	7.2 (1) 2.2 (8)	1.9 (5) 10.6 (9)	1.7 (2)	0.5-1M $CDCl_3$	F 3
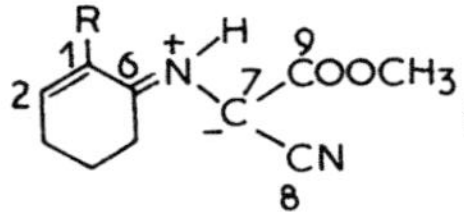 (b)	21.6 (6) 21.3 (7)	2.0 (8)	3.3 (9)	3.0 (2)	0.5-1M $CDCl_3$	F 3
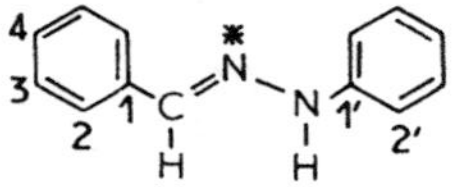	7.2	7.9 (1)	5.2 (1')	3.4 (2) 2.8 (2')	$CDCl_3$ ^{5}J = 1.0 (4)	L 12
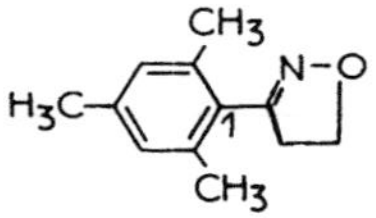		7 ± 1 (1)			Dioxane	C 14

Table 7.45 (continued)

Compounds with -C=N- Bonds	^1J	^2J	^3J	Conditions	Ref.
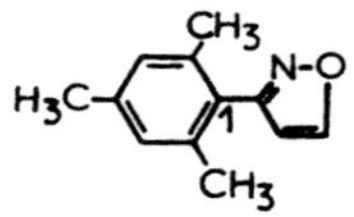	4.0 ± 0.5	7.0 ± 0.5 (1)		Dioxane	C 14
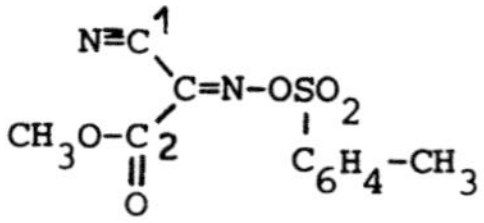		10.5 (2) 1.0 (1)		0.5-1M $CDCl_3$	F 3
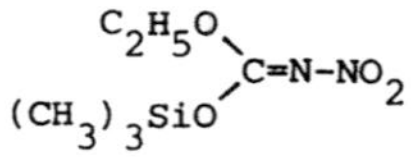	4.4 ± 0.2			CH_2Cl_2	I 1
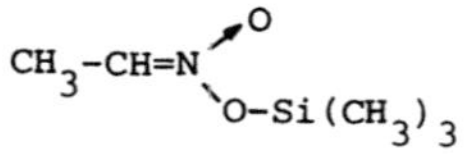	27.4				I 6
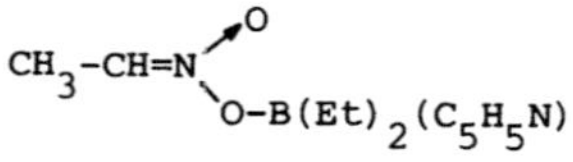	27.5				I 6
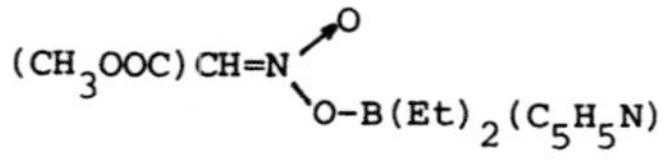	27.4				I 6
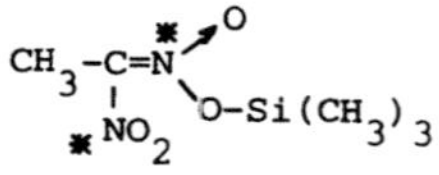	21.4 and 39.7				I 6

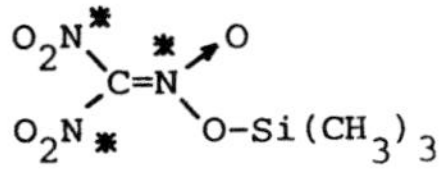		33.6		I 6
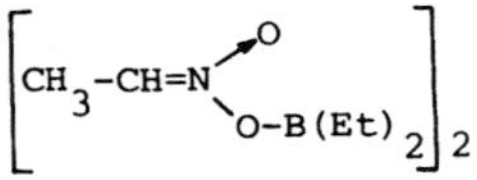		27.5		I 6
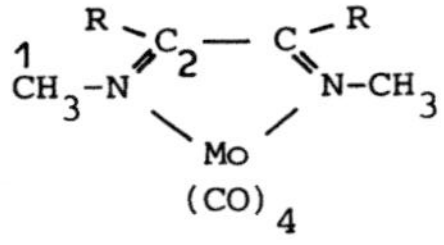	R=H	6.4 (1)	$CDCl_3$	S 13
		10.9 (2)		
	$R=CH_3$	5.5 (1)	$CDCl_3$	S 13
		3.9 (2)		

- Obtained in ^{13}C-NMR except (a) measured using ^{15}N spectroscopy

(b) R = morpholino

Ref. K 42 gives coupling constants for salicylidene-aniline in $CDCl_3$

Table 7.46. $^nJ_{^{15}N-^{13}C}$ Coupling constants of amides, ureas and imides

Amides, Ureas and Imides	^1J	^2J	^3J	Conditions	Ref.
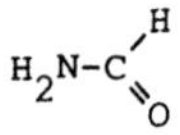	13.9			neat	H 6
	14			/	E 2
	-14.8			H_2O	W 10
	13.8			0.9M H_2O	H 7
$H_2N-CO-CH_3$	-14.4	-8.5		/	D 9
		14		/	E 2
	≃15	9.5		/ (a)	L 8
	< 15			H_2O (a)	B 24
 CO-NH$_2$	13.4	7.4	(b)	DMSO-d_6	L 14
 CO-NH$_2$	13.4	6.9	(b)	DMSO-d_6	L 14
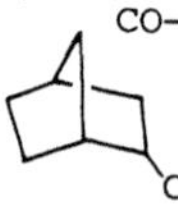 CO-NH$_2$	14.0	7.8	(b)	DMSO-d_6	L 14
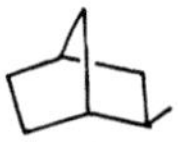 CO-NH$_2$	14.3	8.2	(b)	DMSO-d_6	L 14
 CO-NH$_2$	14.5	8.0	(b)	DMSO-d_6	L 14
 CO-NH$_2$	15.1	10.3	(b)	DMSO-d_6	L 14
$CH_2=CH-CO-NH_2$	15.0	9.0	(b)	DMSO-d_6	L 14
$\overset{4}{(CH_3)_2}\overset{3}{CH}-NH-\overset{1}{\underset{\underset{O}{\Vert}}{C}}-\overset{2}{C}(CH_3)_3$	13.2 (1)	6.5 (2)		$CDCl_3$	B 19
	9.5 (3)	0.4 (4)			
id, HCl	18.8 (1)	2.3 (2)		$CDCl_3$	B 19
	6.3 (3)				
$C_6H_5-\overset{1}{NH}-\overset{2}{CO}-CH_3$	13.0$\pm$1.5 (1)			CH_3OH (a)	B 24
	-14.3$\pm$0.5 (1)	9$\pm$0.3 (2)		20% $CDCl_3$	W 10

Table 7.46 (continued)

Amides, Ureas and Imides	^{1}J	^{2}J	^{3}J	Conditions	Ref.
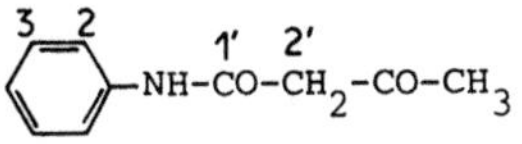	14.4 (1) 15.1 (1')	<0.5 (2) 7.2 (2')	1.4 (3)	$CDCl_3$	N 1
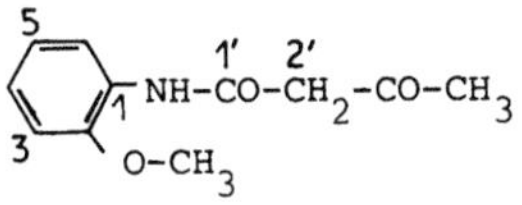	14.9 (1) 15.4 (1')	7.7 (2')	<0.5 (3) 1.0 (5)	$CDCl_3$	N 1
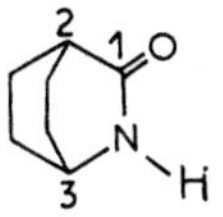	12.1 (1) 7.5 (3)	5.1 (2)		$CDCl_3$	B 19
	13.2 (1) 7.0 (3)	4.2 (2)		0.1M D_2O	K 1
id, HCl	16.2 (1) 6.3 (3)	2.5 (2)		$CDCl_3$	B 19
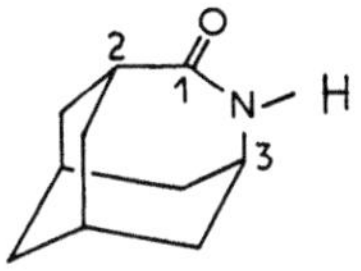	12.3 (1) 8.3 (3)	7.0 (2)		$CDCl_3$	B 19
id, HCl	15.1 (1) 7.2 (3)	3.5 (2)		$CDCl_3$	B 19
$(CH_3)_2$-N-CHO	13.4			/	L 8
H_2N-CO-NH_2	20.2 19 ± 1.5			0.1M H_2O D_2O - H_2O	L 6 S 6
3 1 CH_3-NH-CO-NH-CH_3	22±1.5 (1) 12±1.5 (3)			D_2O - H_2O	S 6
C_6H_5-NH-CS-NH-CH_3	<15 (1)			C_2H_5OH (a)	B 24
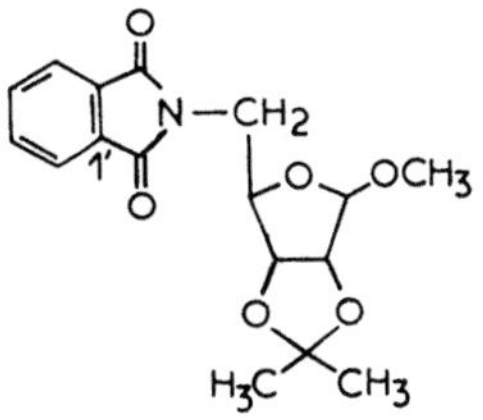	13.4 (CO) 9.8 (CH_2)	8.5 (1')		/	C 19

Tables 7.46 (continued)

Amides, Ureas and Imides	^{1}J	^{2}J	^{3}J	Conditions	Ref.
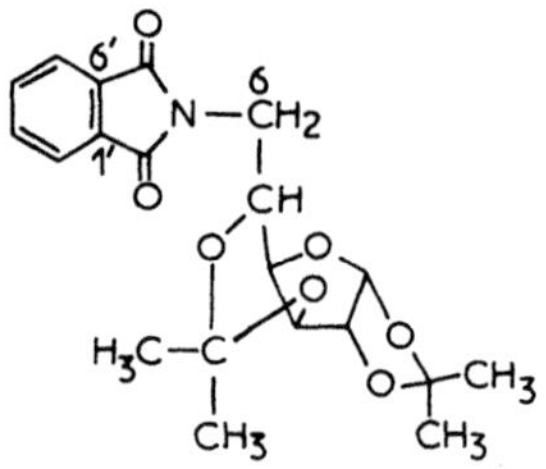	14.6 (CO) 9.8 (6)	7.3 (1')		/	C 19
	12.9 (CO) 9.9	11.4 (1') 11.4 (6')		C_6D_6	C 6

- Obtained in ^{13}C-NMR except (a) measured using ^{15}N spectroscopy

(b) Not observed

Footnote to Table 7.47

- Obtained in ^{13}C-NMR except (a) measured using ^{15}N spectroscopy

(b) $^{3}J_{N\text{-}CO\text{-}C\text{-}C}$ is < 1 Hz (see also G 23)

(c) pH-dependent

(d)

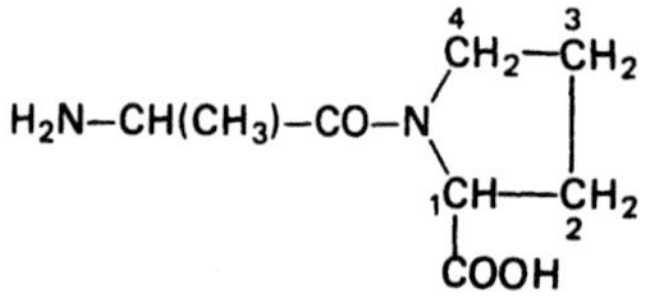

(e) Value for deuterated amide

Cross-peptide bond coupling constants are also reported for tBoc[U-^{13}C] Ala[^{15}N]Gly-OMe and the corresponding cyclic diketopiperazine (N 12)

Table 7.47. $^{n}J_{^{15}N-^{13}C}$ Coupling constants of small peptides and derivatives

Small Peptides and Derivatives	^{1}J	^{2}J	Conditions	Ref.
Ac-Ala^{15}N-NH-CH_3	14.5 (CO) 11.0 (CH)	7.2 (CH_3-CO)	0.1M D_2O	K 1
Ac-L-Phe^{15}N-NH-CH_3	14.4 (CO) 11.0 (CH)		CD_3OH	G 7
Ac-L-Phe^{15}N-ND-CH_3		1.8 (CO) <1 (CH_2)	CD_3OD	G 7
Ac-Pro-NH_2^{15}N			H_2O (b)	D 7
Gly^{15}N-Gly^{15}N	17.7 ± 0.15 to 18.9 ± 0.15		H_2O (c) (a)	I 4
cyclo(Gly^{15}N-Pro)			H_2O (b)	D 7
cyclo(Gly^{15}N-Phe)			H_2O (b)	D 7
Ala-Pro^{15}N (d) cis	9.8 (1) 9.3 (4)	2.2 (2) 1.7 (3)	H_2O pH 6 (a)	B 20
trans	10.2 (1) 9.0 (4)	2.4 (2) 1.7 (2)	H_2O pH 6 (a)	B 20
Gly^{1}-L-Pro-Gly^{2} Gly^{2}-L-Pro-Gly^{1}				W 12
Gly^{1}(^{15}N)	15.90 17.7 18.6 16.25 (e)		DMSO CF_3 - CD_2OH CF_3COOH DMSO	
Gly^{2}(^{15}N)	14.76 15.7 16.5 14.79 (e)		DMSO CF_3 - CD_2OH CF_3COOH DMSO	
Boc-Gly-L-Pro-Gly-OBz Gly^{2}(^{15}N)	14.3 14.7 16.3		$CDCl_3$ DMSO CF_3 - CD_2OH	W 12
Boc-Gly-Gly-OMe Gly^{2}(^{15}N)	15.2 14.6 16.3		$CDCl_3$ DMSO CF_3 - CD_2OH	W 12

Table 7.48. $^{n}J_{^{15}N-^{13}C}$ Coupling constants of compounds with C≡N bond

Nitriles and Derivatives	^{1}J	^{2}J	^{3}J	Conditions	Ref.
$^{-}$C≡N	5.4			H_2O pH 12.5	L 32
	$\pm$5.9			H_2O	M 5
$\overset{2}{C}H_3-\overset{1}{N}$≡C	(+)5.8 (1) -10.7 (2)			/	N 11 M 46
CH_3-C≡N	$\mp$17.5$\pm$0.4	$\pm$3.0$\pm$0.4		/ (a)	M 28
	-17.5			/	M 5
	17.5$\pm$0.2			neat (a)	B 24
C_2H_5-C≡N	(-)16.4			/	M 5
i.C_3H_7-C≡N	(-)15.4			/	M 5
t.C_4H_9-C≡N	(-)15			/	M 5
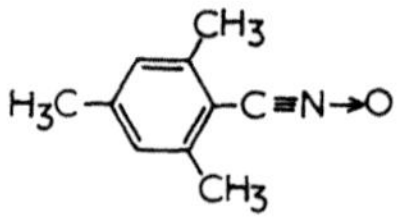	77.5$\pm$0.5	2.2$\pm$0.1	1.0$\pm$0.1	CH_2Cl_2 (b)	C 14
Complex C_2H_5-N≡C + myoglobin	18.5			5mM Phosphate Buffer	M 2
C_2H_5-N≡C free	9.7			id.	M 2
Complex n.C_3H_7-N≡C + myoglobin	19.6			5mM Phosphate Buffer	M 2
n.C_3H_7-N≡C free	9.7			id.	M 2
Complex 2C_2H_5-N≡C + Fe^{II} tetraphenyl porphyrin	22.4			40mM C_5H_5N	M 2
C_2H_5-N≡C free	6.9			id.	M 2

- Obtained in ^{13}C-NMR except (a) measured using ^{15}N spectroscopy
(b) Other couplings measured $^{4}J < 0.5$ Hz; $^{5}J = 0.8 \pm 0.1$ Hz
$^{4}J_{CH_3}$ and $^{6}J_{CH_3}$ < 0.5 Hz

Table 7.49. $^{n}J_{^{15}N-^{13}C}$ Coupling constants of compounds with N-N bonds - Obtained in ^{13}C NMR

Compounds with N-N Bonds		^{1}J	^{2}J	^{3}J	Conditions	Ref.
$(CH_3)_2$-N-NH_2		10.1 (1)	<1		/	L 8
C_6H_5-NH-NH-CO-CH_3 (1 2 3)	Z	10.1 (1) 12 (2)	12 (2-3)		DMSO	E 2
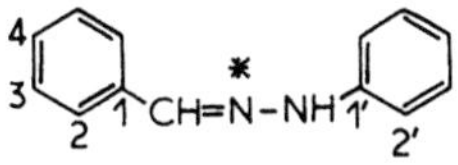		7.2	7.9 (1) 5.2 (1')	3.4 (2) 2.8(2')	$CDCl_3$ (a)	L 12
$(CH_3)_2$-N-N=O			7.5 (b) 1.4 (c)		/	L 8
CH_3-NH-NO_2		8.5 ± 0.5			CH_2Cl_2	I 1
$(CH_3)_2$-N-NO_2		9.1 ± 0.2			CH_2Cl_2	I 1
CH_3(N-NO_2)-C(=O)-OCH_3		8.0±0.2 (CH_3) 21.0±0.2 (CO)			CH_2Cl_2	I 1
CH_3(N-NO_2)-CO-OC_2H_5		8.1±0.2 (CH_3)			CH_2Cl_2	I 1
CH_3(N-NO_2)-Si$(CH_3)_3$		6.4 ± 0.2			CH_2Cl_2	I 1
C_2H_5O-CO(N-NO_2)-Si$(CH_3)_3$		17.6 ± 0.2			CH_2Cl_2	I 1
CH_3-N$(NO_2)_2$		(d)			CH_2Cl_2	I 1
C_2H_5O(C=N-NO_2)-OSi$(CH_3)_3$		4.4 ± 0.2			CH_2Cl_2	I 1
CH_3-N=N*(→O)-O-CH_3 (1, 2)	E		2.8±0.2 (1) 1.4±0.2 (2)		CH_2Cl_2	I 1
	Z		2.3±0.2 (1) 1.8±0.2 (2)		CH_2Cl_2	I 1

Table 7.49 (continued)

Compounds with N-N Bonds	^1J	^2J	^3J	Conditions	Ref.
$\overset{1}{CH_3}-N{=}\overset{*}{N}(\to O)-O-Si(CH_3)_3$ Z		2.0±0.5 (1)		243 K CH_2Cl_2	I 1
$\overset{1}{CH_3}O-CO-N{=}\overset{*}{N}(\to O)-\overset{2}{OCH_3}$		4.5±0.5 (1) 3 ± 0.5 (2)		CH_2Cl_2	I 1
$\overset{2'}{C_6H_5}-N(\to O){=}N-\overset{1'}{C_6H_5}$ (O←N 2, N 1)	1.3 (1-1') 12.5 (2-2')	2.5 (1-2') <0.5 (2-1')		253 K $CDCl_3$	S 9
$C_6H_5-N(\to O){=}N-C_6H_5$ (O←N 2, N 1)	3.7 (1-1') 18.2 (2-2')	9.4 (1-2') <0.5 (2-1')		253 K $CDCl_3$	S 9
$\overset{2'}{C_6H_5}-N(=N-R){=}N-\overset{1'}{C_6H_5}$ (R-N 2, N 1) R=Phthalimido	2.0 (1-1') 12.3 (2-2')	2.0 (1-2') <0.5 (2-1')		253 K $CDCl_3$	S 9
$C_6H_5-N(=N-R){=}N-C_6H_5$ (R-N 2, N 1)	3.0 (1-1') 13.6 (2-2')	6.9 (1-2') 1.5 (2-1')		253 K $CDCl_3$	S 9
$\overset{+}{N}{\equiv}N-\overset{-}{C}H-COOC_2H_5$ (1 2)	21.2	1.0		1-2M $CDCl_3$	A 12
	20.8 and 23.2 (e)	3.2 (1) / (2)		223 K 1M HCl	L 2
	21.36	3.66(1) 1.2 (2)		293 K 1M HCl	L 2
$H_2C{=}\overset{+}{N}{=}\overset{-}{N}$	-20.2 ± 1.8			$CDCl_3$	R 2
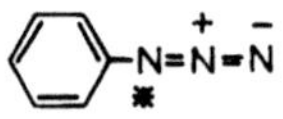	0.0 0.0	4.74 4.74	2.45 2.02	$CDCl_3$ (f) Acetone-d_6 (g)	M 24 M 24
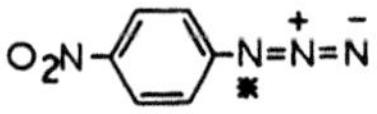	0.0	3.7	1.5	DMSO-d_6 (g)	M 24
$N{\equiv}C-\overset{*}{N}{=}\overset{*+}{N}{=}\overset{*-}{N}$ (h)	0.9	4.7	2.4	238 K CD_3CN	M 24

Table 7.50. $^{n}J_{^{15}N-^{13}C}$ Coupling constants of compounds with N-O bonds - Obtained in ^{13}C-NMR

Compounds with N-O Bonds		^{1}J	^{2}J	Conditions	Ref.
$(CH_3)_2$-CH-O-N=O			< 2	/	D 2
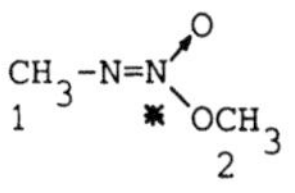	E		2.8 ± 0.2 (1) 1.4 ± 0.2 (2)	CH_2Cl_2	I 1
	Z		2.3 ± 0.2 (1) 1.8 ± 0.2 (2)	CH_2Cl_2	I 1
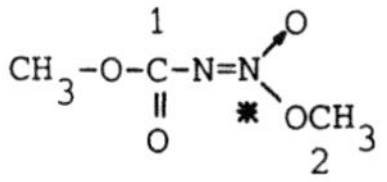			4.5 ± 0.5 (1) 3 ± 0.5 (2)	CH_2Cl_2	I 1
CH_3-NO_2		-10.5		/	W 2 A 13 B 49
$CH_3-CH_2-NO_2$		9.2		/	I 6
$CH_3-CH(NO_2)_2$		10		/	I 6
$C_6H_5-NO_2$		-14.57	-1.67	70% Acetone-d_6 (a)	E 1
		14.5		/	N 1 W 13

(a) Other couplings measured ^{3}J = -2.32 Hz and ^{4}J = ±0.6 Hz

Footnote to Table 7.49

(a) Other coupling measured $^{5}J_{C_4}$ = 1.0 Hz

(b) C is anti with respect to N = O

(c) C is syn with respect to N = O

(d) Not observed

(e) Due to Z and E forms

(f) Other coupling measured ^{4}J = 0.32 Hz

(g) $^{4}J \simeq 0$ Hz

(h) The product contains N≡C-Br

Table 7.51. $^{1}J_{^{15}N-^{15}N}$ Coupling constants of diaza derivatives - Obtained in ^{15}N-NMR

Diaza Derivatives with $^{15}N-^{15}N$ Bonds	^{1}J	Conditions	Ref.
C_6H_5-NH-NH_2	6.7	DMSO	B 7
C_6H_5-NH-NH-$COCH_3$ E	3.0	DMSO-d_6	E 2
Z	3.6	DMSO-d_6	E 2
C_6H_5-NH-N=CH-C_6H_4-NO_2p.	10.7	DMSO	B 7
$(C_6H_5)_2$N-N=O	22	DMSO	A 11
$(C_6H_5-CH_2)_2$N-N=O	19.0	DMSO	B 7
O_2N-N=O	11.7	CH_2Cl_2 173 K	S 1
$CH_2(N(O^-)-N=O)_2$ $2Na^+$	17.5	D_2O	S 1
CH_3-NH-NO_2	4.9 ± 0.2	CH_2Cl_2	I 1
$(CH_3)_2$N-NO_2	6.7 ± 0.5	CH_2Cl_2	I 1
	4.9		A 11
	4.9	$CDCl_3$	B 7
CH_3O-CO-NH-NO_2	4.4 ± 0.5	CH_2Cl_2	I 1
CH_3O-CO-N(CH_3)-NO_2	6.2 ± 0.3	CH_2Cl_2	I 1
C_2H_5O-CO-N(CH_3)-NO_2	5.9 ± 0.3	CH_2Cl_2	I 1
$(CH_3)_3$Si-N(CH_3)-NO_2	7.3 ± 0.3	CH_2Cl_2 243 K	I 1
$(CH_3)_3$Si-N($COOC_2H_5$)-NO_2	6.7 ± 1.0	CH_2Cl_2	I 1
CH_3-N-$(NO_2)_2$	12.2 ± 0.4	CH_2Cl_2	I 1
(CH_3-$\bar{N}$-NO_2) NH_4^+	12.2 ± 0.5	H_2O	I 1
(CH_3O-CO-$\bar{N}$-NO_2) NH_4^+	16.4 ± 0.5	H_2O	I 1
$(CH_3)_3$SiO-C(OC_2H_5)=N-NO_2	14.0 ± 1.2	CH_2Cl_2	I 1
O=N-NO_2	11.7	CH_2Cl_2 173 K	S 1
$(CH_2-N-NO_2)_3$	8.9	DMSO	B 7
$(CH_2-N-NO_2)_4$	4.5	DMSO	B 7
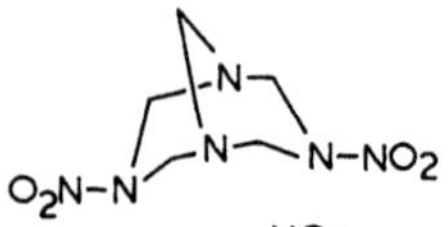	8.5	DMSO	B 7
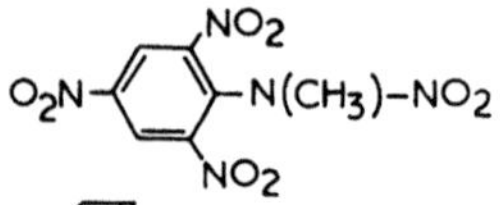	4.9	DMSO	B 7
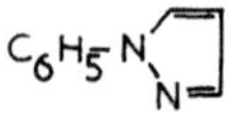	12.8 ± 1.2	0.9M $CDCl_3$	H 4
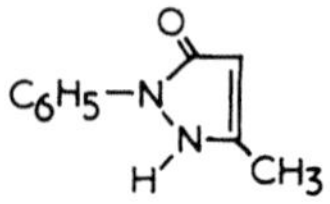	12.0 ± 0.3	0.7M $CDCl_3$	H 4
C_6H_5-N=N-C$(C_6H_5)_3$ trans	17		P 7

Table 7.51 (continued)

Diaza Derivatives with $^{15}N-^{15}N$ Bonds		^1J	Conditions	Ref.
$C_6H_5-N=N-C(CH_3)_2C\equiv N$	trans	17		P 7
	cis	21		P 7
$C_6H_5-N=N-C(CH_3)_2-C_6H_5$	trans	17		P 7
	cis	21		P 7
$C_6H_5-N=N-C_6H_4-OH$ p.		15 $\pm$ 1	Acetone	B 15
$C_6H_5-N(\rightarrow O)=N-C_6H_5$	trans	13.7	$CDCl_3$ 303K(a)	S 9
		12.5	$CDCl_3$ 253K	S 9
		14 $\pm$ 1	Ether	L 15
		13.7 $\pm$ 0.8	Ether	B 24
		12.5	Ether	S 1
	cis	12.5	$CDCl_3$ 253K	S1,S9
$CH_3-N=N(\rightarrow O)(OCH_3)$	trans	14.0 $\pm$ 0.5	CH_2Cl_2	I 1
	cis	12.8 $\pm$ 0.5	CH_2Cl_2	I 1
$CH_3-N=N(\rightarrow O)(O-Si(CH_3)_3)$	trans	11.0 $\pm$ 0.5	CH_2Cl_2 243 K	I 1
$CH_3-O-CO-N=N(\rightarrow O)(OCH_3)$		13.9 $\pm$ 0.3	CH_2Cl_2	I 1
$C_2H_5O-CO-N=N(\rightarrow O)(OCH(CH_3)_2)$		14.4 $\pm$ 0.3	CH_2Cl_2	I 1
$^-O-N=NO_2^-$ $2Na^+$		16.9	D_2O	S 1
$^-O-N=N(O^-)(SO_3)$ $2K^+$		21.6	D_2O	S 1
$^-N=\overset{+}{N}=O$		9.16 $\pm$ 0.3	CCl_4	B 31
		8.9	gaz	J 5
$N\equiv N-CH-COOC_2H_5$		5.65	1 M HCl 293 K	L 2
		5.62 (E)	1 M HCl 223 K	L 2
		5.13 (Z)		L 2
$[C_5(CH_3)_5]_2Ti-N\equiv N$		7 $\pm$ 2	$C_6D_5CD_3$ 212 K	B29,B30
$\{[C_5(CH_3)_5]_2Zr\}_2$ $N\equiv N$		6.2	$C_6D_5CD_3$ 245 K	M 8
trans $[Mo(N_2)_2dppe_2]$	(b)	4.4	THF	C 1
trans $[W(N_2)_2dppe_2]$	(b)	5.4	THF	C 1
cis $[Mo(N_2)_2(P\ Me_2Ph)_4]$		6.3	THF	C 1
cis $[W(N_2)_2(P\ Me_2Ph)_4]$		6.2	THF	C 1

Table 7.51 (continued)

Diaza Derivatives with $^{15}N-^{15}N$ Bonds	^{1}J	Conditions	Ref.
trans $[MoF(N\text{-}NH_2)dppe_2]BF_4$	(b) (c)	CH_2Cl_2	C 21
trans $[WF(N\text{-}NH_2)dppe_2]BF_4$	(b) (d)	CH_2Cl_2	C 21
$[Mo(N\text{-}NH_2)(8\text{-quin})(PMe_2Ph)_3]Cl$	(e) 10 $\pm$ 2	CH_2Cl_2	C 21
$[W(N\text{-}NH_2)(8\text{-quin})(PMe_2Ph)_3]Cl$	(e) 8.8 $\pm$ 1	CH_2Cl_2	C 21
$[MoCl(N\text{-}NH_2)(C_5H_5N)(PMe_2Ph)_3]Cl$	(f) 9.8 $\pm$ 1	CH_2Cl_2	C 21
$[WCl(N\text{-}NH_2)(C_5H_5N)(PMe_2Ph)_3]Cl$	(f) 10.5 $\pm$ 1	CH_2Cl_2	C 21

(a) Without proton decoupling

(b) dppe $Ph_2\text{-}P\text{-}CH_2\text{-}CH_2\text{-}P\text{-}Ph_2$

(c) $|{}^{1}J_{N-N} + {}^{3}J_{N-F}| = 24 \pm 3$ Hz

(d) $|{}^{1}J_{N-N} + {}^{2}J_{N-F}| = 62 \pm 3$ Hz

(e) 8-quin = quinolin-8-olate, O trans with respect to NNH_2

(f) Cl is trans with respect to NNH_2

Table 7.52. ${}^{n}J_{^{15}N-^{15}N}$ Coupling constants of triazenes and azides - Obtained in ^{15}N-NMR

Triazenes and Azides $^{15}N-^{15}N$		^{1}J		^{2}J	Condition	Ref.
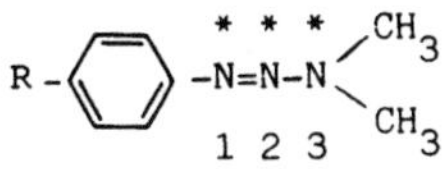		J_{1-2}	J_{2-3}			
R = NO_2		12.8	13.4		$CDCl_3$	A 11
Cl		12.8	13.4		$CDCl_3$	A 11
H		12.8	14.0		$CDCl_3$	A 11
OCH_3		12.2	14.0		$CDCl_3$	A 11
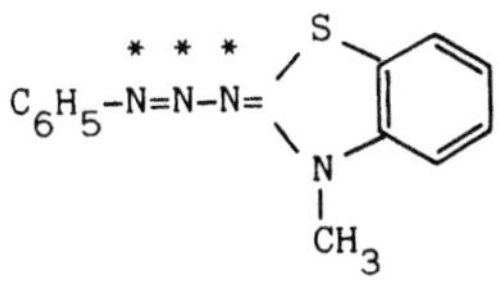	cis	18.9	15.0	0	C_5H_5N	F 1
	trans	13.2	17.7	10.7	C_5H_5N	F 1
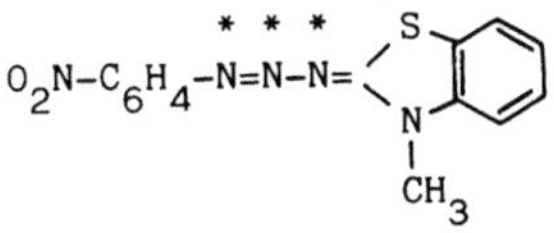	trans	11.8	11.8	5.9	CF_3COOH	F 1
id, 1-2 labelled only		10.3			CF_3COOH	F 1
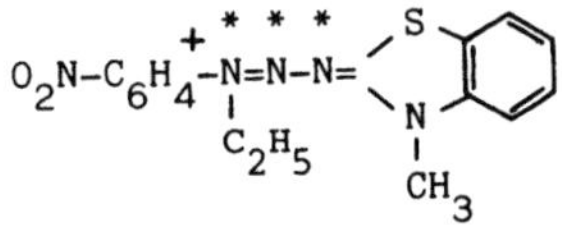	BF_4^-	13.5	11.8	5.1	CF_3COOH	F 1
id, 1-2 labelled only		12.5			CF_3COOH	F 1
$(C_2H_5)_2N-\overset{*}{N}(\overset{+}{\to}O^-)=\overset{*}{N}-O^-$ $(C_2H_5)_2\overset{+}{N}H_2$			16.9		$CDCl_3$ 263K	S 1
$H-\overset{*}{N}=\overset{*}{N}=\overset{*}{N}$ (1 2 3)		13.95	7.20		10% Ether	M 11
Li-N=N=N		11.35				M 10
Na-N=N=N		11.32				M 10
Cl-N=N=N		24.0	7.8		15% CH_2Cl_2	M 11
$(CH_3)_2P(S)\overset{*}{N}=\overset{*}{N}=\overset{*}{N}$		13.6	6.8		Acetone d_6	M 10
$(CH_3)_2P(Se)\overset{*}{N}=\overset{*}{N}=\overset{*}{N}$		14.3	6.8		C_6D_6	M 10
$(CH_3O)_2P(O)\overset{*}{N}=\overset{*}{N}=\overset{*}{N}$		14.4	5.8		CH_3CN	M 10
$[P(\overset{*}{N}_3)_2N]_3$		12.2	5.5		$C_6D_5-CD_3$	M 10

Table 7.52 (continued)

Triazenes and Azides $^{15}N-^{15}N$	^{1}J	^{2}J	Condition	Ref.
$CH_3-N=N=N$ (1 2 3)	14.4	8.2	30% C_6D_6	M 24
$C_6H_5-N=N=N$	13.4	7.8	25% Acetone d_6	M 24
$O_2N-C_6H_4-N=N=N$	13.8	7.4	10% DMSO-d_6	M 24
$(O_2N)_3-C_6H_2-N=N=N$	14.0	6.3	30% DMSO-d_6	M 24
N≡C-N=N=N	16.0	6.0	5% CD_3CN 253K	M 24
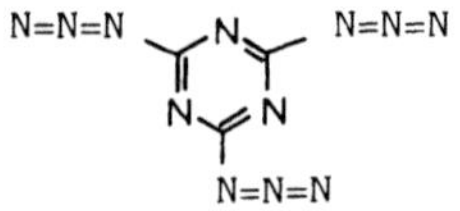	16.1	6.1	10% CD_2Cl_2	M 24
$[(CH_3)_2Al-N=N=N]_3$	9.9	8.8	$C_6D_6CD_3$ 308 K	M 25
	11.5		214 K	
	11.5	5.9	173 K	
$[(CH_3)_2Ga-N=N=N]_3$	11.1	9.8	$C_6D_6CD_3$ 233K	M 25
	12.6	7.3	183K	
$(CH_3)_2As-N=N=N$	12.0	12.0	C_6H_6 341 K	M 25
	12.2	12.2	308 K	
	12.2	12.0	233 K	
	11.9		213 K	
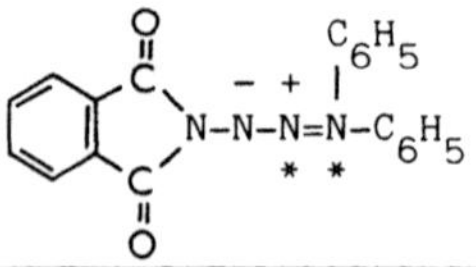 cis	15.0		$CDCl_3$ 253 K	S1,S9
trans	13.4		$CDCl_3$ 253 K	S1,S9

The coupling constants $^{1}J_{N-N}$ of trimethyl tin azide have been measured as a function of temperature (M 52).

Table 7.53. $^{2}J_{^{15}N-^{15}N}$ Coupling constants of polyaza derivatives

Polyaza Derivatives $^{15}N-^{15}N$	^{2}J	Conditions	Ref.
$H_2N-CO-NH_2$	5.1	1:1:1 20mg/ml (a) Acetone:DMSO:TMU	S 6
$CH_3-NH-CO-NH-CH_3$	5.3	40mg/ml DMSO-d_6(a)	S 6
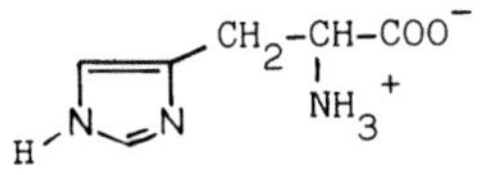	±0.9	pH 1.3 cation	B 35
	±0.6	pH 7.6 amphion	B 35
	±0.9	pH 10.9 anion	B 35
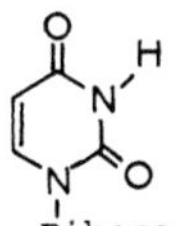 Ribose-3'Phosphate	2.2	0.08 M in H_2O pH 3 and 7	B 35
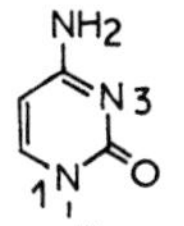 Ribose-3'Phosphate		0.08 M in H_2O	
	1.5(3-NH_2)	pH 3	B 36
	5.8(3-NH_2)	pH 7	B 36
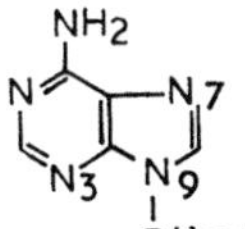 Ribose-3'Phosphate		0.08 M in H_2O	
	1.0(1-NH_2)	pH 3	B 36
	1.5(3-9)		B 36
	5.2(1-NH_2)	pH 7	B 36
	2.2(3-9)		B 36
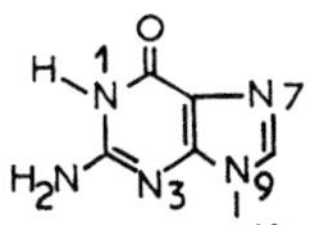 Ribose-3'Phosphate		0.08 M in H_2O	
	2.2(1-NH_2)	pH 3 and 7	B 36
	6.0(3-NH_2)		B 36
	3.7(3-9)		B 36
		0.08 M in H_2O	
	6.0(1-NH_2)	pH 10	B 36
	6.0(3-NH_2)		B 36
	3.7(3-9)		B 36
$(CH_3)_2N-P(\overset{*}{N}H-C_6H_5)_2$	2.2 ± 0.5	C_6H_6	M 26

Table 7.53 (continued)

Polyaza Derivatives $^{15}N-^{15}N$	^{2}J	Conditions	Ref.
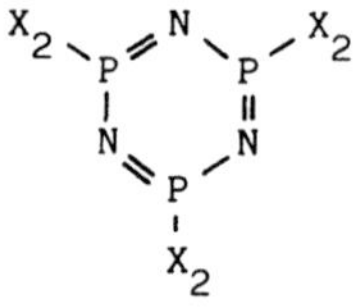 X = Cl or SR	<1	$CDCl_3$	T 4
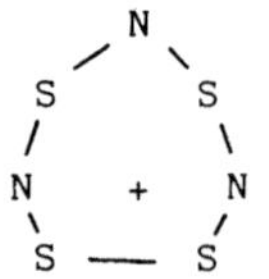 Cl^-	7 ± 1	3.1 M HNO_3 70%	L 4
Pheophytin a 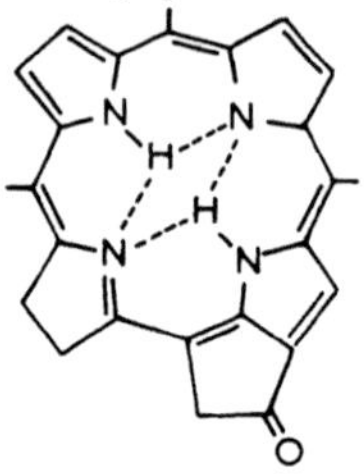	2.0 (1-2) 5.7 (2-3) 1.4 (3-4) 2.5 (1-4)	0.22 M $CDCl_3$	B 32
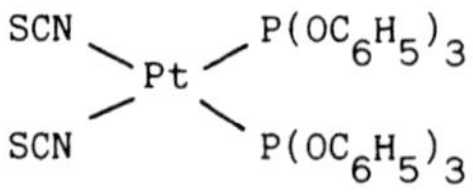	2	CH_2Cl_2	C 18
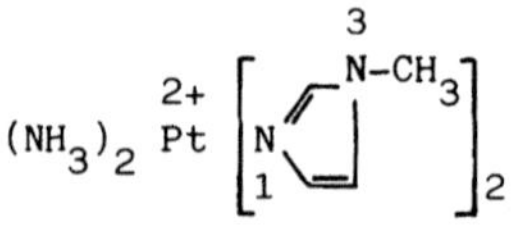 cis	5.4 (b)	H_2O	A 22
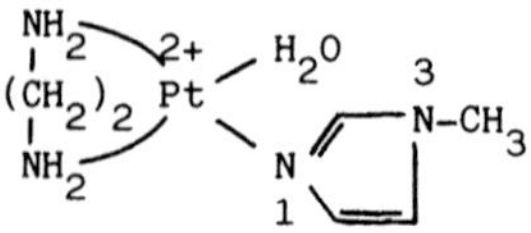	5.1 (b)	H_2O	A 22
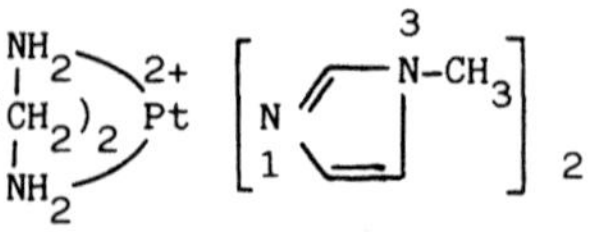	5.4 (b)	H_2O	A 22

- Obtained in ^{15}N-NMR except (a) measured using ^{1}H spectroscopy

(b) Refers to $^{2}J_{N_1-N_{amine}}$

Table 7.54. $^{n}J_{^{15}N-^{19}F}$ Coupling constants

$^{15}N - ^{19}F$	^{2}J	^{3}J	^{4}J	^{5}J	Conditions	Ref.
H_2N-PF_2	-6.4				$TMS:CFCl_3$	A 17
	-6.4				$TMS:C_6D_6$	A 16
$H_2N-P(CF_3)_2$		-1.53				C 12
H_2N-P(Fa)(Fa)(Fe)(Fe)	22.8(Fa)				$CHFCl_2$	C 13
$H-N(SiH_3)-PF_2$	+4.2				$TMS:CFCl_3$	A 17
$H-N(PF_2)_2$	-3.6				$TMS:C_6D_6$	A 16
$H-N(CO-CF_3)-CH_2R$ (e)		0.6			$CDCl_3$ (a)	C 6
$N(PF_2)_3$	±2.5				$TMS:C_6D_6$	A 16
2-Fluoropyridine	-52.64				$TMS:C_6D_6$ (a)	J 1
	52.5				$TMS:C_6D_6$	L 30
	23.1				CF_3COOH	L 30
3-Fluoropyridine		3.6				L 30
		3.1			CF_3COOH	L 30
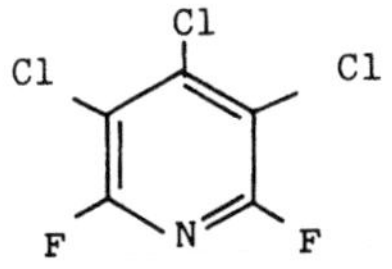	-52.2				$(CFCl_2)_2$ (b)	C7, H2
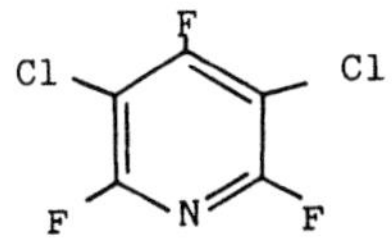	52.6				$(CBrF_2)_2$ (b)	H 2
8-Fluoroquinoline		2.9				L 30
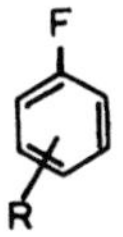 R = $2-NH_2$		0				L 30
$3-NH_2$			0			L 30
$4-NH_2$				1.5		L 30
$2-CH_3,4-NH_2$				1.7		L 30
$2-NHCOCH_3$		1.1				L 30
$3-NHCOCH_3$			0.9			L 30
$4-NHCOCH_3$				0.5		L 30
$2-NHSO_2C_6H_5$		1.9				L 30
$3-NHSO_2C_6H_5$			0.5			L 30

Table 7.54 (continued)

$^{15}N - ^{19}F$	^{2}J	^{3}J	^{4}J	^{5}J	Conditions	Ref.
4-$NHSO_2C_6H_5$				0.8		L 30
2-NH_3^+ CF_3COO^-		1.4				L 30
3-NH_3^+ CF_3COO^-			0.2			L 30
4-NH_3^+ CF_3COO^-				0		L 30
2-$CONH_2$			-7.0		5 % $CDCl_3$	F 2
			-3.2		5 % DMSO-d_6	F 2
$[MoF(N\text{-}NH_2)dppe_2]$ BF_4 (c) trans	77 ± 3(d)				CH_2Cl_2 (a')	C 21
$[WF(N\text{-}NH_2)dppe_2]BF_4$ (c) trans	58 (d)				CH_2Cl_2 (a')	C 21

- Obtained in ^{19}F-NMR except (a) measured using ^{15}N spectroscopy and (a') measured using ^{15}N- and ^{19}F-NMR

(b) From the ^{15}N satellites of the ^{19}F spectrum

(c) dppe $Ph_2\text{-}P\text{-}CH_2\text{-}CH_2\text{-}P\text{-}Ph_2$

(d) trans coupling

(e) R =

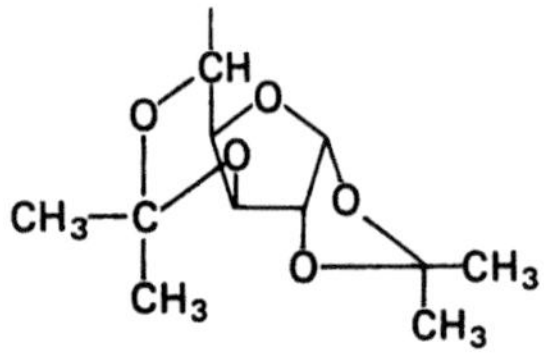

The magnitude of the one-bond coupling constant can be deduced from the value of $^{1}J_{^{14}N-^{19}F}$ which has been shown to vary between 120 and 160 Hz for neutral compounds (F4, H1, M6, R11) but is as high as 231 Hz for NF_4^+ (T1) and 328 Hz for N_2F^+ AsF_6^- (M18, R11). The corresponding values for $^{1}J_{^{15}N-^{19}F}$ can be estimated by multiplying the above data by $\gamma_{^{15}N}/\gamma_{^{14}N}$ (neutral compounds: 170 to 225 Hz, cationic species: 320 and 460 Hz). See also K 45.

Table 7.55. $^{1}J_{^{15}N-^{29}Si}$ Coupling constants - Obtained in ^{15}N-NMR

^{15}N — ^{29}Si	^{1}J	Conditions	Ref.
$(SiH_3)_3N$	+6	$(CH_3)_4Si$	A 17
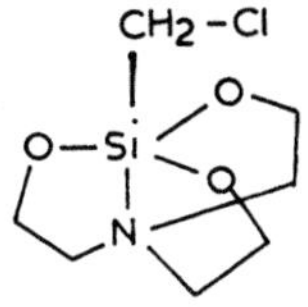	1.5	-	L 42

Table 7.56. $^{1}J_{^{15}N-^{31}P}$ Coupling constants of aminophosphines with P^{III} phosphorus atoms

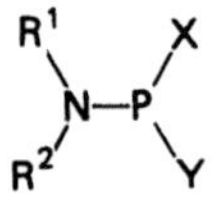

Aminophosphines $^{15}N-^{31}P$ R[1]	R[2]	X	Y	^{1}J	Conditions	Ref.
H	H	CF_3	CF_3	+52.6	(a)	C12,S7
H	H	F	F	+72.5	$(CH_3)_4Si:C_6D_6$ 1:1(a)	A16
				+73.0	$(CH_3)_4Si:CFCl_3$	A17
H	C_6H_5	CH_3	CH_3	+53.0	C_6H_6 (a)	M26,M27
H	C_6H_5	t.Bu	t.Bu	59.6	Mesitylene (a)	M26
H	C_6H_5	$N(CH_3)_2$	$(CH_3)_2N$	53.8	C_6H_6 (a)	M26
H	C_6H_5	$N(CH_3)_2$	C_6H_5NH	52.8	C_6H_6 (a)	M26
H	C_6H_5	$CH_3-N-CH_2-CH_2-N-CH_3$		84.2	C_6H_6 (a)	M26
H	SiH_3	F	F	+69.0	$(CH_3)_4Si:CFCl_3$ (a)	A17,M26
H	PF_2	F	F	+78.9	$(CH_3)_4Si:C_6D_6$ 1:1(a)	A16
CH_3	CH_3	CH_3	CH_3	60.0	neat	G10
CH_3	CH_3	CH_3	Cl	75.0	neat	G10
CH_3	CH_3	C_6H_5	C_6H_5	61.8	neat	G10
CH_3	CH_3	C_6H_5	Cl	76.2	neat	G10
CH_3	CH_3	C_6H_5	$(CH_3)_2N$	75.7	neat	G10
CH_3	CH_3	Cl	Cl	89.4	neat	G5,G10
CH_3	CH_3	Cl	$(CH_3)_2N$	77.6	neat	G10
CH_3	CH_3	Cl	$(Et)_2N$	79.4	neat	G10
CH_3	CH_3	Cl	$(iPr)_2N$	80.9	neat	G10
CH_3	CH_3	OEt	EtO	75.0	neat	G10
CH_3	CH_3	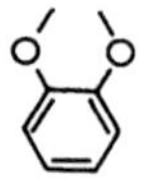		89.0	neat	G5
CH_3	CH_3	OEt	$(CH_3)_2N$	69.0	neat	G10
CH_3	CH_3	$N(CH_3)_2$	$(CH_3)_2N$	60.0	neat	G10
				59.3		C4
				+59.1		G4
CH_3	CH_3	$N(CH_3)_2$	$(Et)_2N$	54.4	neat	G10
CH_3	CH_3	$N(Et)_2$	$(Et)_2N$	50.0	neat	G10
CH_3	CH_3	$N(CH_3)_2$	$(i.Pr)_2N$	51.5	neat	G10
CH_3	CH_3	$N(Et)_2$	$(i.Pr)_2N$	52.9	neat	G10
CH_3	CH_3	$CH_3-N-CH_2-CH_2-N-CH_3$		91.2	neat	G11

Table 7.56 (continued)

Aminophosphines $^{15}N-^{31}P$				^1J	Conditions	Ref.
CH_3	$C_6H_5-CH_2$	C_6H_5	Cl	77.6	neat	G10
CH_3	$C_6H_5-CH_2$	Cl	Cl	90.8	neat	G10
CH_3	PCl_2	Cl	Cl	$+86 \pm 1$		C3
Et	Et	C_6H_5	Cl	79.1	neat	G10
Et	Et	C_6H_5	$(Et)_2N$	75.5	neat	G7
Et	Et	Cl	Cl	92.3	neat	G10
Et	Et	Cl	$(CH_3)_2N$	79.4	neat	G10
Et	Et	Cl	$(Et)_2N$	80.9	neat	G10
Et	Et	Cl	$(i.Pr)_2N$	79.1	neat	G10
Et	Et	$N(CH_3)_2$	$(CH_3)_2N$	72.1	neat	G10
Et	Et	$N(Et)_2$	$(Et)_2N$	61.5	neat	G10
Et	Et	$N(CH_3)_2$	$(i.Pr)_2N$	49.8	neat	G10
Et	Et	$N(Et)_2$	$(i.Pr)_2N$	52.9	neat	G10
Et	Et	$CH_3-N-CH_2-CH_2-N-CH_3$		92.6	neat	G11
$n.C_3H_7$	$n.C_3H_7$	C_6H_5	Cl	80.6	neat	G10
$n.C_4H_9$	$n.C_4H_9$	C_6H_5	Cl	80.6	neat	G10
$C_6H_5-CH_2$	$C_6H_5-CH_2$	C_6H_5	Cl	82.0	neat	G10
$C_6H_5-CH_2$	$C_6H_5-CH_2$	C_6H_5	$C_6H_5-CH_2$ N $C_6H_5-CH_2$	76.7	neat	G7
$i.C_3H_7$	$i.C_3H_7$	C_6H_5	Cl	82.0	neat	G10
$i.C_3H_7$	$i.C_3H_7$	Cl	Cl	95.2	neat	G10
$i.C_3H_7$	$i.C_3H_7$	Cl	$(CH_3)_2N$	83.0	neat	G10
$i.C_3H_7$	$i.C_3H_7$	Cl	$(Et)_2N$	86.4	neat	G10
$i.C_3H_7$	$i.C_3H_7$	Cl	$(i.Pr)_2N$	86.4	neat	G10
$i.C_3H_7$	$i.C_3H_7$	$N(CH_3)_2$	$(CH_3)_2N$	80.9	neat	G10
$i.C_3H_7$	$i.C_3H_7$	$N(CH_3)_2$	$(Et)_2N$	81.0	neat	G10
$i.C_3H_7$	$i.C_3H_7$	$N(Et)_2$	$(Et)_2N$	83.8	neat	G10
$i.C_3H_7$	$i.C_3H_7$	$CH_3-N-CH_2-CH_2-N-CH_3$		96.0	neat	H8,G11
$s.C_4H_9$	$s.C_4H_9$	C_6H_5	Cl	82.0	neat	G10
$s.C_4H_9$	$s.C_4H_9$	Cl	Cl	96.7	neat	G10
C_6H_5	$(CH_3)_3Sn$	CH_3	CH_3	71.9	C_6H_6 (a)	M26
C_6H_5	$(CH_3)_3Sn$	t.Bu	t.Bu	80.0	$C_6H_6 + CH_2Cl_2$ (a)	M26
CH_2 CH_2 $CH_2-CH_2-CH_2$		$CH_3-N-CH_2-CH_2-N-CH_3$		92.7	neat	G11

Table 7.56 (continued)

Aminophosphines $^{15}N-^{31}P$				^{1}J	Conditions	Ref.
PF_2	PF_2	F	F	+87	$(CH_3)_4Si:CFCl_3$ 1:1(a)	A16
$F_3P=$		F	F	+93.76(b)(a)		S7
				-53.20(b)(a)		S7
CH_3	$-CH_2-CH_2-N-CH_3$		C_6H_5	49.8	neat	G7
CH_3	$-CH_2-CH_2-N-CH_3$		Cl	+63.3	neat	B40
				63.2	neat	G11
CH_3	$-CH_2-CH_2-N-CH_3$		s.Bu-O	57.6	neat	G7
CH_3	$-CH_2-CH_2-N-CH_3$		$(CH_3)_2N$	51.5	neat	G11
CH_3	$-CH_2-CH_2-N-CH_3$		$(Et)_2N$	51.5	neat	G11
CH_3	$-CH_2-CH_2-N-CH_3$		$(i-Pr)_2N$	51	neat	H8
CH_3	$-CH_2-CH_2-N-CH_3$		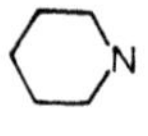	51.5	neat	G11
	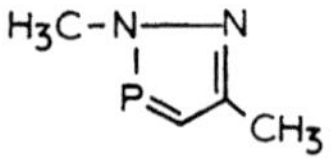			23.5±0.5	80% CCl_4 353 K	N8

- Obtained in ^{15}N-NMR except (a) measured using ^{31}P spectroscopy

(b) The first value (+93.76 Hz) refers to P(III) and the second (-53.2 Hz) to P(V)

Table 7.57. $^{1}J_{^{15}N-^{31}P}$ Coupling constants of compounds with P(V) phosphorus atoms

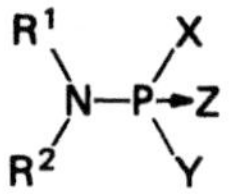

Aminophosphorus PIV Derivatives $^{15}N-^{31}P$					^{1}J	Conditions	Ref.
R^1	R^2	X	Y	Z			
H	C_6H_5	CH_3	CH_3	O	- 0.5	DMSO-d_6	M26,M27
					- 0.5		M10
H	C_6H_5	t.Bu	t.Bu	O	11.5	CH_2Cl_2+Mesitylene(a)	M26
CH_3	CH_3	C_6H_5	Cl	O	≃0	neat	G12
CH_3	CH_3	Cl	Cl	O	13.2	neat	G12
CH_3	CH_3	Cl	$(CH_3)_2N$	O	23.5	neat	G12
CH_3	CH_3	$N(CH_3)_2$	$(CH_3)_2N$	O	29.3	neat	G12
					-26.9		G4
Et	Et	$N(Et)_2$	$(Et)_2N$	O	23.5	neat	G12
Et	Et	CH_3O	CH_3O	O	42.2±0.5	50% C_6D_6	G6
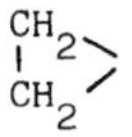		CH_3O	CH_3O	O	9.3±0.5	50% C_6D_6	G6
		CH_3O	CH_3O	O	10.0±0.5	50% C_6D_6	G6
		CH_3O	CH_3O	O	4.4±0.5	50% C_6D_6	G6
		CH_3O	CH_3O	O	4.3±0.5	50% C_6D_6	G6
		CH_3O	CH_3O	O	10.6±0.5	50% C_6D_6	G6
		CH_3O	CH_3O	O	5.0±0.5	50% C_6D_6	G6
		CH_3O	CH_3O	O	⩽ 1±0.5	50% C_6D_6	G6
$(CH_2)_3$		CH_3O	CH_3O	O	20.9±0.5	50% C_6D_6	G6
$(CH_2)_4$		CH_3O	CH_3O	O	41.2±0.5	50% C_6D_6	G6
$(CH_2)_5$		CH_3O	CH_3O	O	39.1±0.5	50% C_6D_6	G6

Table 7.57 (continued)

Aminophosphorus PIV Derivatives $^{15}N-^{31}P$					^{1}J	Conditions	Ref.
$(CH_2)_6$		CH_3O	CH_3O	O	42.7±0.5	50% C_6D_6	G6
$(CH_2)_7$		CH_3O	CH_3O	O	41.8±0.5	50% C_6D_6	G6
$(CH_2)_8$		CH_3O	CH_3O	O	37.6±0.5	50% C_6D_6	G6
N_2		CH_3	CH_3	O	47.9	$CDCl_3$	M10
N_2		Et	Et	O	51.1	C_6D_6	M10
N_2		CH_3O	CH_3O	O	14.9	CH_3CN	M10
H	C_6H_5	CH_3	CH_3	S	+11.3		M10
					+11.3	Dioxane (a)	M26,M27
H	C_6H_5	t.Bu	t.Bu	S	22.2	$CHCl_3$+Mesitylene(a)	M26
H	C_6H_5	$CH_3-N(-)-CH_2-CH_2-N(-)-CH_3$		S	2.6	C_6H_6 + $CHCl_3$ (a)	M26
CH_3	CH_3	$N(CH_3)_2$	$(CH_3)_2N$	S	± 6.0±0.2		G4
					≃ 0	neat	G12
C_6H_5	$(CH_3)_3Sn$	CH_3	CH_3	S	24.8	C_6H_6 (a)	M26
N_2		CH_3	CH_3	S	54.75	Acetone-d_6	M10
N_2		Et	Et	S	58.0	C_6D_6	M10
H	C_6H_5	CH_3	CH_3	Se	+16.5		M10
					16.5	CH_2Cl_2 (a)	M26
H	C_6H_5	t.Bu	t.Bu	Se	27.2	$CHCl_3$+Mesitylene(a)	M26
H	C_6H_5	$CH_3-N(-)-CH_2-CH_2-N(-)-CH_3$		Se	7.8	C_6H_6 (a)	M26
N_2		CH_3	CH_3	Se	57.4	C_6D_6	M10
H	C_6H_5	CH_3	CH_3	Te	36.0	C_6H_6+CH_2Cl_2 (a)	M26
H(b)	C_6H_5	CH_3	CH_3	CH_3	4.1	CH_2Cl_2 (a)	M26
H(b)	C_6H_5	t.Bu	t.Bu	CH_3	- 2.3	DMSO-d_6 (a)	M26
H(b)	C_6H_5	$CH_3-N(-)-CH_2-CH_2-N(-)-CH_3$		CH_3	- 3.3	CH_2Cl_2 (a)	M26
H(b)	C_6H_5	CH_3	CH_3	SCH_3	10.5	$CHCl_3$ (a)	M26
H(b)	C_6H_5	CH_3	CH_3	$SeCH_3$	14.0	CH_2Cl_2 (a)	M26
H(b)	C_6H_5	t.Bu	t.Bu	$SeCH_3$	23.8	DMSO-d_6 (a)	M26
H	H	F	NH_2	F_2	-81.54	(a) (d)	S7

- Obtained in ^{15}N-NMR except (a) measured using ^{31}P spectroscopy

(b) Iodide derivatives

(c) The value of the $^{1}J_{^{15}N-^{31}P}$ coupling constant is 17.0 Hz in CH_2Cl_2 for the meric polymeric compounds $[C_6H_5\ N-P(CH_3)_2BH_2]_n$ (M 26)

(d) P(V) derivative

Table 7.58. $^{1}J_{^{15}N-^{31}P}$ Coupling constants of cyclic amino phosphorus derivatives

Cyclic Aminophosphorus Derivatives $^{15}N-^{31}P$		^{1}J	Conditions	Ref.
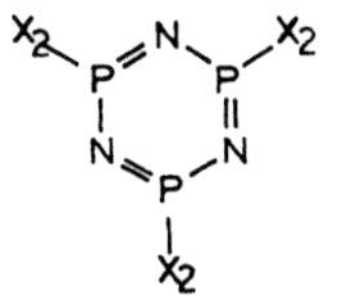	X = F	24.9	$CDCl_3$	T6,T4
	Cl	31.7	$CDCl_3$	T6,T4,T5
	Br	55.8	$CDCl_3$	T6,T4
	$S-C_2H_5$	51.0	$CDCl_3$	T6,T4
	$S-C_6H_5$	53.3	$CDCl_3$	T4
	N_3	17 (N exo)	$C_6D_5CD_3$ (a)	M10
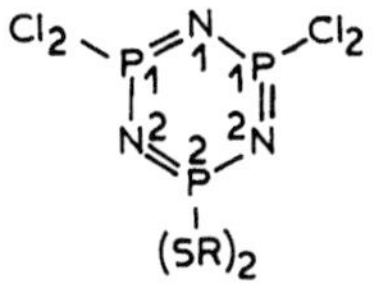	R = C_2H_5	34.1 (P_1-N_1)	$CDCl_3$	T4
		38.8 (P_1-N_2)		
		48.1 (P_2-N_2)		
	C_6H_5	33.7 (P_1-N_1)	$CDCl_3$	T4
		39.4 (P_1-N_2)		
		50.6 (P_2-N_2)		
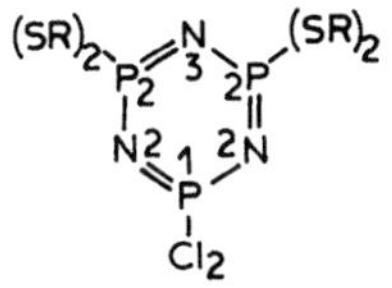	R = C_2H_5	40.3 (P_1-N_2)	$CDCl_3$	T4
		47.6 (P_2-N_2)		
		49.4 (P_2-N_3)		
	C_6H_5	39.7 (P_1-N_2)	$CDCl_3$	T4
		50.9 (P_2-N_2)		
		53.5 (P_2-N_3)		
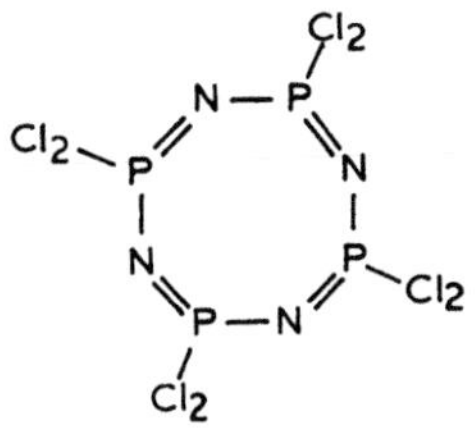		6.9	$CDCl_3$	T5
$(N-P-Cl_2)_5$		2.3	$CDCl_3$	T5

- Obtained in ^{31}P-NMR except (a) measured using ^{15}N spectroscopy

Table 7.59. $^{2}J_{^{15}N-^{31}P}$ and $^{3}J_{^{15}N-^{31}P}$ Coupling constants of various nitrogen-phosphorus derivatives

Various Nitrogen-Phosphorus Derivatives $^{15}N - ^{31}P$		^{2}J	^{3}J	Conditions	Ref.
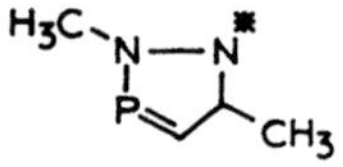		6.0 ± 0.5		80% CCl_4 353 K	N8
 $-N^{*}=N-P^{+}(C_6H_5)_3$		18.5		$CH_2Cl_2-CHCl_3$	D21
$(CH_3)_2P(O)N_3$		4.9		$CDCl_3$	M10
$(CH_3)_2P(S)N_3$		5.05	2.70	Acetone-d_6	M10
$(C_2H_5)_2P(O)N_3$		4.5		C_6D_6	M10
$(C_2H_5)_2P(S)N_3$		5.3		C_6D_6	M10
$[Mo(N_2)_2(PMe_2-Ph)_4]$	cis	5.2 (b)		THF	C1
$[MoF(N-NH_2)dppe_2]BF_4$ (d)	trans	6 ± 3 (c)		CH_2Cl_2	C21
$(\eta^5-C_5H_5)Mo(CO)(PPh_3)(NO)$		5.9		$CHCl_3$	B48
$(n.Bu)_3P-Pd(Cl)_2(NH_2-C_6H_{13})$	trans	50		$CDCl_3$	M7
$(CH_3)(Ph)_2-P-Pd(Cl_2)(NH_2-C_6H_{13})$	trans	54		$CDCl_3$	M7
$(p.tolyl)_3-P-Pd(Cl_2)(NH_2-C_6H_{13})$	trans	54		$CDCl_3$	M7
NCS, SCN $>$Pd$<$ $P(Ph_2)-CH_2$, $P(Ph_2)-CH$ $>$C$-CF_3$		53.71 (b)		213 K (a)	C17
SCN, NCS $>$Pd$<$ $P(Ph)_2-CH_2$, $P(Ph)_2-CH$ $>$C$-CF_3$		53.71 (b)		213 K (a)	C17
$(n.Bu)_3P-Pt(Cl_2)(NH_2-C_6H_{13})$	trans	47		$CDCl_3$	M7
$(CH_3)(Ph)_2P-Pt(Cl_2)(NH_2-C_6H_{13})$	trans	52		$CDCl_3$	M7
$(p.tolyl)_3P-Pt(Cl_2)(NH_2-C_6H_{13})$	trans	50		$CDCl_3$	M7
$Pt(NO_2)_2[P(n.Bu)_3]_2$	cis	61 (b)		$CDCl_3$	M7
		3 (c)			M7
	trans	3 (c)		$CDCl_3$	M7
$Pt(-NCS)_2[P(O-Ph)_3]_2$	cis	≃95 (b)		CH_2Cl_2 (a)	C18
		≃ 7 (c)			
NCS, SCN $>$Pt$<$ $P(O-Ph)_3$, $P(O-Ph)_3$	cis	91 (b)		CH_2Cl_2 (a)	C18
		6 (c)			
NCS, SCN $>$Pt$<$ $P(Ph)_2-CH_2$, $P(Ph)_2-CH$ $>$C$-CF_3$		56.15 (b)		(a)	C17
SCN, NCS $>$Pt$<$ $P(Ph)_2-CH_2$, $P(Ph)_2-CH$ $>$C$-CF_3$		57.37 (b)		(a)	C17

Table 7.59 (continued)

Various Nitrogen-Phosphorus Derivatives $^{15}N-^{31}P$		^{1}J	^{3}J	Conditions	Ref.
$(SCN)(SC\overset{*}{N})Pt[P(Ph)_2-CH_2-C(CF_3)=CH-P(Ph)_2]$		57.3 (b)		(a)	C17
$Pt[P(C_6H_5)_3]_2\ (C_6H_5-N=S=\underline{O})$		25 (b)		0.03 M $CDCl_3$(a)	M19
		2 (c)		243 K	
$Pt[P(C_6H_5)_3]_2\ (C_6H_5-N=S=N-C_6H_5)$		29 (b)		0.03 M $CDCl_3$(a)	M19
		8 (c)		243 K	
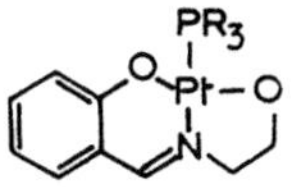	R = n.Bu	50		$CDCl_3$	M7
	p.tolyl	53		$CDCl_3$	M7
	OC_2H_5	76		$CDCl_3$	M7
$[W(N_2)_2dppe_2]$	(d) trans	1.9	0.9	THF	C1
$[W(N_2)_2(PMe_2-Ph)_4]$	cis	16.7 (b)	0.9	THF	C1
		1.2 (c)			
$[W-F(N-NH_2)dppe_2]BF_4$	(d) trans	11±3 (c)		$\cdot CH_2Cl_2$	C21
$[W-Cl(\overset{*}{N}-NH_2)(C_5H_5N)(PMe_2-Ph)_3]Cl$ (e)		5±1 (c)		CH_2Cl_2	C21

- Obtained in ^{15}N NMR except (a) measured using ^{31}P spectroscopy

(b) Trans coupling

(c) cis coupling

(d) dppe = $Ph_2-P-CH_2-CH_2-P-Ph_2$

(e) Cl is trans with respect to NNH_2

Table 7.60. $^{1}J_{^{15}N-^{57}Fe}$ Coupling constants - Obtained in ^{15}N-NMR

$^{15}N-^{57}Fe$	^{1}J	Conditions	Ref.
Bis(pyridine) $[^{15}N_4, ^{57}Fe]$ meso tetraphenylporphyrin	7.8	$C_5D_5N : D_2O$ 5:1	M12
Bis(morpholine) $[^{15}N, ^{57}Fe]$ meso tetraphenylporphyrin	8.0	$CDCl_3$ + 20 % morpholine	M12
Bis(pyrrolidine) $[^{15}N_4, ^{57}Fe]$ meso tetraphenylporphyrin	7.5	$CDCl_3$ + 10 % pyrrolidine	M12

Table 7.61. $^{1}J_{^{15}N-^{59}Co}$ Coupling constants - Obtained in ^{15}N-NMR

$^{15}N-^{59}Co$	^{1}J	Conditions	Ref.
$[Co\ (NH_3)_5(\overset{*}{N}H_3)][Fe(CN)_6]$	62.5 ± 1.0	0.3 M D_2O	Y 15
$[Co\ (NH_2-CH_2-CH_2-NH_2)_3]\ Cl_3$	63.8 ± 1.0	0.3 M D_2O	Y 15

Table 7.62. $^{1}J_{^{15}N-^{103}Rh}$ Coupling constants - Obtained in ^{15}N-NMR

$^{15}N-^{103}Rh$	^{1}J	Conditions	Ref.
Rh(ethylenediamine)$_3Cl_3$	14.6	2M H_2O	B34
Rh(1,2-diaminopropane)$_3Cl_3$	13.5 (a)	2M H_2O	B34
Rh(1,3-diaminopropane)$_3Cl_3$	14.3 (a)	2M H_2O	B34
Rh(2,2-bipyridine)$_3Cl_3$	18.1	2M H_2O	B34
Rh(1,10-phenanthroline)$_3Cl_3$	18.4	2M H_2O	B34
[trans Rh(ethylenediamine)$_2Cl_2$]Cl	13.4	2M H_2O	B34
[trans Rh(pyridine)$_4Cl_2$]Cl	17.1	benzyl alcohol	B34
RhCl[P(iPr)$_3$]$_2$ (C_6H_5NSO)	15.5	0.1 M $CDCl_3$ 0°	M3

(a) Average coupling constants for the isomers present

Table 7.63. $^{1}J_{^{15}N-^{111}Cd}$ Coupling Constants

$^{15}N - ^{111}Cd$	^{1}J	Conditions	Ref.
Cd-EDTA^{2-}	81 ± 1 (b)	0.4 M H_2O (a)	H12
$[^{15}N_4]$Cd octaethyl-porphyrin	151.4 (c)	DMSO-d_6 (a)	K14b
$[^{15}N_4]^{111}$Cd meso tetraphenylporphyrin		$CDCl_3$	D22
with 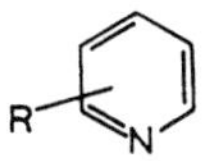 as ligands		8-10 mg/0.5 ml	
R = 4-CN	147.6		
3-Cl	146.4		
4-CH_3OOC	146.3		
4-CH_3CO	146.0		
4-H	142.5		
4-CH_3	141.0		
4-NH_2	137.4		

- Obtained in ^{111}Cd-NMR except (a) measured using ^{15}N spectroscopy

(b) Average coupling with ^{111}Cd or ^{113}Cd

(c) Coupling constant between ^{15}N and ^{111}Cd or ^{113}Cd

Table 7.64. $^{1}J_{^{15}N-^{117}Sn}$ and $^{1}J_{^{15}N-^{119}Sn}$ Coupling constants - Obtained in ^{15}N-NMR

$^{15}N - ^{117}Sn$ / $^{15}N - ^{119}Sn$	^{1}J $^{15}N - ^{117}Sn$	^{1}J $^{15}N - ^{119}Sn$	Conditions	Ref.
$C_6H_5N[P(CH_3)_2]Sn(CH_3)_3$		9.5 ± 3	C_6H_6	M26
$C_6H_5N[PS(CH_3)_2]$ $Sn(CH_3)_3$		-47.5 ± 4	C_6H_6	M26
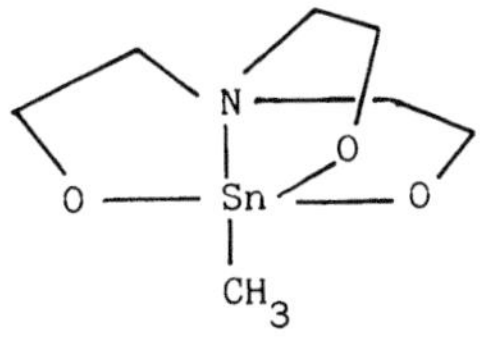	72.4	75.6	$CDCl_3$	J6
	104.6	110.0		J6

Table 7.65. $^{1}J_{^{15}N-^{183}W}$ Coupling constants - Obtained in ^{15}N-NMR

^{15}N — ^{183}W		^{1}J	Conditions	Ref.
$[W(N\text{-}NH_2)(8\text{-quin})(PMe_2Ph)_3]Cl$	(a)	114.0 ± 1	CH_2Cl_2	C21
$[WCl(N\text{-}NH_2)(C_5H_5N)(PMe_2Ph)_3]Cl$	(b)	124.5 ± 1	CH_2Cl_2	C21

(a) 8-quin = quinolin-8-olate, O is trans with respect to NNH_2

(b) Cl is trans with respect to NNH_2

Table 7.66. ${}^{n}J_{{}^{15}N-{}^{195}Pt}$ Coupling constants of R^1R^2-Pt-R^3R^4 and related compounds

Nitrogen-Platinum Derivatives ${}^{15}N$ — ${}^{195}Pt$					^1J	^3J	Conditions	Ref.
R^1	R^2	R^3	R^4					
NH_3	$(CH_3)_2S\underline{O}$	Cl	Cl	cis	336		DMSO	K2
				trans	232		DMSO	K2
H_2N-CH_2-COOH	$(CH_3)_2S\underline{O}$	Cl	Cl	trans	244		H_2O	K2
$n.C_6H_{13}-NH_2$	$(C_4H_9)_3P$	Cl	Cl	trans	138.3		$CDCl_3$ (a')	M7
$n.C_6H_{13}-NH_2$	$(CH_3)-P(C_6H_5)_2$	Cl	Cl	trans	155.9		$CDCl_3$ (a')	M7
$n.C_6H_{13}-NH_2$	$(CH_3-C_6H_4)_3P$	Cl	Cl	trans	158.8		$CDCl_3$ (a')	M7
$n.C_6H_{13}-NH_2$	$(C_4H_9)_3As$	Cl	Cl	trans	183.8		$CDCl_3$ (a')	M7
$n.C_6H_{13}-NH_2$	$CH_3-As(C_6H_5)_2$	Cl	Cl	trans	208.8		$CDCl_3$ (a')	M7
$n.C_6H_{13}-NH_2$	$(CH_3-C_6H_4)_3As$	Cl	Cl	trans	207.4		$CDCl_3$ (a')	M7
$n.C_6H_{13}-NH_2$	C_2H_4	Cl	Cl	cis	295.6		$CDCl_3$ (a')	M7
					296			M13
				trans	283.9		$CDCl_3$ (a')	M7
$(CH_3)_2NH$	C_2H_4	Cl	Cl	trans	299		$CDCl_3$	A20
$(CH_3)_2NH$	$C_2H_4-NH(CH_3)_2$	Cl	Cl	trans	107	51	$CDCl_3$	A20
$(CH_3)_2NH$	$(C_6H_5)_3P$	Cl	Cl	trans	171		$CDCl_3$	A20
C_5H_5N	C_2H_4	Cl	Cl	trans	409		$CDCl_3$ 233K	A20
C_5H_5N	$C_2H_4-C_5H_5N$	Cl	Cl	trans	162	33	$CDCl_3$ 233K	A20
$H_2N-CH_2-CO\underline{O}^-$		Cl	Cl		317		H_2O	K2
$H_2N-CH_2-CO\underline{O}^-$		$(CH_3)_2SO$	Cl		226 (b)		H_2O	K2
					330 (c)			K2
NH_3	NH_3	Cl	Cl	cis	312.2		DMSO	K2

Table 7.66 (continued)

Nitrogen-Platinum Derivatives $^{15}N-^{195}Pt$					^1J	^3J	Conditions	Ref.
NH_3	NH_3	$(CH_3)_2SO$	Cl^+	cis	340 (c)		DMSO	K2
					233.8 (b)			K2
				trans	287		DMSO	K2
n.C_6H_{13}-NH_2	n.C_6H_{13}-NH_2	Cl	Cl	cis	336		DMSO-d_6 (a')	M7
				trans	286.8		$CDCl_3$ (a')	M7
Dodecyl-NH_2	Dodecyl-NH_2	Cl	Cl	cis	351		$CHCl_3$ (a) saturated	P5
				trans	290		$CHCl_3$ (a) saturated	P5
Dodecyl-NH_2	Dodecyl-NH_2	Br	Br	cis	334		$CHCl_3$ (a) saturated	P5
				trans	279		$CHCl_3$ (a) saturated	P5
NH_3	NH_3	NH_3	Cl^+		317 (c)		DMSO	K2
					278 (b)			K2
NH_3	NH_3	NH_3^{2+}	$(CH_3)_2SO$		288 (d)		DMSO	K2
					232 (e)			K2
NH_3	NH_3	H_2O	H_2O	cis	388.6		H_2O (a)	A22
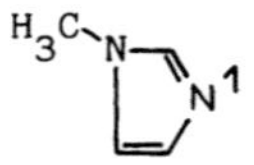	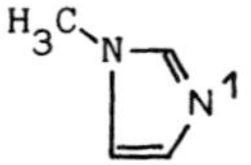	$^2NH_3$	$^2NH_3$	cis	286.5(2)	26.1	H_2O (a)	A22
					436.5(1)			

$H_2N-CH_2-CH_2-NH_2$		H_2O	H_2O		421.4		H_2O (a)	A22
2 2 $H_2N-CH_2-CH_2-NH_2$		H_2O	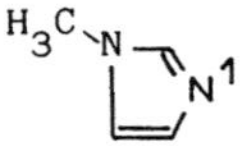		411.1(f) 327 (g) (h)	24.7	H_2O (a)	A22
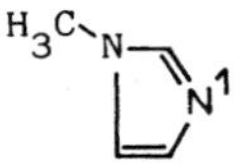	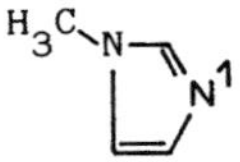	2 2 $H_2N-CH_2-CH_2-NH_2$			318.9(2) 428.6(1)	25.5	H_2O (a)	A22
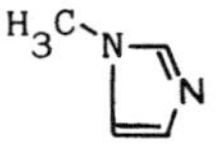	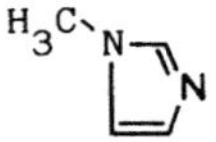	H_2O	H_2O	cis	579.4	32.6	H_2O (a)	A22
$P(C_6H_5)_3$	$P(C_6H_5)_3$	$C_6H_5-N=S=O$			136 $\pm$ 10		0.03 M $CDCl_3$ 243 K	M19
$P(nC_4H_9)_3$	$P(nC_4H_9)_3$	NO_2	NO_2	trans	453			L 30
NO_2	NO_2	NO_2^-	NO_2^-		592			L30
$PtCl_4\left[CH_3-(CH_2)_{11}-NH_2\right]_2$				cis	249		$CHCl_3$ (a) saturated	P5
$PtBr_4\left[CH_3-(CH_2)_{11}-NH_2\right]_2$				cis	223		$CHCl_3$ (a) saturated	P5
$\left\{PtBr\left[CH_3-(CH_2)_{11}-NH_2\right]_3\right\}Br$					342 (c)		$CHCl_3$ (a)	P5
					288 (d)		saturated(a)	P5
$\left\{Pt\left[CH_3-(CH_2)_{11}-NH_2\right]_4\right\}Br_2$					298		$CHCl_3$ (a) saturated	P5
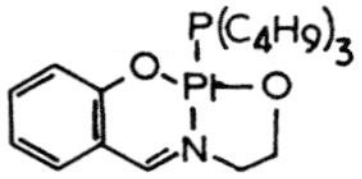					319		(a')	M7
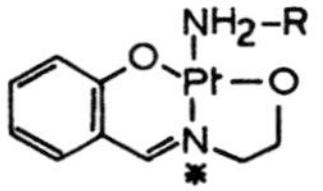					319		(a')	M7

<u>Table 7.66</u> (continued)

Nitrogen-Platinum Derivatives $^{15}N-^{195}Pt$	^{1}J	^{3}J	Conditions	Ref.
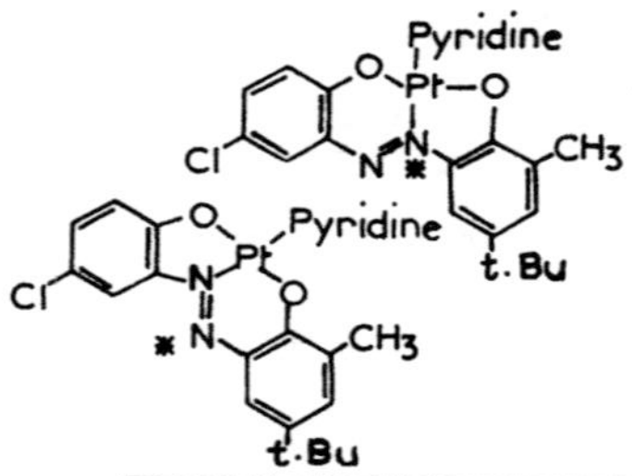	523		$CHCl_3$ (a) saturated + $Cr(acac)_3$	P4
		(i)	$CHCl_3$ (a) saturated + $Cr(acac)_3$	P4

- Obtained in ^{195}Pt-NMR except (a) measured using ^{15}N spectroscopy and (a') measured using ^{195}Pt and ^{15}N spectroscopy

(b) N is cis with respect to Cl
(c) N is trans with respect to Cl or Br
(d) N is trans with respect to N
(e) N is trans with respect to $(CH_3)_2SO$
(f) Doublet of singlets (N_2)
(g) Doublet of 5.1 Hz. Doublet due to $^{2}J_{N_1-N_2}$ (N_2)
(h) $^{1}J_{N_1}$ is not observed
(i) ^{2}J is not observed

Table 7.67. $^{1}J_{^{15}N-^{199}Hg}$ Coupling constants

Nitrogen-Mercury Derivatives $^{15}N - ^{199}Hg$	^{1}J	Ref.
$[^{15}N_4]$ ^{199}Hg meso tetraphenyl porphyrin with pyridine as ligand	340	Y14

Table 7.68. $^{1}J_{^{15}N-^{207}Pb}$ Coupling constants - Obtained in ^{15}N-NMR

Nitrogen-Lead Derivatives $^{15}N - ^{207}Pb$	^{1}J	Conditions	Ref.
cis 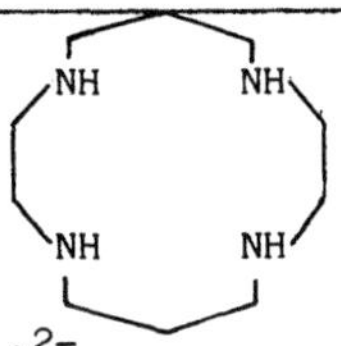 $Pb(NO_3)_2$	207.5(a) 19.8(b)	DMSO	A19a A19b
$Pb(cdta)^{2-}$ (c)	188.3		A19b
$(CH_3)_3Pb-N(CH_3)C_6H_5$	$+261 \pm 5$	20% C_6H_6	K12

(a) Refers to N equatorial

(b) Refers to N axial

(c) cdta $^{4-}$ = trans-1,2-diaminocyclohexane N,N,N',N'-tetracetate ion

Table 7.69. Dipole-dipole coupling constants measured in a nematic phase and used to determine the structure of the molecules

Compounds	Phase		D		Ref.
			$^{15}N-H$	$^{15}N-CH_3$	
CH_3NH_3+	H_2O pH 1.5	53.7%			
	Decyl sulfate $^-$	35.2%	114.19 $\pm$0.13	-27.23 $\pm$ 0.13	R14
	Sodium sulfate	5.6% (b)		or	
	Decanol	5.5%		-25.65 $\pm$ 0.13	
$(CH_3)_2NH_2^+$	Decyl sulfate$^-$	39.4%	111.01	- 8.91	R1
	Sodium sulfate	6.7%			
	Decanol	4.6% (b)			
	H_2O pH 1.5	49.3%			
			$^{15}N-H$	$^{15}N-C-H$	
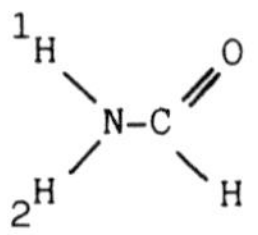	9% weight in				
	Sodium sulfate	7%			
	Sodium decyl sulfate	36% (b)	+130 (1)	+12.6	R10
	Decanol	7%	+116.5(2)		
	H_2O	50%			
			$^{15}N-H$	$^{15}N-^{31}P$	
$CH_3-N(PCl_2)_2$	EBBA (c) (a)		- 86.8 $\pm$ 1	+32$\pm$ 1	C3
			$^{15}N-^{15}N$	$^{15}N-^{31}P$	
$[(CH_3)_2N]_3P$	liquid cristal (a)				
	EBBA 60% MBBA 40% (c)	264.5K	2.2	15.6	C4
		275 K	1.9	14.3	
		280.5K	1.9	14.1	
		291 K	1.7	12.0	
			$^{15}N-^{15}N$		
N=N=O	EBBA (c) (a)		9.16 $\pm$ 0.3		B31

(a) Obtained in ^{15}N-NMR

(b) Weight %

(c) EBBA = p-butyl-N-(p-ethoxybenzylidene)aniline MBBA is the methoxy compound. These results lead to a NPN bond angle value of 97.6 ± 0.3^0 for compound $[(CH_3)_2N]_3P$ and to a PNP bond angle of 117 ± 0.5^0 for compound $CH_3N(PCl_2)_2$

8 Application of ^{15}N Spectroscopy to the Study of Dynamic Processes and Reaction Mechanisms

When dynamic processes are investigated by NMR, the reactions are usually considered as fast or slow according to whether the lifetimes, τ, of the species concerned are short or long with respect to the NMR time scale. In fact, information about molecular dynamics corresponding to very fast motions, is obtained by relaxation time determinations, and this topic has already been discussed in Sect. 2. Here we shall be concerned with chemical exchange processes involving species at equilibrium or with relatively slow reactions in non-equilibrated systems. In terms of lifetimes, such chemical processes may be characterized by τ values down to about 10^{-4}s which correspond to rate constants up to $10^4 s^{-1}$ for first-order reactions. This limit depends, to a certain extent, on the nucleus used as a probe, moreover, it can be displaced towards higher rate constants for second-order reactions investigated in very diluted solutions. Here, we shall consider the exploitation of line-shape modification experiments and then the investigation of equilibration processes by the determination of line intensity variations.

8.1 Exchange Processes Studied by Line-Shape Modification Experiments

In this section we shall be dealing with the effect of temperature, T, and pH on the ^{15}N line shapes. But first it may be useful to recall some general properties of dynamic NMR involving rare isotopes.

8.1.1 Practical Aspects of Dynamic NMR of Rare Isotopes

FT ^{15}N resonance does not present any particuliarity as regards the theoretical aspects of band shape analysis, since it has been generally shown that in suitable experimental conditions FT and CW spectroscopies may lead to equivalent line-shapes (E 7) (K 38). The reader interested in the various methods used to derive lifetimes from line-shapes is referred to general relevant reviews (see for example M 31 and S 34). Indeed, the problems particular to ^{15}N and other rare isotopes mainly concern temperature (regulation and measurement) and the adjustment of the FT parameters (M 44) (M 45).

8.1.1.1 Temperature Measurements

A discussion of the technical conditions of gas flow and temperature control can be found in reference M 31q but it should be pointed out that the difficulties are greater when large diameter cells are used. It is also recognized that errors in measuring the temperature, a major source of uncertainty in the determination of dynamic parameters, are often large in dynamic investigations involving rare isotopes. Indeed, chemical thermometers, such as methanol or glycerol are commonly used in ^{1}H spectroscopy and several mixtures of compounds have been proposed to play this role in ^{13}C-NMR (A 30, M 31q). However, although $\delta^{15}N$ temperature coefficients of at least -2 to 4 10^{-2} ppm K^{-1} (A 5), which may be comparable to, or even larger than, $\delta^{13}C$ coefficients (S 35), are observed, no convenient ^{15}N thermometer has been found yet. An alternative to the use of thermocouples or thermistors is provided by resorting to a broad band NMR probe or a pseudo-dual [^{15}N,^{1}H] probe. In fact, broad-band probes tuned on both the low and high limits of the range of resonance frequencies are not currently available and it is not very convenient to use a ^{13}C or ^{59}Co (L 50) thermometer when observing a ^{15}N spectrum. The use of a pseudo-dual [^{15}N,^{1}H] probe which is constituted from the ^{15}N observing and the ^{1}H decoupling coils may therefore be useful. The proton spectrum of a methanol of glycerol sample observed through the decoupling channel, then provides a convenient chemical thermometer.

8.1.1.2 Adjustment of FT Parameters

Certainly, *sensitivity* is the greatest limitation to ^{15}N-DNMR. Since the intrinsic signal-to-noise ratio decreases as the coalescence temperature is approached, observing coalesced ^{15}N lines makes a considerble demand on spectrometer time and ^{15}N enrichment is often necessary. It should be noted that longitudinal relaxation times and nuclear

Overhauser enhancement factors of exchanging nuclei are not necessarily equal. A careful adjustment of the FT acquisition parameters, taking this possibility into account (M 31r), must therefore be carried out in the slow exchange region. Finally, caution should be exercized when a paramagnetic reagent such as $Cr(acac)_3$, is added to shorten the T_1 values and to quench the nuclear Overhauser effect since selective effects may occur. Indeed, paramagnetic species were shown to interact specifically with basic nitrogen atoms and anomalous line-widths or line intensities are likely to result (see Sect. 2). To conclude, emphasis is laid on the necessity of using a high digital resolution and therefore of selecting narrow sweep-widths since large errors may result from inaccurately defined line-shapes.

8.1.2 Line-Shape Modifications Due to Modulation of Scalar Coupling Interactions by Proton Transfers

Most investigations of chemical dynamic processes at equilibrium are concerned with the exchange of labile hydrogens in various types of NH or NH_2 groups. In such investigations, the coalescence curves are brought about by temperature variations or by changes in the pH of the medium. These experiments often complement results previously obtained using ^1H spectroscopy and in certain instances provide new insight into the mechanisms of exchange processes. Indeed, due to the large values of the $^1J_{N-H}$ coupling constants (Tables 7) the ^{15}N spectrum provides a very sensitive probe for studying proton transfers. Thus, if long range coupling constants are disregarded, the undecoupled nitrogen spectrum of an NH or NH_2 group appears, in the slow exchange limit, as a well-separated doublet or triplet respectively. Due to the intermolecular exchange of the hydrogen atoms, these patterns broaden to give a single line, the width of which decreases as the lifetime becomes shorter. Line-shape analysis at a given temperature can then be carried out by the usual methods of dynamic NMR to derive the lifetimes. If various stages of the coalescence curves are analysed, the activation parameters, for example $\Delta H^{\ddagger}$, $\Delta S^{\ddagger}$, may in principle be determined. However, due to the lack of sensitivity of nitrogen NMR, the coalescence point is not always well-defined and only approximate parameters are sometimes given.

Proton transfers in *aminoacids* which have been extensively studied by ^1H-NMR may also cause modifications in the undecoupled nitrogen spectrum. Thus the ^{15}N signal of glycine at pH 3.13 is *temperature* - de -

pendent and the J structure resolved at low temperatures (J_{NH} = 69 Hz at 275 K) disappears to give a more or less broadened singlet at higher temperatures (L 36). The lifetimes are deduced from the line-widths, $\Delta\nu_{1/2}$, by considering that the decoupled line may be used to deduce the relaxations rates in the absence of exchange. The proton exchange rates are also calculated at 315 K for *different pH* values comprised between 0.56 and 4.13 and, in spite of a limited accuracy the dynamic parameters thus obtained compare favourably with those derived from proton NMR. In this respect, it should be noted that rather complex line-shape behaviour may be expected in certain ranges of pH since both the concentrations of the different forms of glycine and their exchange rates are temperature-dependent. An exchange between the glycine anion and its zwitterion has been postulated in order to explain the line-shape variations above pH 6.4 (C 23). However, it should be kept in mind that variations in the relaxation times and therefore in the line-widths and line intensities may also result from the presence of paramagnetic impurities (I 2). As far as the fast exchange limit is satisfied, the ^{15}N chemical shift variations as a function of pH provide titration curves (§ 6.2). In this respect, the pH titration curves of glycyl-glycine amide and ammonium nitrogens, for example, show that the amide C-N double bond character increases on going from the cationic to the zwitterionic species (I 4):

The exchange of aminoprotons of *nucleotides* with the protons of the solvent (water) can also be studied by the analysis of the nitrogen spectrum (B 36). A spin-spin coupling structure is observed at neutral pH values for ^{15}N - enriched guanosine-3'-phosphate, cytidine-3'-phosphate and adenosine-3'-phosphate. At high pH values, a base catalyzed exchange process transforms the spectra into singlets, similarly for the three compounds. By contrast, the spectral behaviour differs for the three nucleotides at acid pH values and certain mechanisms are proposed to rationalize the rather complex effects which are observed.

Since the rates of proton transfers are base or acid catalyzed, they are very sensitive to the nature of the medium. As a consequence, *comparisons* between the exchange rates of NH or NH_2 groups belonging either to the same molecule or to different species are generally preferred to absolute determinations.

8.1.2.1 Internal Comparison of Exchange Rates

Proton exchange rates are easily compared when the considered NH groups belong to the same molecule. Thus, differential rates of proton transfer are observed for the guanidino NH_2 groups (B 39) (L 6) (Y 8) and NH groups (Y 8) of *arginine* as a function of pH.

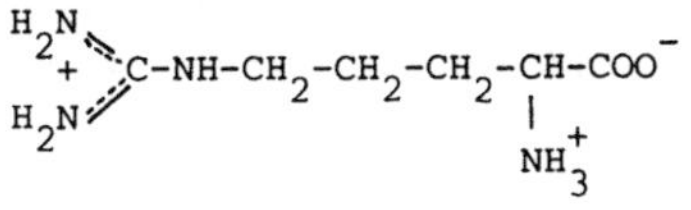

At pH 8 all exchange processes are fast, on the NMR time scale, and three single lines are observed for the three types of nitrogens. As the pH is decreased the spin-spin splittings are resolved: a sharp doublet (J_{NH} = 92 Hz) and a sharp triplet (J_{NH} = 91 Hz) are observed at pH 5 for the NH and NH_2 groups respectively. The signal of the α nitrogen begins to broaden below pH 5. Line-shape analysis of the coalescence curves shows that the exchange phenomenon is about twice as fast for the NH doublet than for the NH_2 triplet, a behaviour which is attributed to the predominant contribution of a resonance structure involving a $>C=\dot{N}<^{+}$ double bond. Similarly, coalescence figures are observed for both the NH doublet and the NH_2 triplet of *phenylhydrazine* (Y 7). At 303 K, the neat compound exhibits resonances at -294.8 ppm (NH) and -320.0 ppm (NH_2) respectively with $^1J_{NH}$ = 88 Hz for the well-resolved NH doublet. On raising the temperature, the broad NH_2 triplet coalesces first and from a line-shape simulation performed at 333 K, rate constants of 2 s^{-1} and 60.0 s^{-1} respectively are determined. The fact that the free energy of activation for proton transfer is higher for NH (80 kJ mol^{-1}) than for NH_2 (64 kJ mol^{-1}), rules out the exclusive operation of a bimolecular exchange process which would simultaneously involve the NH and NH_2 groups.

In the same respect the rates of proton exchanges involving the NH groups of *ureas* have been internally compared (Y 16). In the case of aqueous *N*-methylurea, for example, an NH doublet and an NH_2 triplet are clearly resolved at pH 7.2 but broaden at different rates when the pH is raised, the rate of collapse being three times faster for NH_2 than for NH. The exchanges are also accelerated in acid media and, in this case, the NH_2 triplet collapses 15 times faster than the NH doublet.

8.1.2.2 External Comparison of Exchange Rates

The investigations of the influence of substituents or of stereochemistry on the rates of exchange processes may require the comparison between the line-shapes of ^{15}N spectra of different species. This type of external comparison is more significant if the species concerned are mixed in order to be studied in identical conditions. Various proton transfer phenomena have been investigated in this way. Thus, the base catalyzed NH proton exchange of 1-aza-2-cycloheptanethione (B) is shown to be 1500 times faster at 298 K than

(A) $\overline{(CH_2)_5\text{-NH-CO}}$ — δ = -262.4 ppm, J = 88 Hz

(B) $\overline{(CH_2)_5\text{-NH-CS}}$ — δ = -218.7 ppm, J = 92 Hz

(equimolar solutions 1.5 M in DMSO at 298 K)

that of the corresponding 1-aza-2-cycloheptanone (A) (Y 6). Related to this, the influence of the solvent on the relative rates of proton transfer in various lactams, may also be determined (Y 4). Similarly, it appears that the base catalyzed NH proton exchange of 1 aza-2-cyclononanone is faster for the *cis* (C) than for the *trans* (D) isomer (Y 3).

(C) $\overline{(CH_2)_7\text{-NH-CO}}$ *(cis)* — δ = -259.6 ppm, J = 89 Hz

(D) $\overline{(CH_2)_7\text{-NH-CO}}$ *(trans)* — δ = -262.1 ppm, J = 91 Hz

(2.5 M solution in DMSO at 298 K)

8.1.3 Line-Shape Modifications Involving Chemical Shift Averaging

Although the majority of the dynamic results concerns the effects of the modulation of the scalar coupling interaction J_{NH} through intermolecular proton transfer, coalescence figures resulting from chemical shift averaging may be observed in certain cases.

Owing to their importance in biological systems, *porphyrins* have been the subject of several studies intended to determine the nature and rates of the exchange phenomena involved. Thus, for octaethylporphyrin (OEP) (K 14) and *meso* tetraphenylporphyrin (TPP) (G 8) (Y 17), the two ^{15}N resonances observed at low temperature on the decoupled spectrum coalesce as the temperature is raised, indicating a raprid exchange of the two central hydrogens between the four nitrogen sites.

[^{15}N]OEP 0.019 M in $CDCl_3$ — δ = -253.1 and -143.7 ppm at 220 K

[^{15}N]mTPP 0.011 M in $CDCl_3$ + 10% $(CD_3)_2CO$ — δ = -247.2 and -138.2 ppm at 261 K

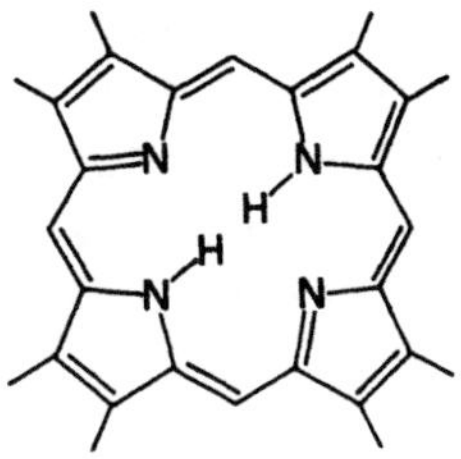

A tautomeric structure in which the two central hydrogens are bound to opposite nitrogens is confirmed on the basis of these results and the dynamic phenomena observed for N-methyloctaethylporphyrin also indicate that the hydrogen is fixed on the nitrogen located opposite the N-methyl group (K 14). It is interesting to note that the proton coupled ^{15}N spectrum of a deuterated pyridine solution of OEP at 343 K exhibits a triplet with the value J_{NH} = 23 Hz which corresponds to the quarter of the $^{1}J_{NH}$ coupling constant measured in proton spectroscopy at lower temperatures. In fact, intramolecular exchange of the inner hydrogens averages but does not collapse the J structure. In the same area, ^{15}N band-shape studies also give some insight into the mechanisms of tautomerism of the hydrogens in protoporphyrin dimethyl ester and coproporphyrin tetramethyl ester (I 5).

Dynamic ^{15}N-NMR was also used to investigate the *double inversion at the two nitrogen* atoms of bicyclic hydrazines (N 10).

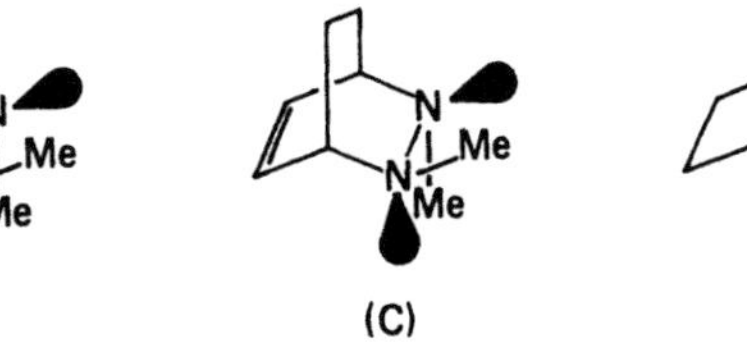

(A)

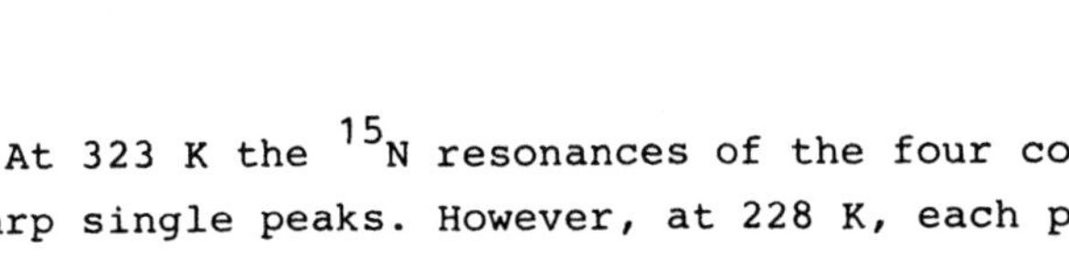

(B) (C) (D)

At 323 K the ^{15}N resonances of the four compounds (A-D) appear as sharp single peaks. However, at 228 K, each peak of A, B, C splits into sharp lines of equal intensities. This behaviour is in accordance with *trans-trans* double inversion at the two nitrogen atoms. (For compound D, the two nitrogens remain unexpectedly equivalent at low temperature). The activation parameters $\Delta G^{\ddagger}$, determined at the coalescence temperature are in good agreement with the results from ^{1}H and ^{13}C dynamic NMR.

	$\delta^{15}N$ (323 K) ppm/CH_3NO_2	$\Delta\nu$ (228 K) Hz (10.09 MHz)	T_c K	$\Delta G^{\ddagger}_{T_c}$ (kJ mol^{-1}) (a)	(b)	(c)
A	-269.5	10.90	269	58.4	55.2	55.2
B	-277.7	8.58	281	61.6	61.6	
C	-282.6	21.90	252	53.1	51.4	50.4

(a) from ^{15}N-NMR
(b) from ^{1}H-NMR
(c) from ^{13}C-NMR

Finally, we shall mention that two ^{15}N signals associated with the bound and terminal nitrogens of *tropylium azide*

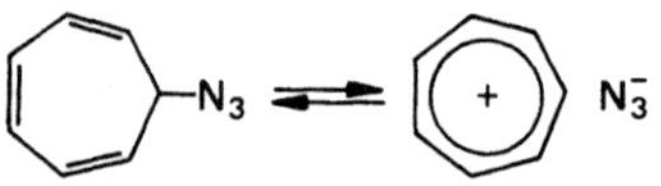

are detectable at low temperature and coalescence occurs when the temperature is raised (F 12). A rough estimation of the free energy of activation at 273 K is in good agreement with the results of ^{13}C spectroscopy interpreted on the basis of 1,3-shift.

8.2 CIDNP Experiments

Chemical Induced Dynamic Nuclear Polarization is a useful source of information about chemical mechanisms involving free radicals and it has been largely exploited in ^{1}H and ^{13}C spectroscopies. Due to large values of the hyperfine coupling constants and of the longitudinal relaxation times, nitrogen 15 is an interesting mechanistic probe and high coefficients of nuclear polarization may be observed (L 41). Several types of reactions have been investigated by means of ^{15}N-CIDNP. Thermal decomposition of ^{15}N-enriched diazoaminobenzene was studied by ^{1}H, ^{13}C and ^{15}N spectroscopies (L 41) (L 55). In addition to the polarized species already identified in ^{13}C-NMR, namely diazoaminobenzene aniline and diphenylamine, the presence of a weak amount of polarized *p*-aminoazobenzene is detected in the course of ^{15}N-NMR experiments performed in cycloheptanone at 443 K. The polarization coefficient is calculated to be E $\simeq$ 10 000 for the nitrogen of aniline and is probably higher for diphenylamine. ^{15}N-CIDNP was also used to identify intermediates in the decomposition of unsymmetric azo compounds. It is noted that the ^{15}N-CIDNP spectrum for *cis* C_6H_5-N=N-C$(CH_3)_2C_6H_5$ is more easily detected than the corresponding ^{1}H or ^{13}C spectra (P 17). The results give further support for the intermediacy of diazenyl radicals R - N = N*. Moreover, they agree with the conclusion drawn from selective labelling experiments and show that no detectable rearrangement of the phenyldiazenyl radical occurs (P 18). ^{15}N-CIDNP is also more suitable than ^{1}H-CIDNP for detecting the presence of the 1-norbornyldiazenyl radical in the thermal decomposition of the *cis* azo compound A (G 20). More generally

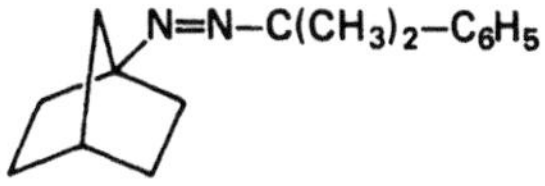

A

^{15}N-CIDNP investigations of the decomposition of several unsymmetric diazenes and radical trapping experiments both confirm a mechanism involving initial cleavage of one carbon-nitrogen bond (P 7). In the case of phenylazotriphenylmethane, for example this cleavage leads to the phenyldiazenyl-triphenylmethyl radical pair

$$\mathrm{Ph{-}N=N{-}CPh_3} \rightleftarrows \mathrm{PhN=N^{*}\ \ ^{*}CPh_3}$$

In addition, the polarized signal observed at about 66 ppm from NO_3^- can be attributed to the formation of molecular nitrogen, N_2 (P 17) (B 15).

8.3 Application of ^{15}N Spectroscopy to the Study of Chemical and Biochemical Mechanisms

In principle, reaction mechanisms can be studied by following the evolution of reactants mixed either outside or in the NMR probe. ^{1}H and ^{13}C-NMR, for example, are thus often used to identify intermediates and/or to determine the kinetic parameters by means of line intensity measurements as a function of time. Due, in particular, to its unfaforable sensitivity, ^{15}N NMR is not well suited to such kinetic experiments and no quantitative kinetic determination has been carried out. In fact, the investigation of reaction mechanisms is therefore based usually on an identification of the reaction products at equilibrium. However, two formal cases may be distinguished according to whether the species concerned are in fast or slow equilibrium with respect to the NMR time scale:

8.3.1 Tautomerism in Nitrogen Derivatives

Numerous tautomeric equilibria are fast with respect to NMR and the observed parameters then represent weighted averages over the δ and J values of the individual species. As already mentioned in Sect. 5.3.3 the values of the measured parameters, and their behaviour as a function of solvent and temperature, may provide information on the nature and the relative proportions of the intervening tautomers. In fact, most investigations of tautomeric equilibria involving an NH group have been carried out using ^{1}H-NMR and we have given various examples in the course of the discussion devoted to the behaviour of $^{1}J_{N-H}$. Here we have limited ourselves to experiments involving ^{15}N spectroscopy, however several references to ^{1}H investigations of typical compounds are given for the benefit of the reader:

Formamide (D 6)	Porphyrins (K 14b)
N-substituted anilines (B 22, L 24)	Schiff's bases (D 20)
Enamines (D 3, D 5, D 11, F 15)	Hydrazo derivatives (B 43, B 51, Y 1)

Indeed, ^{15}N resonance may be used in an analogous way and several problems of tautomerism have been examined. The determination of the favored tautomeric form of the free base of a novel acylguanidine diuretic, amiloride (B, C), was greatly facilitated by the use of

^{15}N-NMR (S 27). Evidence of a tautomeric form (B, C) is given, in particular, by the observation of a triplet structure which can be attributed to two equivalent NH_2 groups at -296.6 ppm and a singlet at -203.6 ppm corresponding to a N = C<(N)(N) nitrogen. Tautomeric forms of the type

R-CO-NH-C(=NH)(NH$_2$) or R(HO)C = NC(=NH)(NH$_2$)

are therefore excluded. In addition, the equivalence of the terminal amino groups implies that they are interconverting rapidly via conformational equilibrium.

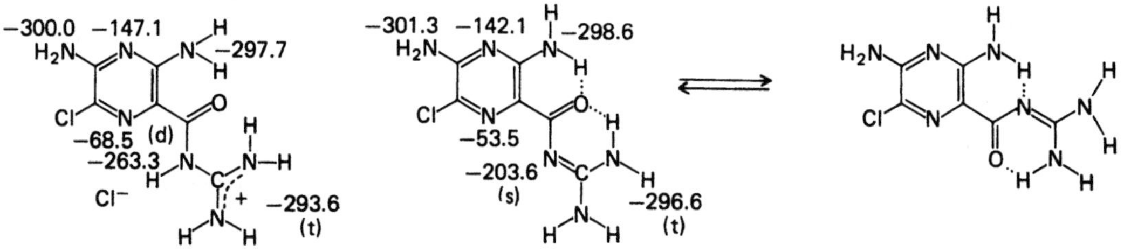

A B C

(s) singlet (d) doublet (t) triplet (ppm/CH_3NO_2)

It can also be seen, from the ^{15}N spectra, that the amiloride, A, preferentially assumes the acylamino tautomer form shown. Another example is provided by the investigation of ^{15}N labelled histidine in aqueous solution (B 35). As previously mentioned ^{15}N-NMR is useful for specifying the tautomeric equilibrium of the deprotonated imidazole ring.

8.3.2 Application of ^{15}N Spectroscopy to the Study of Chemical Mechanisms

To the extent that ^{15}N-NMR constitutes a good method of structure elucidation, it may be of much help in the identification of the intermediate and final products of various reactions, and it has a wide range of applications. Some examples of the use of such information in clarifying mechanistic problems will be given here.

Elucidation of the stereochemistry of the photochemical rearrangement of 2-ethoxypyrrolin-5-one, for example, was rendered possible by the examination of the ^{15}N shifts and these results have provided information on the stereospecificity and the stereoselectivity of the reaction (B 61) (H 18). In this work, the existence of a ^{15}N-γ steric effect previously established (N 4) was exploited to identify

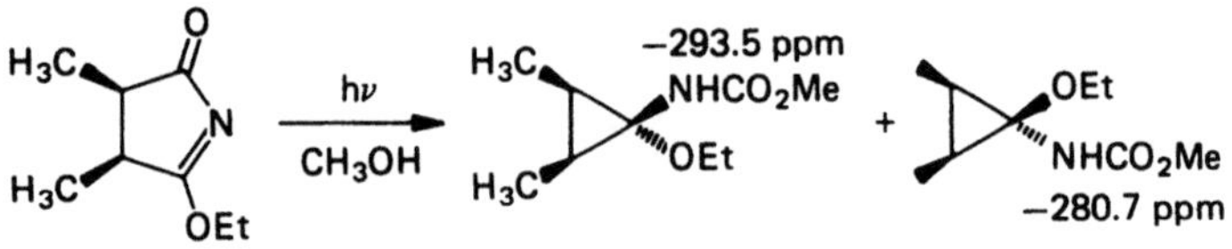

the stereochemistry of the N *cis* - (2,3 dimethyl-1 ethoxy cyclopropyl) carbamates. Another example of structural application is related to the industrial synthesis of sulfamic acid derivatives from the reaction of sulfuric anhydride with ammonia. This reaction was known to yield undesired quantities of ammonium sulphate and other nitrogen derivatives. A ^{15}N study has shown (N 6) that various sulfonamides, formed during the synthesis, are readily identified from their ^{15}N chemical shifts.

	$NH_2SO_3NH_4$	$NH(SO_3NH_4)_2$	$N(SO_3NH_4)_3$	
$\delta^{15}N$	-288.6	-226.3	-154.8	(ppm/CH_3NO_2)

Selective nitrogen *labelling experiments* may also provide a good means of elucidating certain reaction mechanisms. The reaction of 4-chloro-2-phenylquinazoline with ^{15}N labelled potassium azide was shown to lead to tautomer species, A, B, in which the position of the label is easily identified. Thus, in trifluoroacetic acid solution, the azide tautomers of type B

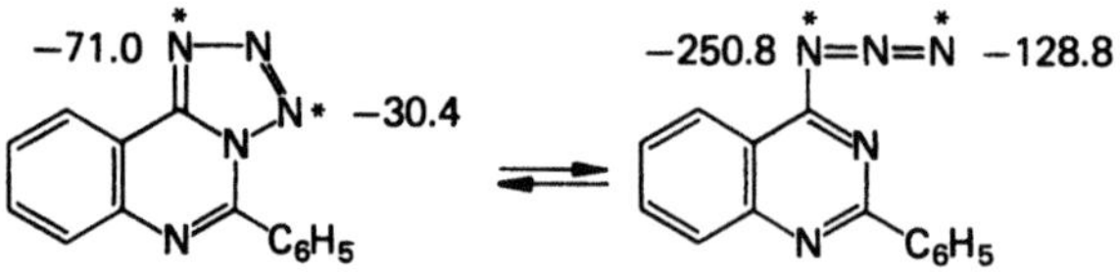

A B

(in phenylazide the α, β and γ nitrogen atoms are respectively characterized by the following shifts: -286.0, -135 and -194 ppm)

are exclusively present whereas only the tetrazoles A are detected in dimethylsulfoxide (T 7). These results corroborate a direct nucleophylic substitution mechanism. An example of the effective *observation of intermediate steps* in the course of chemical reactions is provided by the study of the Fisher indolization reaction (D 27). Two peaks which can be attributed to proton-bearing nitrogens appear simultaneously at -242.7 ppm and -185.2 ppm and their intensities decrease slowly with time. These peaks are characteristic of an amidic and an iminium nitrogen and the intervention of an intermediate of structure C in the course of the formation of indomethacin is therefore confirmed.

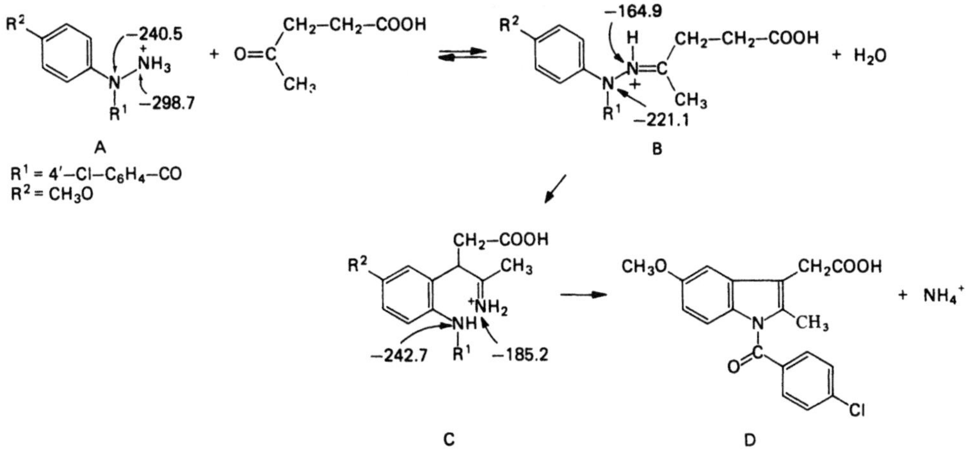

8.3.3 Application of ^{15}N Spectroscopy to the Study of Biochemical Mechanisms

In Sect. 6.2, it was emphasized that ^{15}N-NMR is particularly attractive for the structural study of biomolecules. Its potential use in the determination of biological interactions can be illustrated by the following examples.

A careful analysis of the chemical shift changes observed for various solutions of a model cyclic pentapeptide (W 25) reveals that the glycine (1)

	Gly(1) -	Pro(1) -	Gly(2) -	d-Ala -	Pro(2)	
CH_3OH	-274.6	-244.0	-272.9	-258.9	-244.0	ppm/CH_3NO_2
92% $CHCl_3$ 8% acetone	-276.2	-244.5	-275.4	-261.1	-244.5	
H_2O	{-270.8 -273.2	/	-271.3 -271.8	-255.5	/	

residue is probably involved in the peptide-bond izomerization. In fact the *cis, trans diastereoisomerism* is clearly evidenced by the dual resonance of the Gly (1) and Gly (2) fragments observed in aqueous solutions.

The sensitivity of the ^{15}N shifts to hydrogen bonding interactions (see Sect. 5) can also be used to study *base pairing*. Thus, when 5'-O-acetyl-2', 3'-O-isopropylidene adenosine is added to 3-^{15}N-labelled 2', 3', 5'-tri-O-benzoyluridine, using chloroform as a solvent, the $\delta^{15}N$ value of the 3-nitrogen of uridine moves from -228 ppm to higher frequencies (-222 ppm). This change is explained by a A - U base pairing phenomenon, which is also apparent in ^{13}C and ^{1}H spectra (P 6).

The ^{15}N shift can also be used as a probe for following the change in the *redox state* of coenzymes. In this respect, the N-1 position of [1-^{15}N] nicotinamide dinucleotide (NAD^+) undergoes an upfield shift of about 100 ppm on reduction to NADH (O 10).

-Oxidized Forms (pD = 7)

nicotinamide adenine dinucleotide (NAD^+): -153.9 ppm/CH_3NO_2

mononucleotide (MNN^+): -152.8

- Reduced forms (pD = 7)

(NADH): -263.1 ppm/CH_3NO_2

(MNNH): -263.0

From another point of view, ^{15}N spectroscopy provides a mean for *detecting the eventual formation of nitroso derivatives* when nitrites are allowed to react with organic nitrogen-containing compounds. An attempt to detect nitrosation in situ was performed by treating peptide models with sodium [^{15}N] nitrite at pH ca.3 (B 67). With N-acetylglycylglycine and N-acetylmethionine, no new signal was detected. By contrast, new signals were observed in a range of chemical shifts + 160 to + 180 ppm / CH_3NO_2 , for N-acetyltryptophan, N-acetyltryptophylglycine ethyl ester and propylglycine. These shifts are characteristic of N-nitroso derivatives. Thionitrite derivatives are, for example, detected for N-acetylcysteine of N-acetylpenicillamine.

In principle, ^{15}N-NMR constitues a non-degradative method for the study *of the metabolism of nitrogen-containing compounds.* In this area, the incorporation of l[^{15}N]- valine into penicillin G was examined (B 62). A penicillin G potassium salt, A, was prepared and standards containing three different amounts of ^{15}N labelling were obtained. 130 000 pulses, i e 18h of accumulation, were necessary in order to observe the ^{15}N spectrum at 10.14 MHz for a standard containing 0.5 atom % excess. It could then be shown that the incorporation of [^{15}N] valine introduces a measurable enhancement of the N(a) signal whereas the intensity of N(b) is not modified. The quantitative interpretation of these results shows that little deamination of l-valine occurs before incorporation.

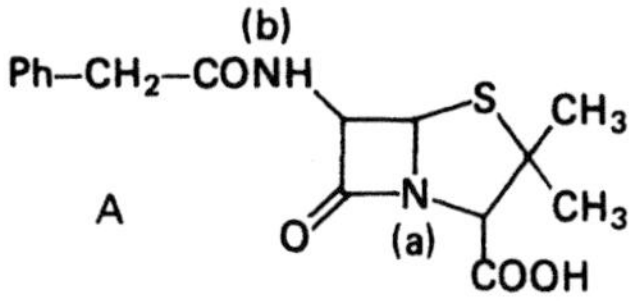

A

Finally, we shall mention that ^{15}N-NMR has proved to be an unequivocal method for detecting and identifying intermediates in *enzyme mechanisms* (B 55).

References

A

1 Axenrod, T. et al.: J. Am. Chem. Soc. *91*, 3681 (1969)

2 Ahlbrecht, H., Papke, G.: Tetrahedron *30*, 2571 (1974)

3 Axenrod, T.: Spectrosc. Lett. *3*, 263 (1970)

4 Axenrod, T. et al.: J. Am. Chem. Soc. *93*, 6536 (1971)

5 Alei, M. et al.: J. Phys. Chem. *75*, 932 (1971)

6 Axenrod, T., Wieder, M.J.: J. Am. Chem. Soc. *93*, 3541 (1971)

7 Axenrod, T. et. al.: J. Am. Chem. Soc. *92*, 6066 (1970)

8 Axenrod, T., Wieder, M.J., Milne, G.W.A.: (a) Tetrahedron Lett. *1969*, 401; (b) Tetrahedron Lett. *1969*, 1397

9 Axenrod, T., Pregosin, P.S., Milne, G.W.A.: (a) Tetrahedron Lett. *1968*, 5293; (b) J. Chem. Soc. Chem. Commun. *1968*, 702

10 Ahlbrecht, H., Papke, G.: Tetrahedron Lett. *1972*, 4443

11 Axenrod, T., Mangiaracina, P., Pregosin, P.S.: Helv. Chim. Acta *59*, 1655 (1976)

12 Albright, T.A., Freeman, W.J.: Org. Magn. Res. *9*, 75 (1977)

13 Axenrod, T.: Nitrogen NMR. In: Webb, G.A., Witanowski, M. (eds.), p. 261. New York: Plenum Publish 1973

14 Axenrod, T. et al.: Org. Magn. Res. *12*, 1 (1979)

15 Axenrod, T., Watnick, C.M., Wieder, M.J.: Org. Magn. Res. *12*, 476 (1979)

16 Arnold, D.E.J., Rankin, D.W.H.: J. Chem. Soc. Dalton *1975*, 889

17 Anderson, D.W.W., Bentham, J.E., Rankin, D.W.H.: J. Chem. Soc. Dalton *1973*, 1215

18 Axenrod, T.: NMR Spectroscopy of nuclei other than proton. In: Axenrod, T., Webb, G.A. (eds.). New York: Wiley 1974

19 Alcock, N.W., Herron, N., Moore, P.: (a) Inorg. Chim. Acta *32*, 25 (1979) (b) J. Chem. Soc. Dalton *1979*, 1486

20 Al-Najjar, I.M. et al.: J. Chem. Research (S) *1979*, 206

21 Axenrod, T., Wieder, M.J.: Org. Magn. Res. *8*, 350 (1976)

22 Alei, M., Vergamini, P.J., Wageman, W.E.: J. Am. Chem. Soc. *101*, 5415 (1979)

23 Alei, M., Florin, A.E., Litchman, W.M.: J. Phys. Chem. *75*, 1758 (1971)

24 Ando, I. et al.: Bull. Chem. Soc. Jap. *52*, 2240 (1979)

25 Alei, M., Wageman, W.E.: Tetrahedron Lett. *1979*, 667

26 Alei, M., Florin, A.E., Litchman, W.M.: J. Am. Chem. Soc. *92*, 4828 (1970)

27 Alei, M., Morgan, L.O., Wageman, W.E.: Inorg. Chem. *17*, 2288 (1978)

28 Adler, G., Lichter, R.L.: J. Org. Chem. *39*, 3547 (1974)

29 Anet, F.A.L., Ghiaci, M.: J. Am. Chem. Soc. *101*, 6857 (1979)

30 Anet, F.A.L.: Variable temperature ^{13}C-NMR. In: Topics in ^{13}C-NMR spectroscopy, Levy, G.C. (ed.), vol. 3, p. 70. New York: Wiley 1979

31 Appel, R., Halstenberg, M., Knoll, F.: Z. Naturforsch. *32b*, 1030 (1977)

B

1 Bramwell, M.R., Randall, E.W.: J. Chem. Soc. Chem. Commun. *1969*, 250

2 Bernheim, R.A., Batiz-Hernandez, H.: J. Chem. Phys. *40*, 3446 (1964)

3 Baldeschwieler, J.D.: J. Chem. Phys. *36*, 152 (1962)

4 Bourn, A.J.R., Randall, E.W.: J. Mol. Spectrosc. *13*, 29 (1964)

5 Becker, E.D., Bradley, R.B., Axenrod, T.: J. Magn. Res. *4*, 136 (1971)

6 Blanchard, M.L., et al.: Bull. Soc. Chim. France *1967*, 2677

7 Bulusu, S., Autera, J.R., Axenrod, T.: J. Chem. Soc. Chem. Commun. *1973*, 602

8 Bystrov, V.F., Gavrilov, Y.D., Solkan, V.N.: J. Magn. Res. *19*, 123 (1975)

9 Briggs, J.M., Rahkamaa, E., Randall, E.W.: J. Magn. Res. *12*, 40 (1973)

10 Briggs, J.M., Rahkamaa, E., Randall, E.W.: J. Magn. Res. *11*, 416 (1973)

11 Berger, S, Roberts, J.D.: J. Am. Chem. Soc. *96*, 6757 (1974)

12 Bundgaard, T., Jakobsen, H.J., Rahkamaa, E.J.: J. Magn. Res. *19*, 345 (1975)

13 Buchanan, G.W., Dawson, B.A.: Can. J. Chem. *54*, 790 (1976)

14 Braun, S.: Org. Magn. Res. *8*, 273 (1976)

15 Bubnov, N.N., et al.: J. Chem. Soc. Chem. Commun. *1972*, 1058

16 Bystrov, V.F.: Progress in NMR spectroscopy. In Emsley, J.W., Feeney, J., Sutcliffe, L.H. (eds.), vol. 10, p. 41. London, 1976

17 Buchanan, G.W., Dawson, B.A.: Can. J. Chem. *55*, 1437 (1977)

18 Buchanan, G.W., Dawson, B.A.: Can. J. Chem. *56*, 2200 (1978)

19 Berger, S.: Tetrahedron *34*, 3133 (1978)

20 Blomberg, F., et al.: Org. Magn. Res. *11*, 598 (1978)

21 Binsch, G., Roberts, J.D.: J. Phys. Chem. *72*, 4310 (1968)

22 Bottin-Strzalko, T., Pouet, M.J., Simonnin, M.P.: Org. Magn. Res. *8*, 120 (1976)

23 Bungaard, T., Jakobsen, H.J.: Tetrahedron Lett. *1976*, 1621

24 Binsch, G., et al.: J. Am. Chem. Soc. *86*, 5564 (1964)

25 Bose, A.K., Kugajevsky, I.: Tetrahedron *23*, 1489 (1967)

26 Bourn, A.J.R., Randall, E.W.: Mol. Phys. *8*, 567 (1964)

27 Bourn, A.J.R., Gillies, D.G., Randall, E.W.: Tetrahedron *20*, 1811 (1964)

28 Batterham, T.J., Bigum, C.: Org. Magn. Res. In press

29 Bercaw, J.E., Rosenberg, E., Roberts, J.D.: J. Am. Chem. Soc. *96*, 612 (1974)

30 Bercaw, J.E.: J. Am. Chem. Soc. *96*, 5087 (1974)

31 Bhattacharyya, P.K., Dailey, B.P.: J. Chem. Phys. *59*, 5820 (1973)

32 Boxer, S.G., Closs, G.L., Katz, J.J.: J. Am. Chem. Soc. *96*, 7058 (1974)

33 Bojesen, I.N., et al.: Acta Chem. Scand. *25*, 2739 (1971)

34 Bose, K.S., Abbott, E.H.: Inorg. Chem. *16*, 3190 (1977)

35 Blomberg, F., Maurer, W., Rüterjans, H.: J. Am. Chem. Soc. *99*, 8149 (1977)

36 Büchner, P., Maurer, W., Rüterjans, H.: J. Magn. Res. *29*, 45 (1978)

37 Burows, E.: Private communication quoted in C13

38 Bradbury, A.F., et al.: FEBS Letters *42*, 179 (1974)

39 Blomberg, F., Maurer, W., Ruterjans, H.: Proc. Natl. Acad. Sci USA, *73*, 1409 (1976)

40 Barlos, K., Noth, H., Wrackmeyer, B.: Z. Naturforsch. *33b*, 115 (1978)

41 Becker, E.D., Miles, H.T., Bradley, R.B.: J. Am. Chem. Soc. *87*, 5575 (1965)

42 Birchall, T., Jolly, W.L.: J. Am. Chem. Soc. *87*, 3007 (1965)

43 Bekarek, V., et al.: Tetrahedron Lett. *1968*, 3711

44 Bachovchin, W.W., Roberts, J.D.: J. Am. Chem. Soc. *100*, 8041 (1978)

45 Botto, R.E., Roberts, J.D.: J. Org. Chem. *42*, 2247 (1977)

46 Briggs, J.M., Farnell, L.F., Randall, E.W.: J. Chem. Soc. Chem. Commun. *(1971)*, 680

47 Botto, R.E., Westerman, P.W., Roberts, J.D.: Org. Magn. Res. *11*, 510 (1978)

48 Botto, R.E., et al.: Inorg. Chem. *18*, 2049 (1979)

49 Becker, E.D.: Private communication quoted in W2

50 Bogdanov, V.S., et al: Izv. Akad. Nauk. SSSR, Ser. Khim *12*, 2783 (1967)

51 Bekarek, V., et al.: Collect. Czech. Chem. Commun. *35*, 1406 (1970)

52 Bloom, S.M., Dudek, G.O.: J. Org. Chem. *36*, 235 (1971)

53 Botto, R.E., Roberts, J.D.: J. Org. Chem. *44*, 140 (1979)

54 Brewer, D. et al.: J. Chem. Soc. Perkin I, *1978*, 1248

55 Burton, G., et al.: J. Am. Chem. Soc. *101*, 3114 (1979)

56 Berger, S., Kaletsch, H.: Org. Magn. Res. *8*, 438 (1976)

57 Bertrand, R.D., et al.: J. Magn. Res. *32*, 465 (1978)

58 Briggs, J.M., Randall, E.W.: Mol. Phys. *26*, 699 (1973)

59 Brownstein, S., Bornais, J.: J. Magn. Res. *38*, 131 (1980)

60 Burden, A.G., Collier, G., Shorter, J.: J. Chem. Soc. Perkin II *1976*, 1627

61 Burns, J.M., et al.: J. Am. Chem. Soc. *99*, 6924 (1977)

62 Booth, H., et al.: J. Chem. Soc. Chem. Commun. *1976*, 110

63 Becker, E.D.: J. Magn. Res. *4*, 142 (1971)

64 Bradley, C.H., et al.: J. Am. Chem. Soc. *97*, 1958 (1975)

65 Bertrand, R.D., et al.: J. Am. Chem. Soc. *100*, 5227 (1978)

66 Bouchet, P., et al.: Bull. Soc. Chim. Fr. *1972*, 2264

67 Bonnett, R., et al.: J. Chem. Soc. Perkin I, *1975*, 2261

68 BIO-RAD Laboratories, Product information 2020, Richmond 1976

69 Becker, E.D.,: Private communication quoted in W23

70 Barfield, M., Geahart, H.L.,: Mol. Phys. *27*, 899 (1974)

71 Bodenhausen, G., Ruben, D.J.: Chem. Phys. Letters *69*, 185 (1980)

72 Buglass, A.J.: J. C. S. Chem. Comm. *313* (1974)

C

1 Chatt, J., et al.: J. Chem. Res. (S) *1979*, 44

2 Crepaux, D., Lehn, J.M.: Org. Magn. Res. *7*, 524 (1975)

3 Colquhoun, I.J., Mac Farlane, W.: J. Chem. Soc. Faraday II, *73*, 722 (1977)

4 Cogne, A., Robert, J.B., Wiesenfeld, L.: Chem. Phys. Lett. *57*, 627 (1978)

5 Clippard, P.H., Taylor, R.C.: Inorg. Chem. *8*, 2802 (1969)

6 Coxon, B. Johnson, L.F.: Carbohydrate Res. *20*, 105 (1971)

7 Cunliffe, A.V., Harris, R.K.: Mol. Phys. *15*, 413 (1968)

8 Chuck, R.J., Gillies, D.G., Randall, E.W.: Mol. Phys. *16*, 121 (1969)

9 Crepaux, D., Lehn, J.M.: Mol. Phys. *14*, 547 (1968)

10 Crepaux, D., Lehn, J.M., Dean, R.R.: Mol. Phys. *16*, 225 (1969)

11 Cowley, A.H., Schweiger, J.R.: J. Chem. Soc. Chem. Commun. *1970*, 1492

12 Cowley, A.H., Schweiger, J.R., Manatt, S.L.: J. Chem. Soc. Chem. Commun. *1970*, 1491

13 Cowley, A.H., Schweiger, J.R.: J. Am. Chem. Soc. *95*, 4179 (1973)

14 Christl, M. et al.: J. Am. Chem. Soc. *95*, 4392 (1973)

15 Christl. M.: Org. Magn. Res. *7*, 349 (1975)

16 Coxon, B., Hough, L.: Carbohydrate Res. *73*, 47 (1979)

17 Carty, A.J.: Inorg. Chem. *15*, 1956 (1976)

18 Carty, A.J. Jacobson, S.E.: J. Chem. Soc. Chem. Commun. *1975*, 715

19 Coxon, B.: Pure and Applied Chem. *49*, 1151 (1977)

20 Cain, A.H., Sullivan, G.R., Roberts, J.D.: J. Am. Chem. Soc. *99*, 6423 (1977)

21 Chatt, J. et al.: J. Chem. Res. *1979*, 322

22 Coburn, R.A., Dudek, G.O.: J. Phys. Chem. *72*, 1177 (1968)

23 Cooper, R.A., Lichter, R.L., Roberts, J.D.: J. Am. Chem. Soc. *95*, 3724 (1973)

24 Chastrette, M.: Tetrahedron Lett. *1979*, 1441

25 Chingas, G.C., et al.: J. Am. Chem. Soc. *101*, 4058 (1979)

26 Charton, M.: J. Org. Chem. *29*, 1222 (1964)

27 Clark, J., Perrin, D.D.: Quart. Rev. *18*, 295 (1964)

28 Chingas, G.C., et al.: J. Magn. Res. *35*, 283 (1979)

29 Craig, R.A., Harris, R.K., Morrow, R.J.: Org. Magn. Res. *13*, 229 (1980)

30 Coxon, B.: Carbohydrate Res. *35*, Cl (1974)

D

1 Danoff, A. et al.: Org. Magn. Res. *12*, 83 (1979)

2 Detsina, A.N. Mamatyuk, V.I.: Bull. Acad. Sci. USSR *22*, 2279 (1973)

3 Dudek, G.O., Dudek, E.P.: J. Am. Chem. Soc. *86*, 4283 (1964)

4 Dudek, G.O., Dudek, E.P.: Tetrahedron *23*, 3245 (1967)

5 Dudek, G.O., Dudek, E.P.: J. Am. Chem. Soc. *88*, 2407 (1966)

6 Drakenberg, T., Forsen, S.: J. Phys. Chem. *74*, 1 (1970)

7 Diblasi, R., Kopple, K.D.: J. Chem. Soc. Chem. Commun. *1975*, 33

8 De Marco, A., Llinas, M., Wuthrich, K.: Biopolymers *17*, 2727 (1978)

9 De Marco, A., Llinas, M.: Org. Magn. Res. *12*, 454 (1979)

10 Digioia, A.J., Lichter, R.L.: J. Magn. Res. *27*, 431 (1977)

11 Dudek, G.O., Dudek, E.P.: J. Chem. Soc. Chem. Commun. *1965*, 464

12 Duthaler, R.O., Roberts, J.D.: J. Am. Chem. Soc. *100*, 3889 (1978)

13 Duthaler, R.O., et al.: J. Am. Chem. Soc. *99*, 8406 (1977)

14 Duthaler, R.O., Roberts, J.D.: J. Am. Chem. Soc. *100*, 3882 (1978)

15 Dorie, J., Mechin, B., Martin, G.J.: Org. Magn. Res. *12*, 229 (1979)

16 Digioia, A.J. et al.: J. Phys. Chem. *82*, 1644 (1978)

17 Duthaler, R.O., Roberts, J.D.: J. Am. Chem. Soc. *100*, 4969 (1978)

18 Dorie, J., et al.: Org. Magn. Res. *13*, 126 (1980)

19 Danilewizc, J.C., Sewell, M.J., Thurman, J.C.: J. Chem. Soc. (C) *1971*, 1704

20 Dudek, G., Dudek, E.P.: J. Chem. Soc., (B) *1971*, 1356

21 Duthaler, R.O., Forster, H.G., Roberts, J.D.: J. Am. Chem. Soc. *100*, 4974 (1978)

22 Dominguez, D.D., King, M.M., Yeh, H.J.C.: J. Magn. Res. *32*, 161 (1978)

23 von Dreele, P.H.: J. Am. Chem. Soc. *98*, 1270 (1976)

24 Dorman, D.E., Paschal, J.W., Merkel, K.E.: J. Am. Chem. Soc. *98*, 6885 (1976)

25 Dimroth, K., et al.: Lieb. Ann. Chem. *661*, 1 (1963)

26 Duthaler, R.O., Roberts, J.D.: J. Magn. Res. *34*, 129 (1979)

27 Douglas, A.W.: J. Am. Chem. Soc. *101*, 5676 (1979)

28 Depaire, H., et al.: Tetrahedron Lett. *1977*, 1401

E

1 Ernst, L., Lustig, E., Wray, V.: J. Magn. Res. *22*, 459 (1976)

2 Elguero, J., et al.: Org. Magn. Res. *9*, 145 (1977)

3 Erashko, V.I., et al.: Izv. Akad. Nauk SSSR, Ser Khim *1970*, 958

4 Evans, D.F.: J. Chem. Soc. *1963*, 5575

5 Exner, O.: Correlation analysis chemistry, recent advances. In: Chapman, N.B., Shorter, J., (eds.), p. 439. New York: Plenum Press 1978

6 Ebraheem, K.A.K., Webb, G.A.: Org. Magn. Res. *9*, 248 (1977)

7 Ernst, R.R.: J. Chem. Phys. *59*, 989 (1973)

F

1 Fanghänel, E., et al.: J. Prakt. Chem. *320*, 168 (1978)

2 Fritz, H., Winkler, T., Kung, W.: Helv. Chim. Acta *58*, 1822 (1975)

3 Fritz, H., Clerin, D., Fleury, J.P.: Org. Magn. Res. *8*, 269 (1976)

4 Fox, W.B., et al.: Inorg. Chem. *7*, 2064 (1968)

5 Fischman, A.J., et al.: J. Am. Chem. Soc. *100*, 54 (1978)

6 Fusstetter, H., Noth, H., Wrackmeyer, B.: Chem. Ber. *110*, 3172 (1977)

7 Fenton, D.E., et al.: Private communication quoted in R 11

8 Filleux-Blanchard, M.L., An, N.D.: Org. Magn. Res. *12*, 12 (1979)

9 Fischer, P.B., Kaul, B.L., Zollinger, H.: Helv. Chim. Acta *51*, 1449 (1968)

10 Fraser, R.R., Haque, K.E.: Can. J. Chem. *46*, 2855 (1968)

11 Flygare, W.H.: J. Chem. Phys. *41*, 793 (1964)

12 Feigel, M., et al.: J. Am. Chem. Soc. *101*, 1943 (1979)

13 Fanso-Free, S.N.Y., et al.: J. Am. Chem. Soc. *101*, 1549 (1979)

14 Fujita, T.: The analysis of the ortho effect. Progress in Physical Organic Chemistry. In: Taft, R.W. (ed.), vol. 12, p. 49, New York: Interscience, 1976

15 Filleux-Blanchard, M.L., et al.: J. Mol. Struct. *3*, 351 (1969)

16 Fronza, G., et al.: J.C.S. Perkin II, *1977*, 1746

17 Fung, B.M., et al.: Inorg. Chem. *12*, 1203 (1973)

G

1 Green, R.D.: Can. J. Chem. *47*, 2407 (1969)

2 Gagnaire, D., Ramasseul, R., Rassat, A.: Bull. Soc. Chim. Fr. 415 (1970)

3 Gould, S.J., Chang, C.C.: J. Am. Chem. Soc. *100*, 1624 (1978)

4 Gray, G.A., Albright, T.A.: J. Am. Chem. Soc. *98*, 3857 (1976)

5 Gray, G.A., Albright, T.A.: J. Am. Chem. Soc. *99*, 3243 (1977)

6 Gray, G.A., Buchanan, G.W., Morin, F.G.: J. Org. Chem. *44*, 1768 (1979)

7 Gavrilov, Yu.D., Solkan, V.N., Bystrov, V.F.: Izv. Akad. Nauk. SSSR, Ser, Khim., in press, quoted in B 16

8 Gust, D., Roberts, J.D.: J. Am. Chem. Soc. *99*, 3637 (1977)

9 Gust, D., Dirks, G., Pettit, G.R.: J. Org. Chem. *44*, 314 (1979)

10 Gouesnard, J.P., Dorie, J., Martin, G.J.: Can. J. Chem. *58*, 1295 (1980)

11 Gouesnard, J.P., Dorie, J.: J. Mol. Struct. *67*, 297 (1980)

12 Gouesnard, J.P., Dorie, J.: Unpublished results

13 Gattegno, D., Hawkes, G.E., Randall, E.W.: J. Chem. Soc. Perkin II, *1976*, 1527

14 Gouesnard, J.P., Martin, G.J.: Org. Magn. Res. *12*, 263, (1979)

15 Gust, D., Neal, D.N.: J. Chem. Soc. Chem. Commun. *1978*, 681

16 Giannini, D.D., et al.: J. Am. Chem. Soc. *97*, 3416 (1975)

17 Gust, D., Moon, R.B., Roberts, J.D.: Proc. Natl. Acad. Sci USA *72*, 4696 (1975)

18 Gansow, O.A., et al.: Org. Magn. Res. *13*, 77 (1980)

19 Gutmann, V.: (a) Electrochim. Acta *21*, 661 (1976) (b) Coordinat. Chem. Rev. *18*, 225 (1976)

20 Green, J.G., Dubay, G.R., Porter, N.A.: J. Am. Chem. Soc. *99*, 1264 (1977)

21 Gil, V.M.S., Formosinho-Simoes, S.J.S.: Mol. Phys. *25*, 639 (1968)

22 Geppert, H., et al.: Z. Chem. *17*, 23 (1977)

23 Gibbons, W.A., et al.: Proc. Nat. Acad. Sci. *67*, 239 (1970)

H

1 Hale, W.H., Williamson, S.M.: Inorg. Chem. *4,* 1343 (1965)

2 Harris, R.K., et al.: Mol. Phys. *19,* 145 (1970)

3 Hogeveen, H.: Rec. Trav. Chim. *86,* 1288 (1967)

4 Hawkes, G.E., et al.: J. Chem. Soc. Perkin II, *1977,* 1024

5 Hansen, M., Jakobsen, H.J.: Acta Chem. Scand. *26,* 2151 (1972)

6 Hinton, J.F., Ladner, K.H., Stewart, W.E.: J. Magn. Res. *12,* 90 (1973)

7 Hinton, J.F., Ladner, K.H.: J. Magn. Res. *6,* 586 (1972)

8 Hargis, J.H., et al.: J. Am. Chem. Soc. *99,* 8090 (1977)

9 Haber, A., Johnson, R.D., Rinehart, K.L.: J. Am. Chem. Soc. *99,* 3541 (1977)

10 Hawkes, G.E., Randall, E.W., Hull, W.E.: J. Chem. Soc. Perkin II, *1977,* 1268

11 Happe, J.A., Morales, M.: J. Am. Chem. Soc. *88,* 2077 (1966)

12 Hagen, R., et al.: J. Am. Chem. Soc. *95,* 5712 (1973)

13 Hawkes, G.E., Litchman, W.M., Randall, E.W.: J. Magn. Res. *19,* 255 (1975)

14 Hawkes, G.E., Randall, E.W., Hull, W.E.: J. Chem. Soc. Chem. Commun. *1977,* 546

15 Harrison, J.F.: J. Chem. Phys. *47,* 299, (1967)

16 Harris, R.K., Kimber, B.J.: J. Magn. Res. *17,* 174 (1975)

17 Hargis, J.H., et al.: J. Am. Chem. Soc. *102,* 13 (1980)

18 Haltiwanger, R.C., et al.: J. Am. Chem. Soc. *100,* 5110 (1978)

19 Harris, R.K.: Personal Communication 1980

20 Hartmann, S.R., Hahn, E.L.: Phys. Rev. *128,* 2042 (1962)

21 Hawkes, G.E., Randall, E.W., Bradley, C.H.: Nature, (London) *257,* 767 (1975)

22 Heilbronner, E., Schmelzer, A.: Nouv. J. Chim. *4,* 23 (1980)

I

1 Ioffe, S.L., Blyumenfeld, A.L., Shaskov, A.S.: Izv. Akad. Nauk. SSSR, Ser. Khim. *1978,* 246

2 Irving, C.S., Lapidot, A.: J. Am. Chem. Soc. *97,* 5945 (1975)

3 Irving, C.S., Lapidot, A.: Proc. Int. Conf. Stable Isotopes *3rd,* 307 (1979)

4 Irving, C.S., Lapidot, A.: J. Chem Soc. Chem. Commun. *1976,* 43

5 Irving, C.S., Lapidot, A.: J. Chem. Soc. Chem. Commun. *1977,* 184

6 Ioffe, S.L., et al.: Izv. Akad. Nauk. SSSR, Ser. Khim. *27,* 1009 (1978)

7 Ishol, L.M., Scott, T.A., Goldblatt, A.: J. Magn. Res. *23,* 313 (1976)

8 IUPAC Recommendations for nuclei other than ^{1}H: Pure Appl. Chem. *45,* 219 (1976)

J

1 Jakobsen, H.J., Brey, W.S.: J. Chem Soc. Chem. Commun. *1979,* 478

2 Jennings, W.B., et al.: J. Am. Chem Soc. *94,* 8501 (1972)

3 Jacobsen, J.P., et al.: J. Magn. Res. *10,* 130 (1973)

4 Jerina, D.M., et al.: Tetrahedron Lett. *1970,* 1483

5 Jameson, C.J., et al.: J. Chem. Phys. *68,* 2861 (1978)

6 Jurkschat, K., et al.: J. Organometal. Chem. *171,* 301 (1979)

7 Jones, A.J., Mc Nab, H., Hanisch, P.: Aust. J. Chem. *31,* 1005 (1978)

8 Jacobsen, J.P., Schaumburg, K., Nielsen, J.T.: J. Magn. Res. *13*, 372 (1974)

9 Jakobsen, H.J., Brey, W.S.: J. Am. Chem Soc. *101,* 774 (1979)

10 Jones, R.G., Wilkins, J.M.: Org. Magn. Res. *11,* 20 (1978)

11 Jackman, L.M., Trewella, J.C.: J. Am. Chem. Soc. *101,* 6428 (1979)

12 Johannesen, R.B.: J. Chem. Phys. *48,* 1414 (1968)

K

1 Kopple, K.D., Ahsan, A., Barfield, M.: Tetrahedron Lett. *1978,* 3519

2 Kerrison, S.J.S., Sadler, P.J.: J. Chem. Soc. Chem. Commun. *1977,* 861

3 Kreissl, F.R., et al.: J. Organometal. Chem. *108,* C 29 (1976)

4 Kainosho, M.: J. Am. Chem. Soc. *101,* 1031 (1979)

5 Kawazoe, Y., Ohnishi, M., Kataoka, N.: Chem. Pharm. Bull. Tokyo *13*, 396 (1965)

6 Kurkovkaya, L.N., et al.: Zhur. Strukt. Khim. *16,* 139 (1975)

7 Kintzinger, J.P., Lehn, J.M.: J. Chem. Soc. Chem. Commun. *1967,* 660

8 Kintzinger, J.P., Lehn, J.M.: Mol. Phys. *14,* 133 (1968)

9 Kalabin, G.A., et al.: Izv. Akad. Nauk. SSSR, Ser. Khim. *1969,* 2627

10 Kawazoe, Y., et al.: Chem. Pharm. Bull. Tokyo *18*, 381 (1970)

11 Karplus, S., Karplus, M.: Proc. Natl. Acad. Sci. USA, *69,* 3204 (1972)

12 Kennedy, J.D., Mc Farlane, W., Wrackmeyer, B.: Inorg. Chem. *15,* 1299 (1976)

13 Kricheldorf, H.R., Schilling, G.: Makromol. Chem. *179,* 2667 (1978)

14 Kawano, K., et al.: (a) J. Chem. Soc. Chem. Commun. *1977,* 226 (b) J. Chem. Soc. Perkin II, *1978,* 1319

15 Kainosho, M., et al.: Biochem. Biophys. Res. Commun. *64*, 425 (1975)

16 Klose, G., Uhlemann, E.: Tetrahedron *22,* 1373 (1966)

17 Kricheldorf, H.R., Hull, W.E.: Org. Magn. Res. *12,* 607 (1979)

18 Kricheldorf, H.R.: Makromol. Chem. *179*, 2687 (1978)

19 Kricheldorf, H.R., Hull, W.E.: Makromol. Chem. *180,* 161 (1979)

20 Kricheldorf, H.R.: Makromol. Chem. *180,* 147 (1979)

21 Kozlov, E.S., et al.: Zhur. Obshch. Khim. SSSR *42,* 748 (1972)

22 Kawazoe, Y., et al.: private communication, quoted in C 9

23 Kamei, H.: Bull. Chem. Soc. Japan *41,* 2269 (1968)

24 Kricheldorf, H.R., Hull, W.E.: J. Macromol. Sci. Chem. *A 12,* 51 (1978)

25 Kricheldorf, H.R., Hull, W.E., Formacek, V.: Biopolymers *16,* 1609 (1977)

26 Kricheldorf, H.R.: Org. Magn. Res. *13,* 52 (1980)

27 Krynicki, K., Rahkamaa, E.J., Powles, I.G.: Mol. Phys. *29,* 539 (1975)

28 Kukolich, S.G., Wofsy, S.C.: J. Chem. Phys. *52,* 5477 (1970)

29 Kricheldorf, H.R.: Angew. Chem. I.E. *17,* 442 (1978)

30a Kosower, E.M.: J. Am. Chem. Soc. *80*, 3253 (1958)

b Kosower, E.M., Mohammad, M.: J. Am. Chem. Soc. *90*, 3271 (1968); *93*, 2713 (1971)

31 Kricheldorf, H.R.: Org. Magn. Res. *12*, 414 (1979)

32 Kricheldorf, H.R.: Makromol. Chem. *179*, 2675 (1978)

33 Koppel, I.A., Palm, V.A.: The influence of the solvent on organic reactivity. In: Advances in linear free energy relationships. Chapman, N.B., Shorter, J. (eds.), p. 203, New York: Plenum Press 1972

34 Korver, P.D., et al.: Org. Magn. Res. *2*, 295 (1970)

35 Karplus, M., Das, T.P.: J. Chem. Phys. *34*, 1683 (1961)

36 Karplus, M., Pople, J.A.: J. Chem. Phys. *38*, 2803 (1963)

37 Knudson, S.K., Idoux, J.P.: J. Org. Chem. *44*, 520 (1979)

38 Kaplan, J.I.: J. Chem. Phys. *59*, 990 (1973)

39 Khin, T., Webb, G.A.: Org. Magn. Res. *11*, 487 (1978)

40 Khin, T., Webb, G.A.: J. Magn. Res. *33*, 159 (1979)

41 Kanamori, K., Cain, A.H., Roberts, J.D.: J. Am. Chem. Soc. *100*, 4979 (1978)

42 Kurkovskaya, L.N.: J. Struct. Chem. *19*, 665 (1978)

43 King, M.M., Yeh, H.J.C., Dudek, G.O.: Org. Magn. Res. *8*, 208 (1976)

44 Kricheldorf, H.R., Hull, W.E.: Makromol. Chem. *178*, 253 (1977)

45 Khin, T., Duangthai, S., Webb, G.A.: Org. Magn. Res. *13*, 240 (1980)

L

1 Lichter, R.L., Roberts, J.D.: J. Org. Chem. *35*, 2806 (1970)

2 Lichter, R.L., et al.: J. Chem. Soc. Chem. Commun. *1977*, 366

3 Lichter, R.L.: Determination of organic structures by physical methods. In: Zuckerman, J.J., Nachod, F. (eds.), Vol.4, New York: Academic Press 1971

4. Logan, N., Jolly, W.L.: Inorg. Chem. *4*, 1508 (1965)

5 Lecoustumer, G., Mollier, Y.: Bull. Soc. Chim. France *1970*, 3076

6 London, R.E., et al.: Org. Magn. Res. *9*, 598 (1977)

7 Liler, M.: J. Magn. Res. 5, 333 (1971)

8 Lichter, R.L., Roberts, J.D.: J. Am. Chem. Soc. *93*, 5218 (1971)

9 Lambert, J.B., Oliver, W.L., Roberts, J.D.: J. Am. Chem. Soc. *87*, 5085 (1965)

10 Litchman, W., Alei, N., Florin, A.E.: J. Chem. Phys. *50*, 1897 (1969)

11 Llinas, J.R., Vincent, E.J., Peiffer, G.: Bull. Soc. Chim. France *1973*, 3209

12 Lichter, R.L., Dorman, D.E., Wasilishen, R.: J. Am. Chem Soc. *96*, 930 (1974)

13 Liler, M.: J. Chem. Soc. Perkin II, *1972*, 816

14 Lichter, R.L., et al.: J. Chem. Soc. Chem. Commun. *1974*, 114

15 Lambert, J.B., Binsch, G., Roberts, J.D.: Proc. Nat. Acad. Sci. U.S.A. *51*, 735 (1964)

16 Llinas, M., Horsley, W.J., Klein, M.P.: J. Am. Chem Soc. *98*, 7554 (1976)

17 Lichter, R.L., Roberts, J.D.: Spectrochim. Acta *26 A*, 1813 (1970)

18 Lipnick, R.L., Fissekis, J.P.: J. Org. Chem. *44*, 1627 (1979)

19 Lapidot, A., Irving, C.S.: J. Am. Chem. Soc. *99,* 5488 (1977)

20 Levy, G.C., Godwin, A., Holloway, C.E.: J. Magn. Res. *34,* 327 (1979)

21 Levy, G.C., et al.: Org Magn. Res. *8,* 643 (1976)

22 Live, D.H., et al.: J. Am. Chem Soc. *101,* 474 (1979)

23 Lestina, G.J., Regan, T.H.: J. Org. Chem. *34,* 1685 (1969)

24 Lestina, G.J., et al.: J. Org. Chem. *33,* 3336 (1968)

25 Lichter, R.L., Roberts, J.D.: J. Am. Chem Soc. *94,* 2495 (1972)

26 Lichter, R.L., Roberts, J.D.: Org. Magn. Res. *6,* 636 (1974)

27 Levy, G.C., et al.: Org. Magn. Res. *8,* 643 (1976)

28 Lichter, R.L., Roberts, J.D.: J. Am. Chem. Soc. *94,* 4904 (1972)

29 Liler, M.: (a) J. Chem Soc. Chem. Commun. *1971,* 115; (b) J. Chem. Soc. Perkin II, *1974,* 71

30 Levy, G.C., Lichter, R.L.: Nitrogen ^{15}NMR Spectroscopy. New York: Wiley, J., 1979

31 Leonard, N.J., Henderson, T.R.: J. Am. Chem. Soc. *97,* 4990 (1975)

32 Loftus, P., Bearden, W.H., Roberts, J.D.: Nouv. J. Chim. *1,* 283 (1977)

33 Litchman, W.M., Alei, M.: J. Chem. Phys. *56,* 5818 (1972)

34 Levy, G.C., Dechter, J.J., Kowalewski, J.: J. Am. Chem. Soc. *100,* 2308 (1978)

35 Levy, G.C., et al.: J. Magn. Res. *29,* 553 (1978)

36 Leipert, T.K., Noggle, J.H.: J. Am. Chem. Soc. *97,* 269 (1975)

37 Lambert, J.B., Netzel, D.A.: J. Magn. Res. *20,* 575 (1975)

38 Lippmaa, E., Saluvere, T., Laisaai, S.: Chem. Phys. Lett. *11,* 120 (1971)

39 Lambert, J.B., Netzel, D.A.: J. Magn. Res. *25,* 531 (1977)

40 Lapidot, A., Irving, C.S.: Proc. Natl. Acad. Sci. USA *74,* 1988 (1977)

41 Lippmaa, E., et al.: Org. Magn. Res. *5,* 429 (1973)

42 Liepinsh, E., et al.: Zhur. Obshch. Khim. *49,* 1537 (1979)

43 Landolt-Bornstein: Numeral tables and functions, vol. II, 6th edit. Heidelberg, Berlin, New York: Springer-Verlag 1967

44 Litchman, W.M., Alei, M., Florin, A.E.: J. Am. Chem. Soc. *91,* 6574 (1969)

45 Lichter, R.L.: J. Magn. Res. *18,* 367 (1975)

46 Lippmaa, E., et al.: Org. Magn. Res. *4,* 153 (1972)

47 Lippmaa, E., et al.: Org. Magn. Res. *4,* 197 (1972)

48 Lynch, B.M.: Tetrahedron Lett. *1969,* 1357

49 Lehninger, A.L.: Biochemistry. 2nd ed. New York: Worth 1975

50 Levy, G.C., Bailey, J.T., Wright, D.A.: J. Magn. Res. *37,* 353 (1980)

51 Ley, B.E., Foti, A.E., Rothstein, S.M.: J. Am. Chem. Soc. *97,* 2030 (1975)

52 Lee, K., Anderson, W.A.: Handbook of chemistry and physics. 53rd ed., E 57. Cleveland: R.C. Weast Ed. Chem. Rubber Co. 1972-1973

53 Lapidot, A., Irving, C.S., Malik, Z.: J. Am. Chem. Soc. *98,* 632 (1976)

54 Lichter, R.L., Dorman, D.L.: J. Org. Chem. *41,* 582 (1976)

55 Lippmaa, E., et al.: Org. Magn. Res. *5,* 441 (1973)

56 Lichter, R.L., Roberts, J.D.: J. Am. Chem. Soc. *93,* 3200 (1971)

57 Litchman, W.M., Alei, M., Florin, A.E.: J. Chem. Phys. *50*, 1031 (1969)
58 LLinas, M., Wüthrich, K.: Biochim. Biophys. Acta *532*, 29 (1978)
59 Llinas, M., et al.: Nature *257*, 817 (1975)
60 Lapidot, A., Irving, C.S.: Proc. Intern. Conf. Stable isotopes *427* (1975)

M

1 Momenteau, M., et al.: Can. J. Chem. *56*, 2598 (1978)
2 Morishima, I., et al.: J. Chem Soc. Chem. Commun. *1979*, 483
3 Meij, R.D.J., et al.: J. Organometal. Chem. *164*, 353 (1979)
4 Marchal, J.P., Canet, D.: J. Am. Chem. Soc. *97*, 6581 (1975)
5 Maciel, G.E., et al.: J. Am. Chem. Soc. *92*, 11 (1970)
6 Muetterties, E.L., Phillips, W.D.: J. Am. Chem. Soc. *81*, 1084 (1959)
7 Motschi, H., Pregosin, P.S., Venanzi, L.M.: Helv. Chim. Acta *62*, 667 (1979)
8 Manriquez, J.M., et al.: J. Am. Chem. Soc. *100*, 3078 (1978)
9 Martin, M.L., et al.: Org. Magn. Res. *13*, 396 (1980)
10 Muller, J., Schröder, H.Fr.: Z. Anorg. Allg. Chem. *450*, 149 (1979)
11 Muller, J.: Z. Naturforsch. *33b*, 993 (1978)
12 Morishima, I., Inubushi, T., Sato, M.: J. Chem. Soc. Chem. Commun. *1978*, 106
13 Motschi, H., Sze, S.N., Pregosin, P.S.: Helv. Chim. Acta. *62*, 2086 (1979)
14 Miles, H.T., Bradley, R.B., Becker, E.D.: Science, *142*, 1569 (1963)
15 Mester, L., et al.: Tetrahedron Lett. *1968*, 4053
16 Mester, L., Stephen, A., Parello, J.: Tetrahedron Lett. *1968*, 4119
17 Mester, L., Vass, G.: Tetrahedron Lett. *1969*, 3847
18 Moy, D., Young, A.R.: J. Am. Chem. Soc. *87*, 1889 (1965)
19 Meij, R., et al.: J. Organometal. Chem. *155*, 323 (1978)
20a Martin, G.J., Guglielmetti, R., unpublished work
b Martin, M.L, Poignant, S., unpublished work
c Martin, G.J., Dorie, J., Gouesnard, J.P., unpublished work
d Martin, G.J., Rabiller, C., Gouesnard, J.P.: unpublished work
e Martin, G.J., Cousseau, J.: unpublished work
f Martin, G.J., Chen, P.C.: unpublished work
21 Markowski, V., Sullivan, G.R., Roberts, J.D.: J. Am. Chem. Soc. *99*, 714 (1977)
22 Martin, G.J., et al.: J. Am. Chem. Soc. *99*, 1381 (1977)
23 Markowski, V., et al.: Proc. Natl. Acad. Sci, USA, *74*, 1308 (1977)
24 Müller, J.: Z. Naturforsch. *34b*, 437 (1979)
25 Müller, J.: Z. Naturforsch. *34b*, 531 (1979)
26 Mc Farlane, W., Wrackmeyer, B.: J. Chem. Soc. Dalton *1978*, 2351
27 Mc Farlane, W., Wrackmeyer, B.: Inorg. Nucl. Chem. Lett. *11*, 719 (1975)
28 Mc Farlane, W.: Mol. Phys. *10*, 603 (1966)
29 Morishima, I., Inubushi, T.: J. Chem. Soc. Chem. Commun. *1977*, 616

30 Morishima, I., Inubushi, T.: J. Am. Chem. Soc. *100,* 3568 (1976)

31 Martin, M.L., Delpuech, J.J., Martin, G.J.: Practical NMR spectroscopy London: Heyden 1980. See the section corresponding to the page indicated and the references cited.

a) p. 400, b) p. 177, c) P.231, d) p.216 remembering that $\gamma^{15}N$ is negative, whereas the discussion refers to the case of positive gyromagnetic ratios
e) p. 209, f) p.235, g) p. 356, h) p.134, i) p. 222, j) p. 224, (fig. 6-10-e-f)
k) p. 287, l) p. 281, m) p.163, n) p. 206, o) p. 396, p) p. 377 q) p. 330
r) p. 359

32 Mason, J., Vinter, J.G.: J. Chem. Soc. Dalton *1975,* 2522

33 Mc Farlane, W.: Ann. Rep. NMR Spectroscopy, Mooney, E.F. (ed.), vol.1, p. 135 New York: Academic Press 1968

34 Mc Clellan, A.L.: Tables of experimental dipole moments. London: Freeman 1963

35 Martin, M.L.: Ann, Phys. *7,* 36 (1962)

36 Martin, G.J., et al.: J. Chem. Res. *1979*, 408

37 Morris, D.G., Murray, A.M.: J. Chem. Soc. Perkin II, *1976,* 1579

38 Morris, G.A.: J. Am. Chem. Soc. *102,* 428 (1980)

39 Morris, G.A., Freeman, R.: J. Am. Chem Soc. *101,* 760 (1979)

40 Maudsley, A.A., Muller, L., Ernst, R.R.: J. Magn. Res. *28,* 463 (1977)

41 Murphy, P.D., et al.: J. Am. Chem. Soc. *101,* 4055 (1979)

42 Murray, A., Williams, D.L.: Organic Syntheses with Isotopes, Part II, New York: Interscience 1958

43 Morishima, I., et al.: J. Chem. Soc. Chem. Commun. *1970*, 1321

44 Martin, G.J.: La RMN dynamique du Carbone, Informat. Chim. *1975,* 127

45 Mann, B.E.: Dynamic NMR spectroscopy. In: Progress in NMR spectroscopy. Emsley, J.W., Feeney, J., Sutcliffe, L.H. (eds.), vol. 11, p.95, Oxford: Pergamon Press 1977

46 Mc Farlane, W.: J. Chem. Soc. A *1967,* 1660

47 McInnes, A.G., et al.: J. Chem. Soc. Chem. Comm. *281* (1974)

48 McInnes, A.G., Taylor, A., Walter, J.A.: J. Am. Chem. Soc. *98*, 6741 (1976)

49 Morishima, I., et al.: Biochem. Biophys. Res. Comm. *78*, 739 (1977)

50 Morishima, I., Inubushi, T.: Febs. Lett. *81*, 57 (1977)

51 Morishima, I., Inubushi, T.: Biochem. Biophys. Res. Comm. *80*, 199 (1978)

52 Muller, J.: Z. Naturforsch. *34b*, 536 (1979)

N

1 Nadzan, A.M., Rinehart, K.L.: J. Am. Chem. Soc. *99,* 4647 (1977)

2 Negrebetskii, V.V., Bogdanov, V.S., Kessenikh, A.V.: Zhur. Strukt. Khim. *12,* 716 (1971)

3 Negrebetskii, V.V., et al.: Izv. Akad. Nauk. SSSR, Ser. Khim B, *1971*, 2613

4 Naulet, N., Beljean, M., Martin, G.J.: Tetrahedron Lett. *1976*, 3597

5 Naulet, N., Martin, G.J.: Tetrahedron Lett. *1979,* 1493

6 Naulet, N., et al.: J. Chem. Res. (S), *1980*, 158

7 Neunhoeffer, H., Fruhauf, H.W.: Liebigs Ann. Chem. *760*, 102 (1972)

8 Negrebetskii, V.V., et al.: Zhur. Strukt. Khim. *19*, 532 (1978)

9 Noth, H., Wrackmeyer, B.: NMR Spectroscopy of boron compounds in NMR basic principles and progress. Diehl, P., Fluck, E., Kosfeld, R. (eds.), vol. 14, p. 104. Berlin, Heidelberg, New York: Springer Verlag 1978

1o Nomura, Y., Takeuchi, Y.: J. Chem. Soc. Chem. Commun. *1979*, 295

11 Nakatsuji, H. et al.: Bull. Chem. Soc. Japan *44*, 2010 (1971)

12 Niu, C., et al.: J. Biochem. and Biophys. methods *1*, 135 (1979)

O

1 Oatis, J.E., Schultz, H.P.: Org. Magn. Res. *11*, 40 (1978)

2 Ogg, R.A., Ray, J.D.: J. Chem. Phys. *26*, 1515 (1957)

3 Ohtsuru, M., Tori, K.: Tetrahedron Lett. *1970*, 4043

4 Olah, G.A., Kiovsky, T.E.: J. Am. Chem. Soc. *90*, 4666 (1968)

5 Olah, G.A., White, A.M.: J. Am. Chem. Soc. *90*, 6087 (1968)

6 Olah, G.A., Schlosberg, R.H.: J. Am. Chem Soc. *90*, 6464 (1968)

7 Olah, G.A., Balaram Gupta, B.G., Narang, S.C.: J. Am. Chem Soc. *101*, 5317, (1979)

8 Ozaki, Y., et al.: J. Chem. Soc. Chem. Commun. *1979*, 76

9 Olesker, A., et al.: J. Chem. Soc. Chem. Commun. *1978*, 577

10 Oppenheimer, N.J., Davidson, R.M.: Org. Magn. Res. *13*, 14 (1980)

11 Oppenheimer, N.J., Davidson, R.M.: J. Labell. Comp. Radiopharm. *15*, 191 (1978)

P

1 Paolillo, L., Becker, E.D.: J. Magn. Res. *3*, 200 (1970)

2 Paolillo, L., Becker, E.D.: J. Magn. Res. *2*, 168 (1970)

3 Pregosin, P.S., Randall, E.W., White, A.I.: J. Chem. Soc. Perkin Trans. II, *1972*, 1

4 Pregosin, P.S., Steiner, E.: Helv. Chim. Acta *59*, 376 (1976)

5 Pregosin, P.S., Omura, H., Venanzi, L.M.: J. Am. Chem. Soc. *95*, 2047 (1973)

6 Poulter, C.D., Livingston, C.L.: Tetrahedron Lett. *1979*, 755

7 Porter, N.A., Dubay, G.R., Green J.G.: J. Am. Chem.Soc. *100*, 920 (1978)

8 Pregosin, P.S., Randall, E.W., White, A.I.: J. Chem. Soc. Chem. Commun. *1971*, 1602

9 Paschal, J.W., et al.: J. Org. Chem. *43*, 2013 (1978)

10 Posner, T.B., et al.: J. Chem. Soc. Chem. Commun. *1975*, 769

11 Poignant, S., Gauvreau, J.R., Martin, G.J.: Can. J. Chem. *58*, 946 (1980)

12 Pestunovich, V.A., et al.: Izv. Akad. Nauk. SSSR, Ser. Khim. *1978*, 2653

13 Paschal, J.W., Dorman, D.E.: Org. Magn. Res. *11*, 632 (1978)

14 Pregosin, P.S., Randall, E.W., White, A.I.: J. Chem. Soc. Perkin II, *1972*, 513

15 Pines, A., Gibby, M.G., Waugh, J.S.: J. Chem. Phys. *59,* 569 (1973)

16 Psota, L., et al.: Org. Magn. Res. *11,* 401 (1978)

17 Porter, N.A., Green, J.G., Dubay, G.R.: Tetrahedron Lett. *1975,* 3363

18 Porter, N.A., Green, J.G.: Tetrahedron Lett. *1975,* 2667

19 Pichon, R.: Dissertation Brest 1978

20 Pestunovich, V.A., et al.: Izv. Akad. Nauk. SSSR, Ser. Khim. *1979,* 2159

R

1 Reeves, L.W., Tracey, A.S.: J. Am. Chem. Soc. *96,* 7176 (1974)

2 Runge, W., Firl, J.: Z. Naturforsch. *316,* 1515 (1976)

3 Randall, E.W., Ellner, J.J., Zuckerman, J.J.: J. Am. Chem. Soc. *88*, 622 (1966)

4 Roberts, B.W., Lambert, J.B., Roberts, J.D.: J. Am. Chem. Soc. *87,* 5439 (1965)

5 Rahkamaa, E.: Mol. Phys. *19,* 727 (1970)

6 Rogers, M.T., Laplanche, L.A.: J. Phys. Chem. *69,* 3648 (1965)

7 Randall, E.W., Zuckerman, J.J.: J. Am. Chem. Soc. *90*, 3167 (1968)

8 Rahkamaa, E.: Z. Naturforsch. *24A,* 2004 (1969)

9 Riddell, F.C., Lehn, J.M.: J. Chem. Soc. (B) *1968*, 1224

10 Reeves, L.W., et al.: Mol. Phys. *25,* 9 (1973)

11 Randall, E.W., Gillies, D.G.: Progress in NMR spectroscopy. In: Emsley, J.W., Feeney, J., Sutcliffe, L.H. (eds.), vol. 6, p. 119. Oxford: Pergamon Press 1971

12 Rosenberg, E., Williamson, K.L., Roberts, J.D.: Org. Magn. Res. *8,* 117 (1976)

13 Rabiller, C., et al.: Nouv. J. Chimie *4,* 35 (1980)

14 Reeves, L.W., Tracey, A.S.: J. Am. Chem. Soc. *96,* 1198 (1974)

15 Reichardt, C., Grahn, W.: (a) Tetrahedron *27,* 3745 (1971) (b) Chem. Ber. *103,* 1065 (1970)

16 Rabiller, C., et al.: Org. Magn. Res. *14*, 479 (1980)

17 Ray, J.D.: J. Chem. Phys. *40,* 3440 (1964)

18 Reichardt, C.: (a) Solvent effects in Organic Chemistry; Weinheim: Verlag Chemie 1969 (b) Angew. Chem. Internat. Edit. Engl. *18,* 98 (1979)

19 Rykowski, A., Van Der Plas, H.C.: Rec. Trav. Chim. *94,* 204 (1975)

20 Rasmussen, C.A., Van Der Plas, H.C.: Rec. Trav. Chim. *96,* 1o1 (1977)

21 Ritchie, C.D.: Interactions in dipolar aprotic solvents. In: Solute-Solvent Interactions. Coetzee, J.F., Ritchie, C.D. (eds.), p. 219. New York: Marcel Decker 1969

22 Rüterjans, H., Blomberg, F., Maurer, W.: Ber. Bunsenges *80*, 1240 (1976)

23 Radeglia, R., et al.: Z. Phys. Chem. Leipzig *261*, 502 (1980)

S

1 Schultheiss, H., Fluck, E.: Z. Naturforsch. *32b,* 257 (1977)

2 Shoup, R.R., Becker, E.D., Miles, H.T.: Biochem. Biophys. Res. Commun. *43,* 1350 (1971)

3 Stefaniak, L., Witanowski, M. : Bull. Acad. Polon. Sci. XXV, *1977,* 261

4 Sunners, B., Piette, L.H., Schneider, W.G.: Can. J. Chem. *38,* 681 (1960)

5 Sogn, J.A., Gibbons, W.A., Randall, E.W.: Biochemistry *12,* 2100 (1973)

6 Stilbs, P., Forsen, S.: Org. Magn. Res. *8,* 384 (1976)

7 Schweiger, J.T., et al.: J. Am. Chem. Soc. *96,* 7122 (1974)

8 Swenson, C.A., Koob, L.: J. Phys. Chem. *74,* 3376 (1970)

9 Schwotzer, W., et al.: Org. Magn. Res. *9,* 382 (1977)

10 Schwotzer, W., Von Philipsborn, W.: Helv. Chim Acta. *60,* 1501 (1977)

11 Sullivan, G., Roberts, J.D.: J. Org. Chem. *42,* 1095 (1977)

12 Schuster, I.I., Doss, S.H., Roberts, J.D.: J. Org. Chem. *43,* 4693 (1978)

13 Staal,L.H., Stufkens, D.J., Oskam, A.: Inorg. Chim. Acta. *26,* 255 (1978)

14 Sibi, M.P., Lichter, R.L.: J. Org. Chem. *42,* 2999 (1977)

15 Schwotzer, W., et al.: Helv. Chim Acta. *61,* 2108 (1978)

16 Sibi, M.P., Lichter, R.L.: J. Org. Chem. *44,* 3017 (1979)

17 Stefaniak, L.: Tetrahedron *33,* 2571 (1977)

18 Stilbs, P.: Tetrahedron Lett. *1972,* 227

19 Suzuki, T., Yamaguchi, T., Imanari, M.: Tetrahedron Lett. *1974,* 1809

20 Shitkin, V.M., et al.: Izv. Akad. Nauk. SSSR, Ser. Khim. *1977,* 2266

21 Schuster, I.I., Dyllick-Brenzinger, C., Roberts, J.D.: J. Org. Chem. *44,* 1765 (1979)

22 Schweitzer, D., Spiess, H.W.: J. Magn. Res. *15,* 529 (1974)

23 Schweitzer, D., Spiess, H.W.: J. Magn. Res. *16,* 243 (1974)

24 Srinivasan, P.R., Lichter, R.L.: J. Magn. Res. *28,* 227 (1977)

25 Stadeli, W., et al.: Helv. Chim. Acta *63,* 504 (1980)

26 Schneider, R.L.: Eastman Organic Chemical Bull. *47,* 1 (1975)

27 Smith, R.L., et al.: J. Am. Chem. Soc. *101,* 191 (1979)

28 Schuster, I.I., Roberts, J.D.: J. Org. Chem. *44,* 3864 (1979)

29 Schuster, I.I., Roberts, J.D.: J. Org. Chem. *45,* 284 (1980)

30 Stothers, J.B.: Carbon-13 NMR Spectroscopy, p. 375. New York: Academic Press 1972

31 Schulman, J.M., Venanzi, T.: J. Am. Chem. Soc. *98,* 4701 (1976)

32 Schulman, J.M., Venanzi, T.: J. Am. Chem. Soc. *98,* 6739 (1976)

33 Schulman, J.M.: J. Magn. Res. *28,* 137 (1977)

34 Steigel, A.: Mechanistic studies of the rearrangements and exchange relations by dynamic NMR spectroscopy. In: NMR basic principles and progress. Diehl, P., Fluck, E., Kosfeld, R. (eds.), vol.15. Berlin, Heidelberg, New York: Springer Verlag 1978

35 Schneider, H.J., Freitag, W.: J. Am. Chem. Soc. *98,* 478 (1976)

36 Schulman, J.M., Ruggio, J., Venanzi, T.J.: J. Am. Chem. Soc. *99,* 2045 (1977)

37 Solkan, V.N., Bystrov, V.F.: Izv. Akad. Nauk. SSSR, Ser. Khim. *1974,* 102

38 Saluvere, T., Lippmaa, E.: Chem. Phys. Letters *7,* 545 (1970)

39 Stenberg, V.I., et al.: J. Heterocyclie. Chem. *14,* 407 (1977)

T

1 Tolberg, W.E., et al.: Inorg, Chem. *6,* 1156 (1967)

2 Tori, K., et al.: J. Am. Chem. Soc. *89,* 2765 (1967)

3 Tarburton, P., et al.: J. Org. Chem. *44,* 2041 (1979)

4 Thomas, B., Grossmann, G.: Z. Anorg. Allg. Chem. *448,* 107 (1979)

5 Thomas, B., et al.: Z. Phys. Chemie, Leipzig. *260,* 225 (1979)

6 Thomas, B., Grossmann, G.: J. Magn. Res. *36,* 333 (1979)

7 Thetaz, C. Wehrli, F.W., Wentrup, C.: Helv. Chim. Acta *59,* 259 (1976)

8 Trierweiler, M., Martineau, H., Martin, M.L.: Results to be published

U

1 Unger, S.H., Hansch, C.: Quantitative models of steric effects. In: Progress in physical organic chemistry. Taft, R.W., (ed.), vol. 12. New York: Interscience 1976

V

1 Vaes, J., Chabanel, M., Martin, M.L.: J. Phys. Chem. *82,* 2420 (1978)

W

1 Wasylishen, R.E., Schaefer, T.: Can. J. Chem. *51,* 3087 (1973)

2 Wasylishen, R.E.: Annual Reports on NMR spectroscopy. In: Webb, G.A. (ed.) vol. 7, p.265, New York: Academic Press 1977

3 Witanowski, M., Stefaniak, L., Webb, G.A.: Annual reports on NMR spectroscopy. Webb, G.A. (ed.), vol. 7, p. 117. New York: Academic Press 1977

4 Wasylishen, R., et al.: Can. J. Chem. *50,* 2575 (1972)

5 Wiemer, D.F., Scopes, D.I.C., Leonard, N.J.: J. Org. Chem. *41,* 3051 (1976)

6 Walter, W., et al.: Ann. Chem. *719,* 210 (1968)

7 Wasylishen, R., Schaefer, T.: Can. J. Chem. *49,* 3627 (1971)

8 Wamsler, T., et al.: J. Magn. Res. *31,* 177 (1978)

9 Witanowski, M.: Pure Appl. Chem. *37,* 225 (1974)

10 Wasylischen, R.E.: Can. J. Chem. *54,* 833 (1976)

11 Wasylischen, R.: NMR Spectroscopy of nuclei other than proton. Axenrod, T., Webb, G.A., (eds,) p. 105, New York: Wiley Interscience 1974

12 Walter, J.A., Wright, J.L.C.: Tetrahedron Lett. *1979,* 3909

13 Weigert, F.J., Roberts, J.D.: Unpublished results quoted in N 1 and S 30

14 Williamson, K.L., Roberts, J.D.: Heterocycles *11,* 121 (1978)

15 Whidby, J.F., Edwards III, W.B., Phil Pitner, T.: J. Org. Chem. *44,* 794 (1979)

16 Westerman, P.W., Roberts, J.D.: J. Org. Chem. *42,* 2249 (1977)

17 Westerman, P.W., Roberts, J.D.: J. Org. Chem. *43,* 1177 (1978)

18 Williamson, K.L., Roberts, J.D.: J. Am. Chem. Soc. *98,* 5082 (1976)

19 Westerman, P.W., Botto, R.E., Roberts, J.D.: J. Org. Chem. *43*, 2590 (1978)

20 Walter, W., Sieveking, M.F., Schaumann, E.: Tetrahedron Lett. *1974*, 839

21 Warren, J.P., Roberts, J.D.: J. Phys. Chem. *78*, 2507 (1974)

22 Witanowski, M., et al.: J. Magn. Res. *28*, 217 (1977)

23 Witanowski, M., Webb, G.A.: Nitrogen NMR. New York: Plenum Press 1973

24 Wrackmeyer, B.: J. Magn. Res. *27*, 345 (1977)

25 Williamson, K.L., Pease, L.G., Roberts, J.D.: J. Am. Chem. Soc. *101*, 714 (1979)

26 Wentrup, C., Thetaz, C.: Helv. Chim. Acta *59*, 256 (1976)

27 Witanowski, M., et al.: Tetrahedron Lett. *1971*, 1653

28 Wüthrich, K.: NMR in biological research: peptides and proteins, p. 310, Amsterdam, London, New York: Elsevier 1976

29 Webb, G.A., Witanowski, M.: Nitrogen NMR. In: Witanowski, M., Webb, G.A., (eds.), p. 1. London: Plenum Press 1973

30 Wasylsishen, R., Schaefer, T.: Can. J. Chem. *50*, 2989 (1972)

Y

1 Yoder, C.H., et al.: J. Org. Chem. *37*, 4121 (1972)

2 Yamada, H., et al.: J. Am. Chem Soc. *100*, 4617 (1978)

3 Yavari, I., Roberts, J.D.: J. Am. Chem. Soc. *100*, 5217 (1978)

4 Yavari, I. Roberts, J.D.: Nouv. J. Chimie *3*, 279 (1979)

5 Yeh, H.J.C., et al.: J. Am. Chem. Soc. *95*, 2741 (1973)

6 Yavari, I. Roberts, J.D.: Tetrahedron Lett. *1978*, 2491

7 Yavari, I. Roberts, J.D.: J. Am. Chem. Soc. *100*, 4662 (1978)

8 Yavari, I., Roberts, J.D.: Biochem. Biophys. Res. Commun. *83*, 635 (1978)

9 Yavari, I., Roberts, J.D.: Org. Magn. Res. *12*, 87 (1979)

10 Yavari, I., Roberts, J.D.: J. Org. Chem. *43*, 4689 (1978)

11 Yavari, I., Staral, J.S., Roberts, J.D.: Org. Magn. Res. *12*, 340 (1979)

12 Yavari, I., Botto, R.E., Roberts, J.D.: J. Org. Chem. *43*, 2542 (1978)

13 Yamaguchi, T., et al.: Bull. Chem. Soc. Japan *49*, 1161 (1976)

14 Yeh, H.J.C.: Private communication

15 Yamasaki, A., et al.: J. Inorg. Nucl. Chem. *41*, 473 (1979)

16 Yavari, I., Roberts, J.D.: Org. Magn. Res. *13*, 68 (1980)

17 Yeh, H.J.C., Morishima, J.: J. Magn. Res. *26*, 365 (1977)

18 Yamasaki, A., et al.: Denki Tsushin Daigaku Gakuho. *28*, 107 (1977)

Index of Compounds

Subject Index

A

B

C

D

NMR

Basic Principles and Progress
Grundlagen und Fortschritte

Editors: P. Diehl, E. Fluck, R. Kosfeld

Volume 12
B. Lindman, S. Forsén

Clorine, Bromine and Iodine NMR

Physico-Chemical and Biological Applications

1976. 74 figures, 29 tables. XIV, 368 pages
ISBN 3-540-07725-1

Contents:
Introductory Aspects. Relaxation in Molecules or Ions with Covalently Bonded Halogens. Shielding Effects in Covalent Halogen Compounds. Scalar Spin Couplings. Relaxation of Chloride, Bromide and Iodide Ions. Shielding of Halide Ions. Quadrupole Splittings in Liquid Crystals. Halide Ions in Biological Systems. Studies of the Perchlorate Ion.

Volume 13

Introductory Essays

Editor: M. M. Pintar
1976. 48 figures. XI, 154 pages
ISBN 3-540-07754-5

Contents:
A. G. Redfield: A. Guide to Relaxation Theory. – *J. Jenner:* Thermodynamics of Spin Systems in Solids. An Elementary Introduction. – *J. S. Waugh:* Coherent Averaging and Double Resonance in Solids. – *E. L. Hahn:* Macroscopic Dipole Coherence Phenomena. – *G. J. Bene:* Nuclear Spins and Non Resonant Electromagnetic Phenomena. – *F. R. McCourt:* Nuclear Spin Relaxation in Molecular Hydrogen. – *R. L. Armstrong:* Longitudinal Nuclear Spin Relaxation Time Measurements in Molecular Gases. – *R. Blinc:* Spin-Lattice Relaxation in Nematic Liquid Crystals Via the Modulation of the Intramolecular Dipolar Interactions by Order Fluctuations. – *S. Clough:* NMR Studies of Molecular Tunneling. – *M. M. Pintar:* Effect of Molecular Tunneling on NMR Absorption and Relaxation in Solids. – *A. G. Redfield:* How to Build a Fourier Transform NMR-Spectrometer for Biochemical Applications.

Volume 14
H. Nöth, H. Wrackmeyer

Nuclear Magnetic Resonance Spectroscopy of Boron Compounds

1978. 1 figure, 96 tables. XII, 461 pages
ISBN 3-540-08456-8

Contents:
Introduction and Scope. Nuclear Magnetic Properties of Boron. ^{11}B NMR or Two-Coordinate Boron. ^{11}B Chemical Shifts of Three Coordinate Boron. ^{11}B NMR of Transition Metal Boron Compounds. ^{11}B NMR of Diborane and Derivatives. ^{11}B NMR of Tetracoordinate Boron. Spin-Spin Coupling Constants $^{n}J(^{11}BX)$. – Tables of ^{11}B NMR Data.

Volume 15

Dynamic NMR Spectroscopy

1978. 77 figures, 22 tables. VI, 214 pages
ISBN 3-540-08784-2

Contents:
A. Steigel: Mechanistic Studies of Rearrangements and Exchange Reactions by Dynamic NMR Spectroscopy. – *H. W. Spiess:* Rotation of Molecules and Nuclear Spin Relaxation.

Volume 16
P. S. Pregosin, R. W. Kunz

^{31}P and ^{13}C NMR of Transition Metal Phosphine Complexes

1979. 26 figures, 37 tables. IX, 156 pages
ISBN 3-540-09163-7

Contents:
Motivation and Methodology: Introduction. Measurement Techniques. Spin-Lattice Relaxation Times. The ^{31}P Nuclear Overhauser Effect. – Coupling Constants: One-Bond Metal-Phosphorus Coupling Constants. Two-Bound Phosphorus-Phosphorus Coupling Constants. One-Bond Metal-Phosphorus Coupling Constants – A Survey. Two-Bond Phosphorus-Phosphorus Coupling Constants-Data. – Chemical Shifts. – Applications. – ^{13}C Studies of Phosphorus Containing Complexes: Studies of Coordinated Phosphorus Ligands. Studies of Non-Phosphorus Ligands in Phosphorus-Complexes. – Appendices. – Tables. – References. – Index.

Inorganic Chemistry Concepts

Editors:
M. Becke
C. K. Jørgensen
M. F. Lappert
S. J. Lippard
J. L. Margrave
K. Niedenzu
R. W. Parry
H. Yamatera

Springer-Verlag
Berlin
Heidelberg
New York

Volume 1
R. Reisfeld, C. K. Jørgensen

Lasers and Excited States of Rare Earths

1977. 9 figures, 26 tables. VIII, 226 pages
ISBN 3-540-08324-3

Contents:
Analogies and Differences Between Monatomic Entities and Condensed Matter. – Rare-Earth Lasers. – Chemical Bonding and Lanthanide Spectra. – Energy Transfer. – Applications and Suggestions.

Volume 2
R. L. Carlin, A. J. van Duyneveldt

Magnetic Properties of Transition Metal Compounds

1977. 149 figures, 7 tables. XV, 264 pages
ISBN 3-540-08584-X

Contents:
Paramagnetism: The Curie Law. – Thermodynamics and Relaxation. – Paramagnetism: Zero-Field Splittings. – Dimers and Clusters. – Long-Range Order. – Short-Range Order. – Special Topics: Spin-Flop, Metamagnetism, Ferrimagnetism and Canting. – Selected Examples.

Volume 3
P. Gütlich, R. Link, A. Trautwein

Mössbauer Spectroscopy and Transition Metal Chemistry

1978. 160 figures, 1 folding plate, 19 tables.
X, 280 pages
ISBN 3-540-08671-4

Contents:
Basic Physical Concepts. – Hyperfine Interactions. – Experimental. – Mathematical Evaluation of Mössbauer Spectra. – Interpretation of Mössbauer Parameters of Iron Compounds. – Mössbauer-Active Transition Metals Other than Iron. – Some Special Applications.

GPSR Compliance

The European Union's (EU) General Product Safety Regulation (GPSR) is a set of rules that requires consumer products to be safe and our obligations to ensure this.

If you have any concerns about our products, you can contact us on ProductSafety@springernature.com

In case Publisher is established outside the EU, the EU authorized representative is:

Springer Nature Customer Service Center GmbH
Europaplatz 3
69115 Heidelberg, Germany

Batch number: 10838062

Printed by Printforce, the Netherlands